Introduction to Computational Fluid Dynamics

Atul Sharma

Introduction to Computational Fluid Dynamics

Development, Application and Analysis

Ane Books Pvt. Ltd.

Atul Sharma
Department of Mechanical Engineering
Indian Institute of Technology Bombay
Mumbai, Maharashtra, India

ISBN 978-3-030-72886-1   ISBN 978-3-030-72884-7   (eBook)
https://doi.org/10.1007/978-3-030-72884-7

Jointly published with ANE Books Pvt. Ltd.
In addition to this printed edition, there is a local printed edition of this work available via Ane Books in
South Asia (India, Pakistan, Sri Lanka, Bangladesh, Nepal and Bhutan) and Africa (all countries in the
African subcontinent).
ISBN of the Co-Publisher's edition: 9789385462535

This Springer imprint is published by the registered company Springer Nature Switzerland AG
The registered company address is: Gewerbestrasse 11, 6330 Cham, Switzerland

This book is dedicated

to

Suhas V. Patankar

Dr. Suhas V. Patankar[1] (born February 22, 1941) is an Indian mechanical engineer. He is a pioneer in the field of computational fluid dynamics (CFD) and finite volume method. He received his B.E. from the University of Pune in 1962, M.Tech. from IIT Bombay in 1964 and Ph.D. from Imperial College, London in 1967.

---

[1] *Source* (for the Biography): Personal Communication.

Dr. Patankar is currently a Professor Emeritus in the Mechanical Engineering Department at the University of Minnesota, where he worked for 25 years (from 1975–2000). Earlier, he held teaching and research positions at IIT Kanpur, Imperial College, and University of Waterloo. He is also the President of Innovative Research, Inc. He has authored or co-authored four books, published over 150 papers, advised 35 completed Ph.D. theses, and lectured extensively in the USA and abroad. His 1980 book "Numerical heat Transfer and Fluid Flow" is considered to be a groundbreaking contribution to CFD. He is one of the most cited authors in science and engineering.

For excellence in teaching, Dr. Patankar received the 1983 George Taylor Distinguished Teaching Award and the 1989–90 Morse-Alumni Award for Outstanding Contributions to Undergraduate Education. For his research contributions to computational heat transfer, he was given the 1991 ASME Heat Transfer Memorial Award and the 1997 Classic Paper Award. He was awarded the 2008 Max Jakob Award, which is considered to be the highest international honor in the field of heat transfer. In 2015, an International Conference on Computational Heat Transfer (CHT-15) held at Rutgers University in New Jersey was dedicated to Professor Patankar.

Dr. Patankar's widespread influence on research and engineering education has been recognized in many ways. In 2007, the Editors of the International Journal of Heat and Mass Transfer wrote, "There is no person who has made a more profound and enduring impact on the theory and practice of numerical simulation in mechanical engineering than Professor Patankar."

Foreword

The subject of computational fluid dynamics (CFD) emerged in the academic context, but in the past two decades, it has developed to a point where it is extensively employed as a design and analysis tool in the industry. The impact of the subject is deeply felt in diverse disciplines ranging from aerospace spreading all the way to the chemical industry. New, compact, and efficient designs are enabled, and it has become that much more convenient to locate optimum conditions for the operation of engineering systems. As it stands, CFD is compulsory learning in several branches of engineering. Many universities now opt to develop theoretical courses around it. The great utility of this tool has influenced adjacent disciplines, and the process of solving complex differential equations by the process of discretization has become a staple for addressing engineering challenges.

For an emerging discipline, it is necessary that new books frequently appear on the horizon. Students can choose from the selection and will stand to gain from it. Such books should have a perspective, of the past and the future, and it is all the more significant when the author has exclusive training in the subject. Dr. Atul Sharma represents a combination of perspective, training, and passion that combines well and delivers a strong text.

I have personal knowledge of the fact that Dr. Atul Sharma has developed a whole suite of computer programs on CFD from scratch. He has wondered and pondered over why approximations work and conditions under which they yield meaningful solutions. His spread of experience over 15-and-odd years includes two- and three-dimensional Navier–Stokes equations, heat transfer, interfacial dynamics, and turbulent flow. He has championed the finite volume method which is now the industry standard. The programming experience is important because it brings in the precision needed for assembling the jigsaw puzzle. The sprinkling of computer programs in the book has no coincidence.

Atul knows the conventional method of discretizing differential equations but has never been satisfied with it. He has felt an element of discomfort going away from a physical reality into a world of numbers connected by arithmetic operations. As a result, he has developed a principle that physical laws that characterize the differential equations should be reflected at every stage of discretization and every stage of approximation, indeed at every scale—whether a single tetrahedron or the sub-domain or the region as a whole. This idea permeates the book where it is shown that discretized versions must provide reasonable answers because they continue to represent the laws of nature.

This new CFD book is comprehensive and has a stamp of originality of the author. It will bring students closer to the subject and enable them to contribute to it.

Prof. K. Muralidhar
Indian Institute of Technology
Kanpur, India

Preface

As a Postgraduate (PG) student in 1997, I got introduced to a course on CFD at the Indian Institute of Science (IISc) Bangalore. During the "introductory" course on CFD, although I struggled hard in the physical understanding and computer programming, it didn't stop my interest and enthusiasm to do research, teach, and write a textbook on an "introductory" course in CFD. Since then, I did my Masters project work on CFD (1997-98); introduced and taught this course to Undergraduate (UG) students at the National Institute of Technology, Hamirpur, H.P. (1999–2000); did my Doctoral work on this subject at the Indian Institute of Technology (IIT) Kanpur (2000–2004); and co-authored a chapter on "finite volume method" in a revised edition of an edited book on computational fluid flow and heat transfer, during my Ph.D. in 2003. After joining as a faculty in 2004, I taught this course to UG as well as PG students of different departments at IIT Bombay as well as to the students of different colleges in India through CDEEP (Center for Distance Engineering Programme), IIT Bombay, from 2007 to 2010. Furthermore, I gave a series of lectures on CFD at various colleges and industries in India. More recently, in 2012, I delivered a 5-day lecture and lab session on CFD (to around 1400 college teachers from different parts of India) as a part of a project funded by the Ministry of Human Resource Development under NMEICT (National Mission on Education through Information and Communication Technology).

The increasing need for the development of the customized CFD software (*app*) and widespread CFD application as well as analysis for the design, optimization, and innovation of various types of engineering systems always motivated me to write this book. Due to the increasing importance of CFD, another motivation is to present CFD simple enough to introduce the first course in the early UG curriculum, rather than the PG curriculum. The motivation got strengthened by my own as well as other's (students, teachers, researchers, and practitioners) struggle on physical understanding of the mathematics involved during the Partial Differential Equation (PDE)-based algebraic formulation (finite volume method), along with the continuous frustrating pursuit to effectively convert the theory of CFD into the computer program.

This led me to come up with an alternate physical law-based (PDE free) finite volume method as well as solution methodology, which are much easier to comprehend. Mostly, the success of the programming effort is decided by the perfect execution of the implementation details which rarely appear in print (existing books and journal articles). The physical law-based CFD development approach and the implementation details are the novelties in the present book. Furthermore, using an open-source software for numerical computations (*Scilab*), computer programs are given in this book on the basic modules of conduction, advection, and convection. Indeed, the reader can generalize and extend these codes for the development of the Navier-Stokes solver and generate the results presented in the chapters toward the end of this book.

I have limited the scope of this book to the numerical techniques which I wish to recommend for the book, although there are numerous other advanced and better methods in the published literature. I do not claim that the programming practice presented here are the most efficient; they are presented here more for the ease in programming (understand as your program) than for the computational efficiency. Furthermore, although I am enthusiastic about my presentation of the physical law-based CFD approach, it may not be more efficient than the alternate mathematics-based CFD approach. The emphasis in this book is on the ease of understanding of the formulations and programming practice, for the introductory course on CFD.

I would hereby acknowledge that I owe my greatest debt to Prof. J. Srinivasan (my M.E. Project supervisor at IISc Bangalore) who did a hand-holding and wonderful job in patiently introducing me to the fascinating world of research in fluid dynamics and heat transfer. I would also acknowledge the excellent training from Prof. V. Eswaran (my Ph.D. supervisor at IIT Kanpur), from whom I learned the systematic way of CFD development, application, and analysis. His inspiring comments and suggestions on my writing style, and elaboration of the first chapter, in this book are also gratefully acknowledged. I also want to record my sincere thanks to Prof. Pradip Datta from IISc Bangalore and Prof. Gautam Biswas from IIT Kanpur; my teachers for CFD course. Special thanks to Prof. K. Muralidhar, IIT Kanpur, for his personal encouragement, academic support, and motivation (for the book writing). He also cleared my dilemma in proper presentation of the novelty in this book. I also thank him for writing the foreword of this book.

I am grateful to Prof. Kannan M Moudgalya and Prof. Deepak B. Phatak from IIT Bombay for giving me the opportunity to teach CFD in a distance education mode. I am also grateful to Prof. Amit Agrawal (my colleague at IIT Bombay) who diligently read through all the chapters of this book and suggested some really remarkable changes in the book. I also thank him for the research interactions, over the last decade, which had substantially improved my understanding of thermal and fluid science. I would also like to thank my Ph.D. student, Namshad T., who did a great job in implementing my ideas on generating the CFD simulation-based "CFD" image, for the cover page and couple of figures in the first chapter (Figs. 1.2 and 1.3). Special mention that the Scilab codes developed (for the CFD course under NMEICT) by Vishesh Aggarwal greatly helped me to develop the codes presented in this book; his contributions are gratefully acknowledged. I thank Malhar Malushte for testing some of the formulations presented in this book.

I thank my research students (Sachin B. Paramane, D. Datta, C. M. Sewatkar, Vinesh H. Gada, Kaushal Prasad, Mukul Shrivastava, Absar M. Lakdawala, Mausam Sarkar, and Javed Shaikh) for the diligence and sincerity they have shown in unraveling some of our ideas on CFD. Apart from these, several other research students, and the students, teachers, and CFD practitioners who took my lectures on CFD in the past decade have contributed to my better understanding of CFD and inspired me to write the book. I thank all of them for their attention and enthusiastic response during the CFD courses. The influence of the interactions with all of them can be seen throughout this book. I am grateful to IIT Bombay, for giving me 1-year sabbatical leave to write this book. The ambiance of academic freedom, considerate support by the faculties, and the friendly atmosphere inside the campus of IIT Bombay has greatly contributed to the solo effort of book writing. I thank all my colleagues at IIT Bombay for shaping my perceptions of CFD.

During this long endeavor, the patient support by Mr. Sunil Saxena, Director, Athena Academic Ltd., UK, and Ane Books Pvt. Ltd., India, is gratefully acknowledged. I also thank John Wiley for coming forward to take this book to international markets. Lastly, I would acknowledge that I am greatly inspired by the writing style of J. D. Anderson, and the book on CFD by S. V. Patankar. Of course, you will find that I have heavily cross-referred the CFD book by the two most acknowledged pioneers on CFD. I dedicate this book to Dr. S. V. Patankar, whose book nicely introduced me to the computational as well as flow physics in CFD and greatly helped me to introduce a physical approach of CFD development in this book.

The culmination of the book writing process fulfills a long-cherished dream. It would not have been possible without the excellent environment, continuous support, and complete understanding of my wife Anubha and daughter Anshita. Their kindness has greatly helped to fulfill my dream; now, I plan to spend more time with my wife and daughter.

Mumbai, Maharashtra   Atul Sharma

Contents

About the Author

Dr. Atul Sharma is a Professor in the Mechanical Department at Indian Institute of Technology (IIT) Bombay, with a teaching and research experience on CFD code-development, its applications and big-data analysis for more than 20 years. He has proved leadership in capacity building in CFD, by providing training programme in more than 100 engineering-institutes, with a vision to make CFD-easy-enough to be taught at an early Under-Graduate curriculum. Most of his research contributions are on Computational Multi-Phase Dynamics, comprising Computational Multi-Fluid Dynamics and Computational Fluid-Structure Interactions. He has graduated 15 Ph.D. students, with two best-Ph.D.-thesis awards and SHELL computational-talent-prizes. His wide-variety of research is published as 79 articles in 29 different international-journals and 13 chapters in 4 edited-books; and appeared in the cover page of Journal of Fluid Mechanics, Physics of Fluids and Langmuir. He was a CFD consultant at Global R&D, Crompton Greaves Limited, Mumbai; served as Secretary, National Society of Fluid Mechanics and Fluid Power INDIA; and presently, associate-editor of a Journal Sadhana—academy proceedings in engineering sciences. He received IIT Bombay "Departmental Award for Excellence in Teaching 2019".

Part I
Introduction and Essentials

Chapter 1.	Chapter 2.	Chapter 3.	Chapter 4.
Introduction	Introduction to CFD Development, Application and Analysis	Essentials of Fluid-Dynamics & Heat-Transfer for CFD	Essentials of Numerical-Methods for CFD

The book on an introductory course in CFD starts with the two introductory chapters on CFD, followed by one chapter each on essentials of two prerequisite courses—fluid-dynamics and heat transfer, and numerical-methods. Chapter 1 on introduction presents an eventful, thoughtful, and intuitive discussion on What is CFD? and Why to study CFD?; whereas, How CFD works? which involves lots of details is presented separately in the Chapter 2. Chapter 3 on essentials of fluid-dynamics and heat transfer presents physical-laws, transport-mechanisms, physical-law based differential formulation, volumetric and flux terms, and mathematical formulation. Chapter 4 on essentials of numerical-methods presents finite difference method, iterative solution of the system of linear algebraic equation, numerical differentiation, and numerical integration. The presentation, for the essential of the two prerequisite courses, is customized to CFD development, application and analysis.

Chapter 1
Introduction

The introduction of any course consists of the three basic questions: What is it? Why to study? How it works? An eventful, thoughtful, and intuitive answer to the first two questions are presented in this chapter. The third question which involves lots of working details on the three aspects of the **C**omputational **F**luid **D**ynamics (CFD) (development, application, and analysis) is presented separately in the next chapter. The novelty, scope, and purpose of this book are also presented at the end of this chapter.

1.1   CFD: What Is It?

CFD is a *theoretical method* of scientific and engineering investigation, concerned with the *development* and *application* of a video camera like tool (a *software*) which is used for a unified cause-and-effect-based *analysis* of a fluid dynamics as well as heat and mass transfer problem.

CFD is a subject where you first learn how to *develop* a product—a software which acts like a virtual video camera. Then, you learn how to *apply* or use this product to generate fluid dynamics movies. Finally, you learn how to *analyze* the detailed spatial and temporal fluid dynamics information in the movies. The analysis is done to come up with a scientifically exciting as well as engineering-relevant story (unified cause-and-effect study) of a fluid dynamics situation in nature as well as industrial applications. With continuous development and a wider application of CFD, the word "fluid dynamics" in the CFD has become more generic as it also corresponds to heat and mass transfer as well as chemical reaction.

Fluid dynamics information are of two types: scientific and engineering. *Scientific-information* corresponds to a structure of the heat and fluid flow, due to a physical phenomenon in fluid dynamics; such as boundary layer, flow separation, wake formation, and vortex shedding. The flow structures are obtained from a temporal variation

© The Author(s), under exclusive license to Springer Nature Switzerland AG 2022   3
A. Sharma, *Introduction to Computational Fluid Dynamics*,
https://doi.org/10.1007/978-3-030-72884-7_1

of flow properties (such as velocity, pressure, temperature, and many other variables which characterize a flow), called scientifically exciting movies. *Engineering-information* corresponds to certain parameters which are important in an engineering application, called *engineering parameters*; for example, lift force, drag force, wall shear stress, pressure drop, and rate of heat transfer. They are obtained as the temporal variation of engineering parameters, called engineering-relevant movies. The engineering parameters are the effects which are caused by the flow structures—their correlation leads to the *unified cause-and-effect study*. A fluid dynamics movie consists of a timewise varying series of pictures which give fluid dynamics information. Each flow property and engineering parameter results in one scientifically exciting and engineering-relevant movie, respectively. When the large number of both the types of movies are played synchronized in time, the detailed spatial as well as temporal fluid dynamic information greatly helps in the unified cause-and-effect study of a fluid dynamics problem.

Alternatively, CFD is a *subject* concerned with the development of *computer programs*, for the *computer simulation* and *study* of a natural or an engineering *fluid dynamics system*. The development starts with the *virtual development* of the system, involving a geometrical information-based computational development of a solid model for the system; the system may be already-existing or yet-to-be-developed physically. Thereafter, the development involves numerical solution of the governing algebraic equations for CFD formulated from the physical laws (conservation of mass, momentum, and energy) and initial as well as boundary conditions corresponding to the system. CFD study involves the analysis of the fluid dynamics and heat transfer results, obtained from the computer simulations of the virtual system which is subjected to certain governing parameters.

1.1.1 CFD as a Scientific and Engineering Analysis Tool

For a better understanding of the definition of the CFD and its application as a scientific and engineering analysis tool, two example problems are presented: first, conduction heat transfer in a plate; and second, fluid flow across a circular pillar of a bridge. The first problem corresponds to an engineering-application for electronic cooling, and the second problem is studied for an engineering design of the structure subjected to fluid dynamics forces; and are shown in Fig. 1.1. Both the problems are considered as two-dimensional and unsteady. However, the results asymptote from the unsteady to a steady state in the first problem and to a periodic (dynamic steady) state in the second problem.

The first problem is taken from the classic book on heat transfer by Incropera and Dewitt (1996), shown in Fig. 1.1a. The figure shows a periodic module of the plate where a constant heat-flux $q_W = 10$ W/cm^2 (dissipated from certain electronic devices) is first conducted to an aluminum plate, and then convected to the water flowing through the rectangular channel; grooved in the plate. The forced convection results in a convection coefficient of $h = 5000$ W/m^2.K inside the channel. The

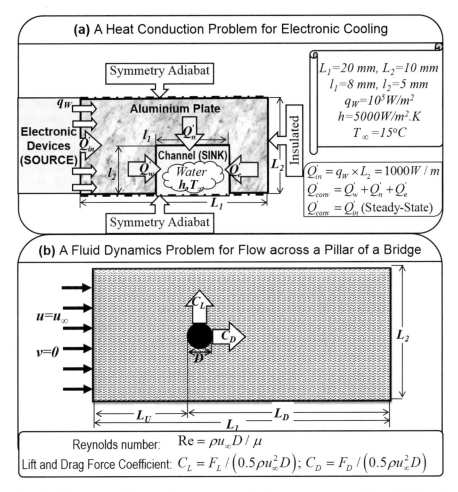

(a) A Heat Conduction Problem for Electronic Cooling

Symmetry Adiabat

q_W

Aluminium Plate

Electronic Devices (SOURCE)

Q_{in}

l_1 | Q_n'

Channel (SINK)

l_2 | Q_w' | *Water* | Q_e'

h, T_∞

L_1

Symmetry Adiabat

Insulated

L_2

L_1=20 mm, L_2=10 mm
l_1=8 mm, l_2=5 mm
q_W=$10^5 W/m^2$
h=$5000 W/m^2.K$
T_∞=15°C

$Q_{in}' = q_W \times L_2 = 1000 W / m$
$Q_{conv}' = Q_w' + Q_n' + Q_e'$
$Q_{conv}' = Q_{in}'$ (Steady-State)

(b) A Fluid Dynamics Problem for Flow across a Pillar of a Bridge

$u=u_\infty$

$v=0$

C_L

C_D

D

L_2

L_U

L_D

L_1

Reynolds number: $\quad Re = \rho u_\infty D / \mu$

Lift and Drag Force Coefficient: $C_L = F_L / (0.5\rho u_\infty^2 D); \quad C_D = F_D / (0.5\rho u_\infty^2 D)$

Fig. 1.1 Schematic and the associated parameters for the two example problems: (**a**) thermal conduction module of a heat sink, and (**b**) the classical flow across a circular cylinder

unsteady heat transfer results in an interesting timewise varying conduction heat flow in the plate and instantaneous convection heat transfer rate Q_{conv}' (per unit width of the plate, perpendicular to the plane shown in Fig. 1.1a) into the water.

The second problem is encountered if you are standing over a bridge and look below to the water passing over a pillar (of circular cross section) of the bridge, shown in Fig. 1.1b as a general case of 2D flow across a cylinder. The fluid dynamics results in a beautiful timewise varying structure of the fluid flow just downstream of the pillar, and certain engineering parameters. The parameters correspond to the fluid dynamic forces acting in the horizontal (or streamwise) and vertical (or transverse) directions, called the drag force F_D and lift force F_L, respectively. The respective force is presented below in a non-dimensional form called the drag coefficient C_D and lift coefficient C_L; equations shown in Fig. 1.1b.

If you use your video camera to capture a movie for the two problems, you will get series of pictures for the plate in the first problem and the flowing fluid in the second problem. Since these movies result in a pictorial information but not flow property-based fluid dynamic information; such movies are not relevant for the fluid dynamics and heat transfer study. Instead, if you use a CFD software for computer simulation, you can capture various types of movies. The various flow properties-based scientifically exciting movies are presented here as the pictures at certain time instant in Fig. 1.2a–d and Fig. 1.3a1–d2. The former figure shows gray-colored flooded contour for the temperature; whereas, the latter figure shows gray-colored flooded contour for the pressure and vorticity and vector plot for the velocity. For the gray-colored flooded contours, a color bar can be seen in the figures for the relative magnitude of the variable. These figures also show line contours of stream function and heat function as streamlines and heatlines, respectively. They are the visualization techniques for a *2D steady* heat and fluid flow. The heatlines are analogous to streamlines and represents the direction of heat flow, under steady-state condition, proposed by Kimura and Bejan (1983); later presented in a heat transfer book by Bejan (1984). The *steady-state heatlines* can be seen in Fig. 1.2d, where the direction for the conduction flux is tangential at any point on the heatlines; analogous to that for the mass flux in a streamline. Thus, for a better presentation of the structure of heat and fluid flow, the scientifically exciting movies are not only limited to flow properties but also correspond to certain flow visualization techniques.

An engineering parameter-based engineering-relevant fluid dynamic movie is presented in Fig. 1.2e, for the convection heat transfer rate per unit width at the various walls on the channel: Q'_w at the west wall, Q'_n at the north wall, Q'_e at the east wall, and their cumulative value Q'_{conv} (refer Fig. 1.1a). Another engineering-relevant movie is presented in Fig. 1.3e, for the lift and drag coefficients. Note that the movie here corresponds to the motion of a filled symbol over a line for the temporal variation of an engineering parameter. Thus, the engineering-relevant movies are one-dimensional lower than the scientifically exciting movies; here, for the 2D problems, the former movie is presented as 1D and the latter as 2D results.

1.1.1.1 Heat Conduction

For the computational setup of the heat conduction problem, shown in Fig. 1.1a, the results obtained from a 2D CFD simulation of the conduction heat transfer are shown in Fig. 1.2. For a uniform initial temperature of the plate as $T_\infty = 15\,°\text{C}$, Fig. 1.2a–d shows the isotherms and heatlines in the plate, as the pictures of the scientifically exciting movies, at certain discrete time instants $t = 2.5, 5, 10$, and 20 sec. For the same discrete time instants, the temporal variation of the position of the symbols over a line plot is presented in Fig. 1.2e. The figure is shown for each of the heat transfer rate and is obtained from an engineering-relevant 1D movie. There is a temporal

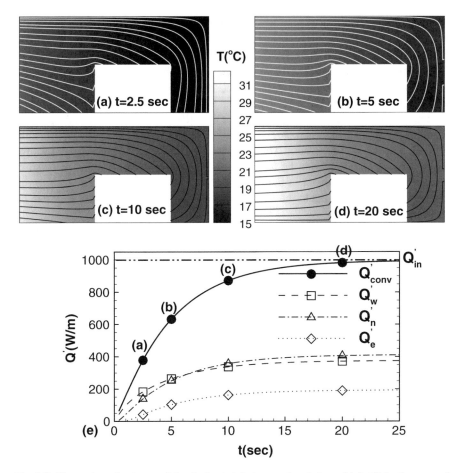

Fig. 1.2 Thermal conduction module of a heat sink: temporal variation of (**a**)–(**d**) isotherms and heatlines, and (**e**) the convection heat transfer rate per unit width at the various walls of the channels and their cumulative total value Q'_{conv}. For the discrete time instant corresponding to heat flow patterns in (**a**)–(**d**), temporal variation of the various heat transfer rates are shown as symbols in (**e**). (**a**)–(**c**) are the transient and (**d**) represents a steady-state result

variation of picture in the scientifically exciting movie and that of the position of symbol (along with the line plot) in the engineering-relevant movie.

According to *caloric theory* of the famous French scientist Antonie Lavoisier, heat is considered as an invisible, tasteless, odorless, weightless fluid, which he called *caloric fluid*. Presently, the calorific theory is overtaken by mechanical theory of heat—heat is not a substance but a dynamical form of mechanical effect (Thomson 1851). Nevertheless, there are certain problems involving heat flow for which Lavoisier's approach is rather useful; such as, the discussion on heatlines here.

The role of heatlines for heat flow is analogous to that of streamlines for fluid flow (Paramane and Sharma 2009). For the fluid flow, the difference of stream function values represents the rate of fluid flow, the function remains constant on a solid wall, the tangent to a streamline represents the direction of the fluid flow (with no flow in the normal direction), and the streamline originate or emerge at a mass source. Analogously, for the heat-flow, the difference of heat function values represents the rate of heat flow, the function remains constant on an adiabatic wall, the tangent to a heatline represents the direction of the heat flow (with no flow in the normal direction), and the heatline originate or emerge at the heat source. For the present problem in Fig. 1.1a, the heat source is at the inlet (or left) boundary and the heat sink is at the surface of the channel. Thus, the *steady-state* heatlines in Fig. 1.2d shows the heat flow emerging from the inlet/source and ending on the channel walls/sink. Furthermore, it can be seen in the figure that the heatlines which are close to the adiabatic walls are parallel to the walls. The source, sink, and adiabatic walls can be seen in Fig. 1.1a.

Figure 1.2e shows an asymptotic timewise increase in the convection heat transfer per unit width at the various walls of the channel and its cumulative value Q'_{conv} $(= Q'_w + Q'_n + Q'_e)$. It can be also be seen that the heat transfer per unit width Q'_w at the west-wall and Q'_n at the north wall are larger than Q'_e at the east wall of the channel. As time progresses, the figure shows that Q'_{conv} increases monotonically and becomes equal to the inlet heat transfer rate $Q'_{in} = 1000W$ under steady-state condition; refer Fig. 1.1a. The difference between the Q'_{in} and Q'_{conv}, under the transient condition, is utilized for the sensible heating and results in the increase in the temperature of the plate. Under the steady-state condition, note from Fig. 1.2d that the maximum temperature is close to $32\,°C$—well within the permissible limit for the reliable and efficient operation of the electronic device.

1.1.1.2   Fluid Dynamics

For the free-stream flow across a circular cylinder/pillar at a Reynolds number of 100, the non-dimensional results obtained from a 2D CFD simulation are shown in Fig. 1.3. In contrast to the previous problem which reaches steady state, the present problem reaches to an unsteady periodic-flow—both flow patterns and engineering parameters start repeating after a certain time period. Thereafter, for four different increasing time instants within one time period of the periodic flow, the results are presented in Fig. 1.3a–d. The figure shows the close-up view of the pictures, of the scientifically exciting movies, for flow properties (velocity, pressure, and vorticity) and streamlines. The velocity vectors are shown as arrows, with a scale shown in the figure for a reference vector as unity. The size of the arrow in the figure represents the relative magnitude and its sense represents the direction of the velocity. Both vorticity and pressure are shown as the flooded color contours. The vorticity is computed from the velocity field and provides an idea of the relative circulation of fluid particles in the flow field. Since the flow is periodic and the time duration between two consecutive figure is one-fourth of the time period, Fig. 1.3a repeats after Fig. 1.3d. For the four time instants corresponding to the flow pattern in Fig. 1.3a–d, the variation of position

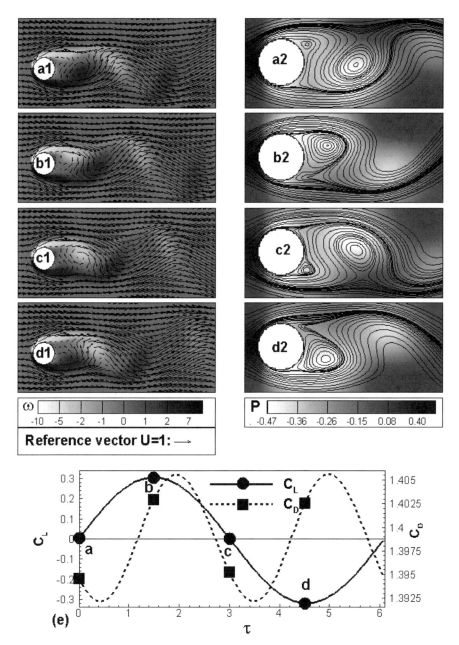

Fig. 1.3 Free-stream flow across a cylinder at a Reynolds number of 100: temporal variation of (a1)–(d1) velocity vector and vorticity contour, and (a2)–(d2) streamlines and pressure contour, for a vortex-shedding flow. For the discrete time instant corresponding to the fluid flow patterns in (**a**)–(**d**), temporal variation of the lift and drag coefficients are shown as symbols in (**e**). Two consecutive figures in (**a**)–(**d**) and (**e**)–(**h**) are separated by a time interval of one-fourth of the time period of the periodic flow

of the symbols over the line plot in Fig. 1.3e represents the temporal variation of the lift and drag coefficients; obtained from the engineering-relevant movie for the C_L and C_D.

The velocity vectors in Fig. 1.3a1–d1 shows a wake region just behind the cylinder, where the streamwise velocity is less than free-stream velocity. The temporal variation of velocity vector does not provide a clear idea about the variation of the size of the vortex in the separated region. The size of the vortex is shown clearly through streamlines in Fig. 1.3a2–d2, as the maximum size of the enclosing streamlines; representing recirculating fluid region. The streamlines in the Fig. 1.3a2–b2 shows the growth of CW (clockwise) vortex at the top of the rear surface of the cylinder followed by its break-up from the surface and moving downstream (Fig. 1.3c2). This is followed by the similar process for the CCW (counter-clockwise) vortex at the bottom of the rear surface of the cylinder, shown in Fig. 1.3c2–d2–a2. This process is called vortex-shedding, with one vortex pair (CW and CCW) shed within one time period of the periodic flow. The time period is used to compute the frequency of vortex shedding. The CW/CCW direction of the shed vortices is clearly seen in the velocity vector plot. These vortices are also seen in the vorticity contours, with a magnitude which is positive for a CCW-vortex and negative for a CW-vortex.

Cause leads to an effect—the periodic variation of the velocity and the pressure results in a periodic variation of the fluid dynamic forces acting on the cylinder. Figure 1.3a1–d2 shows that the pressure as well as velocity behind as compared to ahead of the cylinder is smaller, leading to the fluid dynamics-based drag force on the cylinder. Similarly, the relative variation of the pressure above and below the cylinder leads to the lift force. Both lift and drag forces consist of two components: pressure and viscous, computed from the pressure and shear-stress distribution on the surface of the cylinder, respectively. These instantaneous engineering parameters are calculated from the transient results for the flow properties, called *post-processing*. Within the time-period of the periodic-flow, Fig. 1.3e shows that the periodic variation of the lift and drag coefficients is almost sinusoidal; however, it can be seen that there are two periodic cycles of the drag coefficient within one cycle of the lift coefficient. Thus, the drag-coefficient is oblivious of whether a vortex is shed from the top or the bottom of the rear surface of the cylinder. It is interesting to observe that the time instant for the two maxima in the drag coefficient in Fig. 1.3e almost corresponds to the break-up followed by the vortex-shedding—from the top of the rear surface (Fig. 1.3b2) for the first maxima, and from the bottom of the rear surface for the second maxima (1.3d2). The large size wall vortex slightly before the break-up of the vortex at the top of the rear surface in Fig. 1.3b2 corresponds to the maximum value of positive/upward lift coefficient in Fig. 1.3e; similar wall vortex at the bottom of the rear surface corresponds to the maximum value of the negative/downward lift force. Note from the figure that a vortex corresponds to a minima in the pressure field.

Figure 1.3a1–d2 show flow separation, wake formation, and the classical vortex shedding phenomenon. The drag force tries to drag the pillar while the periodic lift force results in the transverse vibration of the pillar. The undesirable effects of the fluid dynamic forces are prevented by the adequate structural support at the bottom of the pillars. The nature of the flow patterns and the forces depend on certain parameters, called as *governing parameters*; such as the incoming velocity/flow rate, the thermo-physical properties of the fluid, and the shape as well as size of the cross section of the pillars. *Effect study* is the determination of the temporal variation of the lift and the drag forces. Whereas, a *cause study* is to analyze the flow structure/phenomenon near the surface of the cylinder and understand the fluid dynamic reasons for the effect of the shape of the pillar, and the other governing parameters, on the lift and drag force. The cause-and-effect study is done simultaneously by creating a synchronized movie of the flow properties and engineering parameters, with a timewise variation of picture and symbol (over the line plot), respectively.

Scientific investigation corresponds to a story about the interesting fluid dynamics phenomenon (flow separation, wake formation, and vortex-shedding) encountered in this problem. Whereas, the *engineering-investigation* corresponds to a methodology for the structural design of the pillars after obtaining the timewise varying lift and drag forces. A scientific as well as engineering investigation corresponds to identifying a correlation between the forces and flow structure as well as the fluid dynamics phenomenon. The role of a CFD software is not only limited to creating scientifically exciting fluid dynamic movie of flow properties, but also to create engineering-relevant movie of engineering parameters, for a *unified cause-and-effect study* of various heat and fluid flow situations. The above cause-and-effect study for the cylinder is also relevant for the flow across a car and also over an aircraft.

1.1.2 Analogy with a Video Camera

1.1.2.1 CFD as a Finite-Resolution Virtual Video Camera

After discussing the role of CFD-based scientifically exciting and engineering-relevant movies on the investigation of a heat transfer and a fluid dynamics problem, an analogy between a real and virtual (from a CFD software) video camera is presented. One needs to focus to a spatial region of interest to create a movie using a video camera. This is also true in CFD and corresponds to the region involving heat and/or fluid flow—called computational domain. With a video camera, the size of the *focused region* is constraint by its focusing specification. Whereas, in CFD, the size of *computational domain* are taken such that a physically relevant boundary conditions can be applied at the boundary of the domain.

A picture taken from a video camera is made up of the image/color captured at a large number of discrete *pixels*. A picture from a CFD software is made up of the colors-based representation (of a flow property) at the discrete points in the computational domain; the points are called *grid points*. The colors at the grid points correspond to the numbers computed for the associated flow properties, mapped by a graphical software. For example, the flooded gray-colored contour with a color bar is shown in Fig. 1.2a–d for temperature, Fig. 1.3a1–d1 for vorticity, and Fig. 1.3a2–d2 for pressure. The figures correspond to a total number of grid points (analogous to pixels) as 4000 for temperature, 67500 for vorticity, and 32250 for pressure. The quality of a picture obtained from a camera depends on the number of pixels. Whereas, the spatial accuracy of the results (presented as a picture) obtained from a CFD software depends on the number of grid points.

Spatial resolution for a picture is equal to the total number of pixels in a video camera and the number of grid points in CFD. The resolution is uniform in a video camera and mostly non-uniform in CFD. This is because the grid points are non-uniformly spaced to efficiently capture the spatially non-uniform action/dynamics of flow inside the computational domain. A movie is made by a series of pictures obtained at a time interval corresponding to the frame rate of the video camera, and to the time step in case of a CFD software, called as temporal resolution. The spatial and the temporal resolutions controls the quality of the graphical results, judged by the visual appearance in a video camera and the accuracy in a CFD software. The analogy presented above applies not only to CFD but also applicable to computational solid dynamics as well as many other physics-based simulation software in science and engineering.

For almost all the fluid flow and heat transfer situations, CFD provides a numerical solution which is approximate, and at *certain* spatial and temporal coordinates. Thus, *CFD is a finite resolution virtual video camera.*

1.1.2.2 Analytical Solution as an Infinite-Resolution Virtual Video Camera

Analytical solution for an unsteady problem is an infinite resolution virtual video camera. The governing equations in fluid dynamics and heat transfer are **P**artial **D**ifferential **E**quations (PDEs)—with dependent variables as flow properties, and independent variable as spatial and temporal coordinates. The analytical solution of the PDEs is a *functional relationship* of the dependent in terms of the independent variables. The solution results in flow properties at *any* value of spatial and temporal coordinates—infinite resolution virtual video camera.

Unfortunately, the analytical is not yet possible for a general fluid dynamics problem and is one of the most important unresolved problem in science and engineering. The advent of the general purpose analytical solution will lead to closure of all the CFD software companies as well as the existence of this course.

1.2 CFD: Why to Study?

Developments in the engineering simulation methodology has led to a drastic improvements in the design methodology for a product. Now, it is possible to create virtual product in a computer (using a solid modeler), analyze its performance, suggest some design modifications, and confirm enhancement in performance or reduction in cost. It provides an excellent platform for innovations in the development of the new products. Whereas, for the existing products, it leads to a tremendous improvements in the performance and efficiency. Now a days, a computer simulation and analysis have become an integral part of an engineering study.

CFD has led to a more scientific as compared to empirical approach in the design and development of a globally competitive product. Scientists as well as engineers want to use CFD as a design and optimization tool. However, due to the unaffordable computational cost, presently it is more commonly used as a powerful analysis tool. CFD is used for analysis-based improvements in the design of a fluid dynamics and heat transfer systems in various industries; such as aerospace, automobile, turbo-machinery, chemical, electrical, electronics, biomedical, etc. In future, CFD will be more often used as a design and optimization tool.

In academics, CFD is taught as an undergraduate elective and postgraduate core course in Aerospace, Chemical, and Mechanical branches of Engineering. Further-more, it taught as an interdisciplinary course in the Civil, Metallurgical, Electrical, and Biomedical Engineering. The increasing importance of CFD development, appli-cation, and analysis in the various industries as well as research organizations, along with the lack of trained manpower in this subject, has greatly increased the impor-tance of this course.

1.3   Novelty, Scope, and Purpose of This Book

From the fluid dynamics and heat transfer course, we learn to derive governing **P**artial **D**ifferential **E**quations (PDEs) by applying the mass, momentum, and energy conservation laws to an elemental fluid/solid **C**ontrol **V**olume (CV). Whereas, in CFD, the governing PDEs are mathematically operated to derive a system of **L**inear **A**lgebraic **E**quations (LAEs)—called discretization method; such as, **F**inite **V**olume **M**ethod (FVM). The system of LAEs resulting from the algebraic formulation acts as the governing equations for CFD, and consist of the flow properties (velocity, pressure, and temperature) as the unknown field variables. The system of LAEs are solved by a solution methodology. The PDE-based FVM and the mathematical coefficients of the LAEs-based solution methodology, used in almost all the books in CFD, is called here as a *mathematical approach of CFD development*; shown in Fig. 1.4.

The *novelty* in this book is a ***physical approach of CFD development***, proposed here and probably not found in any of the books on CFD. The above PDE-based FVM leads to a question: Why do we switch from the algebraic formulation of the physical laws to the differential formulation to then go back to the algebraic formulation? Why don't we keep the algebraic formulation, and convert the continuous variables in the physical laws to appropriate discrete variables using certain approximations? This led to *physical laws-based (PDE-free) FVM*, shown in Fig. 1.4. The figure also shows that the physical approach of CFD development consist of a *flux (physical quantity)-based solution methodology*. The present book is on physical law-based FVM, and presents both the flux and coefficient of LAEs-based solution methodology. Further

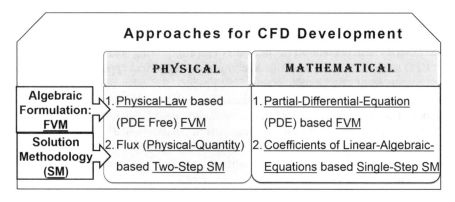

Fig. 1.4 Physical and mathematical approaches of CFD development

details and the difference between the two methodologies will be presented in the next chapter.

With a motivation to increase as well as enrich the learning of this subject which should lead to the development of a customized CFD software as a product—*be a creator than a piece of creation*—the physical approach evolved independently over my 12 years of teaching and 18 years of research experience in CFD. However, much later, the general idea of the differential formulation free discrete algebraic formulation was found in some research papers (recently by Tonti (2014). In the insightful paper, Tonti (2014) has stated:

"The "exact" solution, that is promised by the differential formulation, is hardly ever attained in practice. By contrast, the great technological progress of our days is made possible by the fact of being able to have an approximate solution, in particular a numerical solution to the problems posed by the technique. For our culture, formed on the model of infinitesimal analysis, the term "approximate" sounds like "imperfect." However, we must not forget that the objective of a numerical simulation of physical processes is the agreement with the experimental measurements and not the convergence to an analytical solution, usually not attainable. In addition, the request of reducing the error of an approximate solution does not mean make the error "arbitrarily small", as required by the process of limit, but making the error smaller than a preassigned tolerance. The notion of precision of a measuring device plays the same role as the notion of tolerance in manufacturing and the notion of error in numerical analysis."

Since the advent of computer, there is an enormous development in the discrete mathematics. However, if we change our habits of considering the differential/continuous formulation as indispensable and work on the discrete independent of the continuous mathematics, it will lead to an increase as well as enrichment in the range of applications (Desbrun et al. 2005). This will also lead to a much better understanding as well as developments of the simulation methodology in the various branches of science and engineering.

There are several classics and general purpose books on CFD such as the book by Roache (1976), Peyret and Taylor (1983), Jaluria and Torrance (**?**), Fletcher (1988), Hirch (1989,1990), and Hoffmann and Chiang (2000). Among the several very good books on CFD, would like to mention seven of them: first, by Patankar (1980), a classical and probably the most referred book; second, by Anderson et al. (1984), another classical and probably the first comprehensive book; third, by Anderson (1995), the book with probably the best writing style; fourth, by Versteeg and Malalasekera (1995), an excellent books for beginners; fifth, by Ferziger and Peric (2002), probably the most comprehensive book; sixth, edited by Muralidhar and Sundararajan (2003), another comprehensive book with sufficient details for a beginner; and seventh, by Date (2005), an outstanding book and probably with the largest number of exercise problems. The author of this book is highly inspired by the first, third and fifth book; and has contributed a chapter on the FVM (Sharma and Eswaran 2003) in a revised edition of the sixth book, using the mathematics-based approach of CFD development (Fig. 1.4).

The *scope* of this book is mainly on CFD development than on CFD applications and analysis; however, the applications and analysis are presented concisely in the example and exercise problems. This is motivated by the fact that it is more challenging and satisfying to learn to develop than to use a product. Since the present book is for an introductory course on CFD, the CFD development is limited to the *solution of the Navier-Stokes equations* for a Cartesian as well as complex geometry; the difference between the two geometries presented in the next chapter. Furthermore, the book is mostly on *numerical method* as compared to numerical analysis aspect of CFD. The present book is on *incompressible flow*, commonly encountered in the Chemical and Mechanical Engineering; however, it can be extended to the compressible flow for the Aerospace Engineering. Furthermore, the present book deals with *Newtonian fluids*, commonly encountered in Mechanical and Aerospace engineering; however, it can be extended to non-Newtonian fluids for the Chemical Engineering. FVM is the discretization method which is used in the most of the present day CFD software; thus, the present book is *mainly on FVM*. However, **F**inite **D**ifference **M**ethod (FDM) is also presented here as a topic in the chapter on essentials of numerical methods. Here, the numerical methodology is presented mostly for *2D problems* (which can be easily extended to 3D) and for *unsteady flow and heat transfer*. The present subject involves learning to develop and properly use a product—a CFD software. The details for the CFD development presented here is the main structure of the CFD software; however, the computational methodology for certain finishing touches in the software (such as graphical user interface development for user input and output) are not included here. Thus, a reader should be able to develop CFD codes which can be extended to a CFD software.

A *mind-map* for the material covered in this book is given in Fig. 1.5; inspired by Anderson (1995). The objective of this mind map is to show you a bigger picture on how the material flows in a logical fashion, before going on to the details in the following chapters. Furthermore, localized mind map is included in all the chapters after the first two introductory chapters—to demonstrate the flow of ideas in each chapter. Figure 1.5 shows that the book starts with PART I on INTRODUCTION and

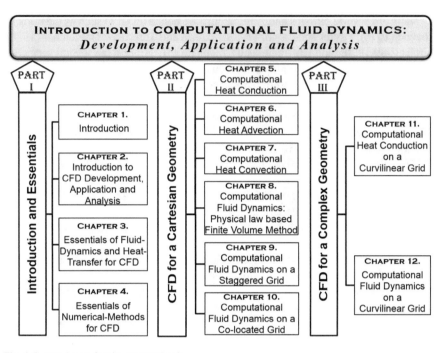

Fig. 1.5 Mind map for the present book

ESSENTIALS, discussed in the present and the next three chapters. The general introduction of CFD is presented in this chapter, and the next chapter presents an overview of CFD development, application, and analysis. Thereafter, the two chapters presents the absolutely essentials of two prerequisite subjects (for a course on CFD): fluid dynamics and heat transfer in Chap. 3, and numerical methods in Chap. 4. After building the foundation of the course in PART I, Fig. 1.5 shows that the book starts with the PART II on CFD FOR A CARTESIAN GEOMETRY, and ends with PART III on CFD FOR A COMPLEX GEOMETRY. For the PART II, it is felt better to introduce the CFD development with the flux based solution methodology (for a better appreciation of the physical law-based FVM), followed by the coefficient of LAEs-based methodology (Fig. 1.4). The PART II presents CFD development for explicit as well as implicit method, uniform as well as non-uniform grid, and staggered as well as co-located grid.

Any development activity involves learning to develop as well as test the different components, before their assembly to obtain the final product. As the product envisaged here is a CFD software, this book is structured with a module-by-module development of the numerical formulation. It presents the methodology of developing as well as testing the different components/modules/solvers of the software: diffusion and advection, and their combination as convection; presented as separate chapters of PART II. Thus, Fig. 1.5 shows that PART II of this book consists of chapters with the title as computational heat conduction (for diffusion), computational heat advection, computational heat convection, and computational fluid dynamics. The computational fluid dynamics corresponds to Navier-Stokes solver as the combination of the diffusion and advection solver, along with a source term. The physical example of the individual solver is more commonly encountered in heat transfer but not in fluid dynamics. Thus, the diffusion, advection and convection solver (presented in Chaps. 5–7 of PART II) are considered here for heat transfer; although they are also encountered (in combination) in fluid dynamics. PART II is on uniform as well as non-uniform Cartesian grid, and PART III is on the curvilinear grid; with the appropriate grid generation presented in the first chapter of PART II and III. PART III on CFD for complex geometry starts with the computational heat conduction and ends with the computational fluid dynamics.

For CFD development, there are five main steps: first, grid generation; second, finite volume method; third, solution methodology; fourth, code development; and fifth, testing. The first chapter of PART II as well as PART III of this book presents the first step. The remaining steps are discussed in the separate sections and the example problems of the various chapters of PART II and III. The computer programs corresponding to the fourth step is presented here for the basic modules on the heat conduction, advection, and convection in Chaps. 5–7; can be extended to the fluid dynamics. The programs are written using SCILAB—an open-source software for numerical computations, with an in-built graphical user interface. CFD application and analysis are presented by certain carefully designed example and exercise problems. With this three-pronged approach of development, application, and analysis, the *purpose* of this book is that the reader will learn to appreciate the theory as well as utilize it for the development of a new CFD software, or an existing one can be used intelligently for the CFD application and analysis. Thus, a reader will be firmly set on the path of becoming a CFD EXPERT.

References

1. Anderson, D. A., Tannehill, J. C., & Pletcher, R. H. (1984). *Computational fluid mechanics and heat transfer* (1st ed.). New York: Hemisphere Publishing Corporation.
2. Anderson, J. D. (1995). *Computational fluid dynamics: The basics with applications*. New York: McGraw Hill.
3. Bejan, A. (1984). *Convection heat transfer*. New York: Wiley.
4. Date, A. W. (2005). *Introduction to computational fluid dynamics*. New York: Cambridge University Press.
5. Desbrun, M., Hirani, A. N., Leok, M., & Marsden J. E. (2005). *Discrete Exterior Calculus*, Technical Report (http://arxiv.org/abs/math/0508341), California Institute of Technology.
6. Ferziger, J. H., & Peric, M. (2002). *Computational methods for fluid dynamics* (3rd ed.). Berlin: Springer.
7. Fletcher, C. A. J. (1988). *Computational techniques for fluid dynamics* (Vol. 1 and 2). New York: Springer.
8. Hirsch, C. (1989). *Numerical computation of internal and external flows* (Vol. 1). Wiley, Chichester: Fundamentals of Numerical Discretization.

9. Hirsch, C. (1990). *Numerical computation of internal and external flows* (Vol. 2). Wiley, Chichester: Computational Methods for Inviscid and Viscous Flows.
10. Hoffmann, K. A., & Chiang, S. T. (2000). *Computational fluid dynamics* (Volume 1, 2 and 3). Kansas: Engineering Education System.
11. Incropera, F. P., & Dewitt, D. P. (1996). *Fundamentals of heat and mass transfer* (4th ed.). New York: Wiley.
12. Jaluria, Y., & Torrance, K. E. (1986). *Computational heat transfer*. New York: Hemisphere Publishing Company.
13. Kimura, S., & Bejan, A. (1983). The heatline visualization of convective heat transfer. *ASME J. Heat Transfer, 105*(4), 916–919.
14. Muralidhar, K., & Sundararajan, T. (Eds.) (2003). *Computational fluid flow and heat transfer* (2nd ed.). New Delhi: Narosa Publishing House.
15. Paramane, S. B., & Sharma, A. (2009). Numerical investigation of heat and fluid flow across a rotating circular Cylinder Maintained at a Constant Temperature in 2-D Laminar Flow Regime. *Int. J. Heat Mass Transfer, 52,* 3205–3216.
16. Patankar, S. V. (1980). *Numerical heat transfer and fluid flow*. New York: Hemisphere Publishing Corporation.
17. Peyret, R., & Taylor, T. D. (1983). *Computational methods for fluid flow*. New York: Springer.
18. Roache, P. J. (1976). *Computational fluid dynamics*. Albuquerque, New Mexico: Hermosa Publishers.
19. Sharma, A., & Eswaran, V. (2003). A Finite Volume Method, Chapter 12. In K. Muralidhar, & T. Sundararajan (Eds.), *Computational fluid flow and heat transfer* (2nd ed.). New Delhi: Narosa Publishing House.
20. Tonti, E. (2014). Why starting from differential equations for computational physics? *J. Comp. Phys., 257,* 1270–1290.
21. Thomson, W. (1851). On the dynamical theory of heat, with numerical results deduced from Mr. Joule's equivalent of a thermal unit, and M. Regnault's observations on steam, *Transactions of the Royal Society of Edinburgh*.
22. Versteeg, H. K., & Malalasekera, W. (1995). *An introduction to computational fluid dynamics: The finite* (Vol. Method). Harlow: Longman Scientific & Technical.

Chapter 2
Introduction to CFD: Development, Application, and Analysis

Computational fluid dynamics involves *development* of a software, its *application* for a fluid dynamics problem (to obtain the scientifically exciting as well as engineering-relevant results), and *analysis* of the results (for a unified cause-and-effect study). A detailed overview of the development, application, and analysis aspects of CFD is presented in this chapter, to answer how CFD works?

2.1 CFD Development

CFD development involves the development of computer programs (also called codes which lead to a software) as well as testing. The development follows a numerical methodology which consists of five main steps: first, grid generation; second, discretization method; third, solution methodology; fourth, computation of engineering parameters; and fifth, testing. They are shown schematically in Fig. 2.1 and presented in separate subsections below.

CFD development has been mainly contributed by two different types of people: fluid dynamicists and computer scientists. Fluid dynamicists developed a new or an improved mathematical model and/or a numerical methodology which can give accurate results in the least computational cost. Whereas, computer scientists developed a new (or an improved) computer system and/or programming paradigm to reduce the computational cost. The computational cost involves an initial cost and a running-cost for a computational facility, where the running cost increases with the computational time.

© The Author(s), under exclusive license to Springer Nature Switzerland AG 2022
A. Sharma, *Introduction to Computational Fluid Dynamics*,
https://doi.org/10.1007/978-3-030-72884-7_2

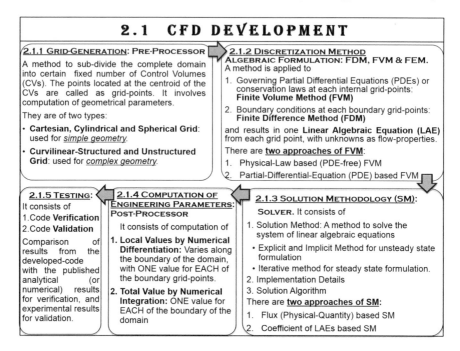

2.1   CFD DEVELOPMENT

2.1.1 GRID-GENERATION: PRE-PROCESSOR

A method to sub-divide the complete domain into certain fixed number of Control Volumes (CVs). The points located at the centroid of the CVs are called as grid-points. It involves computation of geometrical parameters.

They are of two types:

- **Cartesian, Cylindrical and Spherical Grid:** used for *simple geometry*.
- **Curvilinear-Structured and Unstructured Grid:** used for *complex geometry*.

2.1.2 DISCRETIZATION METHOD
ALGEBRAIC FORMULATION: FDM, FVM & FEM.
A method is applied to

1. Governing Partial Differential Equations (PDEs) or conservation laws at each internal grid-points: **Finite Volume Method (FVM)**
2. Boundary conditions at each boundary grid-points: **Finite Difference Method (FDM)**

and results in one **Linear Algebraic Equation (LAE)** from each grid point, with unknowns as flow-properties.

There are **two approaches of FVM**:

1. Physical-Law based (PDE-free) FVM
2. Partial-Differential-Equation (PDE) based FVM

2.1.5 TESTING:

It consists of

1. Code **Verification**
2. Code **Validation**

Comparison of results from the developed-code with the published analytical (or numerical) results for verification, and experimental results for validation.

2.1.4 COMPUTATION OF ENGINEERING PARAMETERS:
POST-PROCESSOR

It consists of computation of

1. **Local Values by Numerical Differentiation:** Varies along the boundary of the domain, with ONE value for EACH of the boundary grid-points.
2. **Total Value by Numerical Integration:** ONE value for EACH of the boundary of the domain

2.1.3 SOLUTION METHODOLOGY (SM):

SOLVER. It consists of

1. Solution Method: A method to solve the system of linear algebraic equations
   - Explicit and Implicit Method for unsteady state formulation
   - Iterative method for steady state formulation.
2. Implementation Details
3. Solution Algorithm

There are **two approaches of SM**:

1. Flux (Physical-Quantity) based SM
2. Coefficient of LAEs based SM

Fig. 2.1 Schematic description of five different steps of CFD development

2.1.1   Grid Generation: Pre-Processor

The first step on grid generation is a discretization procedure of a flow/computational domain, where an infinite number of points in the domain is divided into certain discrete points. The points are joined by straight lines in a certain fashion, such that they form the corners of *contiguous non-overlapping* **C**ontrol **V**olumes (CVs); also called *cells*. Alternatively, Fig. 2.1 presents grid generation as a method to subdivide the domain into certain fixed number of sub-domains/CVs—points located at the centroid of the CVs is called grid points (or cell centers).

The simplest example of grid generation is shown in Fig. 2.2, for a simple Cartesian geometry problem on 2D heat conduction in a square plate. A uniform Cartesian grid generation is shown in the figure as a procedure of drawing equi-spaced horizontal and vertical lines. The lines subdivide the computational domain into sub-domains, called CVs. The grid lines correspond to a constant value of Cartesian coordinate; $x = $ constant for the vertical lines, and $y = $ constant for the horizontal lines. The points located at the centroid of the CVs are called internal grid points. The figure also shows grid points located at the boundary of the domain, called boundary grid points. Similar Cartesian grid generation, for the square plate with a square hole, is shown in Fig. 2.3a. Whereas, for the square plate with a circular hole, a curvilinear (piecewise-linear curve) grid generation is shown in Fig. 2.3b.

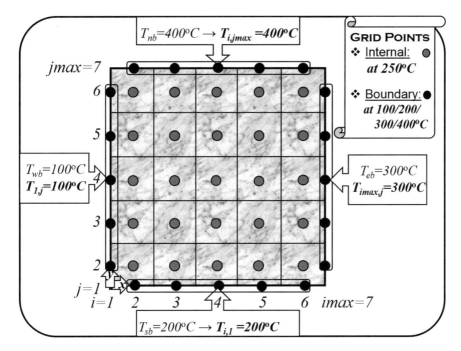

Fig. 2.2 A uniform Cartesian grid generation for 2D heat conduction in a plate. The plate is taken from a furnace at a temperature of 250 °C and subjected to a boundary condition of 100 °C, 200 °C, 300 °C, and 400 °C on the west, south, east, and north boundaries, respectively

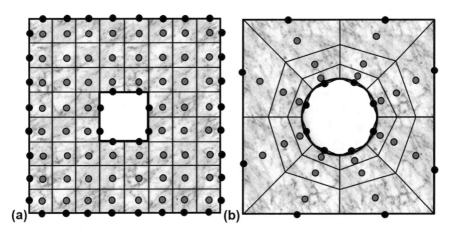

Fig. 2.3 2D heat conduction in a square plate with a (**a**) square hole and (**b**) circular hole: (**a**) Cartesian and (**b**) curvilinear grid generation

One of the classifications of a CFD problem is based on the geometry (or shape) of computational domain: simple and complex. If all the boundaries of the computational domain are aligned along *one* (not any) of the three *standard* coordinate systems: Cartesian, Cylindrical, and Spherical, they are called simple geometry problem; otherwise called complex geometry problem. For the 2D heat conduction in a square plate with a square hole (Fig. 2.3a), the computational domain consists of two boundaries: inner (hole) and outer; both are squares. Since all the boundaries of the domain are aligned along the Cartesian coordinate direction, it becomes a Cartesian geometry problem. Similarly, 2D heat conduction in a circular plate with a circular hole is a Cylindrical geometry problem. Both the Cartesian and Cylindrical geometry problems are simple geometry problem. Whereas, 2D heat conduction in a square plate with a circular hole (Fig. 2.3b), and circular plate with a square hole, are the complex geometry problem. For the two types of geometry of the domain, the procedure for the grid generation is different (Fig. 2.1): Cartesian, Cylindrical, and Spherical grid for the simple geometry, and curvilinear-grid for the complex geometry.

The grid generation procedure is not only limited to pictorial discretization (or subdivision) of computational domain to a certain fixed number of CVs (or grid points). It also involves mathematical computation of coordinates of the vertices of the CVs, width of the CVs, a distance of a cell center from its immediate neighboring cell centers, and surface area as well as volume of the CVs—called as geometrical parameters; their role will be discussed below.

2.1.2   Discretization Method: Algebraic Formulation

Discretization method is an *approximate* mathematical method to convert a governing PDE or a BC into a **L**inear **A**lgebraic **E**quation (LAE); converting the problem of calculus to that of algebra. Considering the applicability of the PDE to the interior of the domain, the method is applied to PDE at each internal grid point; furthermore, the method is applied to BCs at each boundary grid point, as the BCs correspond to the boundary of the domain. The internal and boundary grid points are shown in Fig. 2.2. Thus, the application of a discretization method leads to one LAE for each grid point. This results in a closed system of LAEs, with the number of LAEs equal to a total number of internal and boundary grid points.

There are three types of discretization methods: **F**inite **D**ifference **M**ethod (FDM), **F**inite **V**olume **M**ethod (FVM), and **F**inite **E**lement **M**ethod (FEM). FDM is the oldest while FVM is a relatively recent method in CFD; whereas, FEM is more commonly used in computational solid-mechanics as compared to fluid mechanics. With an advancement in CFD, the FVM as compared to the FDM was proposed as a better alternative for solving the complex geometry problems. However, the FVM is used only for internal grid points and the FDM is still applied to the boundary grid points (Fig. 2.1) in most of the present-day CFD software; and will be presented in this book. For example, considering the grid points in Fig. 2.2, 25 LAEs are obtained for a temperature at 25 internal grid points by applying the FVM for the 2D heat conduction. While applying the BCs using FDM, 20 LAEs are obtained for the temperature at the 20 boundary grid points; as a constant value for the constant wall temperature BC seen in the figure. Note that the coefficients of the FVM-based LAEs consist of the thermo-physical properties and the geometrical parameters calculated in the first step of grid generation.

Mathematical modeling in fluid dynamics is based on mass, momentum, and energy conservation laws, where each conservation law leads to a governing PDE for a particular flow property. For example, energy conservation law leads to derivation of energy equation as the equation for temperature. The law consist of a term as rate of change of energy/enthalpy which indicates that it can be used for CFD-based generation of movie for temperature. Similarly, for each flow property, there a governing PDE/conservation law which is converted into a system of LAEs in CFD; with unknowns as the flow properties. Note that the number of *system* of LAEs is equal to the number of PDEs/conservation laws/flow property. This ensures that an approximate solution for all the flow properties at internal as well as boundary grid points can be obtained, using a solution methodology presented in the next subsection.

Figure 2.1 shows the second step on **F**inite **V**olume **M**ethod (FVM) as a procedure for discrete algebraic representation of either the governing PDEs or the physical (mass, momentum, and energy conservation) laws. The respective methods are called here as *PDE-based FVM* and *physical law-based (PDE-free) FVM*; introduced in Fig. 1.4 earlier and discussed separately below.

2.1.2.1 Partial Differential-Equation-based FVM

Figure 2.4 shows that the governing PDEs are obtained by applying the conservation laws on a CV, dividing by the volume ΔV of the CV, and using the limit $\Delta V \rightarrow 0$; this is done in an undergraduate course on fluid mechanics and heat transfer. Almost all books on FVM (such as Patankar (1980)) start with the PDEs, present FVM by first applying volume-integral to the PDEs, then using Gauss divergence theorem, and finally applying certain approximations; this is called here a PDE-based FVM (Fig. 2.4b).

2.1.2.2 Physical law-based FVM

Using the same approximations, the same LAEs can be directly obtained from an algebraic formulation of the conservation laws applied to a CV; this is called here as a physical law-based FVM (Fig. 2.4a). Note that the mathematical form of the application of the conservation laws are divided by the volume ΔV of the CV to derive the PDEs in the fluid dynamics and heat transfer course; whereas, in a CFD course, the PDEs are somewhat multiplied by the volume (volume integral in the PDE-based FVM; Patankar (1980)). The application of the volume integral and

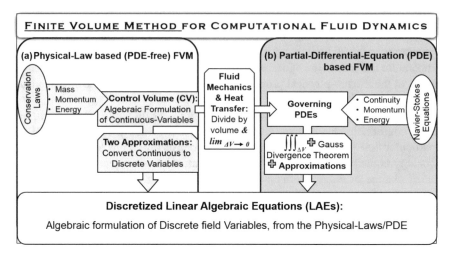

Fig. 2.4 Schematic representation of the two approaches of FVM for CFD

Gauss divergence theorem is avoided in the physical law-based FVM which starts with a picture (CV) as compared to equations (PDEs) in the PDE-based FVM.

Over the last *three centuries*, the tradition to mathematically describe physical laws in terms of differential equations hinders the consideration of any other possible descriptions (Tonti 2014). The *addiction* is such that it was preferred to discretize the differential equations, rather than considering other more convenient tools, for an algebraic formulation of physical laws (required by a computer) since the advent of CFD. However, a beginner in CFD might ask: Why do we divide the algebraic formulation of the conservation law by the volume of the CV, and then take volume integral to go back to the algebraic formulation. Why do we follow this convoluted process? Why don't we keep the algebraic formulation from the outset? This is shown in Fig. 2.4a, as the two-approximations-based direct algebraic formulation of the conservation laws; thus, bypassing the PDE. The approximations are used to convert the continuous physical variables in the laws to a discrete variable; the details on the approximations are presented in the next part of this book.

2.1.3 Solution Methodology: Solver

Figure 2.1 shows that the number of LAEs for a flow property (after the second step) is equal to the number of grid points (in the first step), and is solved using a detailed procedure called solution methodology in the third step of CFD development. The figure shows that the solution methodology consists of three steps: first, *solution method*; second, *implementation details*; and third, *solution algorithm*. The solution method is an *explicit and implicit methods* for an unsteady-state formulation, and an *iterative method* for a steady-state formulation; the different methods will be presented in the following chapters. Implementation details is a detailed procedure to convert the solution method and the algorithm to a computer program—demonstrated with the help of data structures, the programming loops, computational stencil, and figures in the different chapters of this book. Solution algorithm is a step-by-step procedure to develop a computer program.

First picture of a fluid dynamics movie of flow properties is always known—from the initial and the boundary condition. This corresponds to 250 °C for the internal grid points, and 100 °C/200 °C/300 °C/400 °C for the boundary grid points in the example problem (Fig. 2.2). Solution methodology is such that the temperature for the first picture (at a previous or old time instant t) is used to obtain the second picture (at a present or new time instant $t + \Delta t$), with a time interval (or time step) step Δt between the two pictures. For the temperature contours in the conduction problem, the time marching of the picture continues from the second to obtain the third picture, from the third to the fourth picture, and so on—the picture-by-picture generation leads to a movie of the temperature field. The time marching in a movie is stopped if there is a negligible change between two consecutive pictures in case of a steady-state problem; implemented by using certain stopping criterion in an unsteady formulation.

Almost all the books on CFD (such as Patankar (1980)) follows a single-step solution methodology where the final system of LAEs are solved for flow properties (temperature for heat conduction); this is called here as a *coefficients of LAEs-based single-step solution methodology*. For unsteady-state problems solved by an *explicit method*, a *flux (physical quantity)-based two-step solution methodology* is proposed here. It involves a two-step implementation of the physical law-based FVM for a CV; discussed below.

For an *explicit method*, the *implementation* of the two approaches of the solution methodology are shown in Fig. 2.5; presented for any one of the 25 internal grid points considered in the unsteady heat conduction problem (Fig. 2.2). Figure 2.5 shows the implementation for the grid point P of a representative CV, with adjoining north, south, east, and west neighboring grid points as N, S, E, and W, respectively. Furthermore, for each of the two approach, Fig. 2.6 shows three different *computational stencils*. The stencil is a graphical representation of all the grid points involved in the computation of flow properties at a particular grid point. The two types of solution methodology and the associated computational stencils are discussed in separate subsections below.

2.1.3.1 Coefficients of LAEs-based Solution Methodology

For the coefficients of LAEs-based methodology, Fig. 2.5a shows a single-step solution of the LAE for $T_P(t + \Delta t)$ (temperature of the grid point P at the present time instant $t + \Delta t$) which is commonly expressed in terms of the temperature of same grid point (T_P) and its adjacent neighboring grid points (T_E, T_W, T_N, T_S) at previous time instant t. For the 2D heat conduction, note that the discretization leads to a LAE,

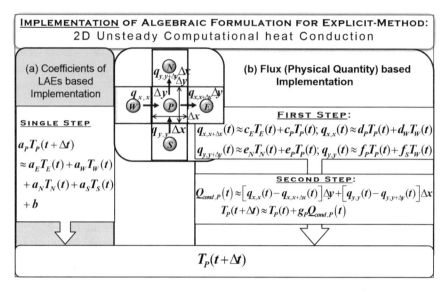

Fig. 2.5 Symbolic representation of the implementation of the explicit method-based FVM, for the two approaches of solution methodology. Coefficients of the algebraic equations, obtained by the FVM, are represented here as a to f with subscripts

where the temperature at a particular grid point is a function of only four adjoining (not all) grid points in the domain. Furthermore, for an explicit method, the LAE consist of only one unknown for each LAE, and each LAE in the system of LAE can be solved independently.

2D matrix is used as the data structure to store the variables encountered for a 2D CFD (conduction here) problem; its element tagged (or addressed) by an integer value i for row and j for column of the matrix. Here, they are called running indices, shown in Fig. 2.6, with increasing i in the x−direction and increasing j in the y−direction. The figure shows the computational stencils for three representative grid points: (i, j) equal to $(2, 2)$, $(4, 4)$, and $(6, 6)$); however, note that the stencil is applicable to all the 25 internal grid points. During implementation of a solution method in a computer program, the subscript P, E, W, N, and S (Fig. 2.5) are replaced by the running indices as (i, j), $(i + 1, j)$, $(i − 1, j)$, $(i, j + 1)$, and $(i, j − 1)$, respectively (Fig. 2.6). Figure 2.6a shows an outer-loop of $j = 2$ to $(jmax − 1)$ and an inner-loop of $i = 2$ to $(imax − 1)$ as well as a generalized equation for the application of the LAE to each of the internal grid points. The recursive calculation along with in-built neighbor information (due to 2D matrix as the data structure) makes it well suited for programming, and then solving in a computer.

The solution methodology is seen in Fig. 2.6 as the temperature from the first picture (at $n = 1$ & $t = 0$) is used to obtain second picture (at $n = 2$ & $t = \Delta t$) for temperature. Here, n is an integer value which represents a running index in the temporal coordinate; analogous to i and j in the spatial Cartesian coordinates. For the coefficients of LAEs-based methodology, Fig. 2.6a shows that the five values of temperature are picked up from the first picture, multiplied by their respective coefficients (denoted by $a's$ in Fig. 2.5a) to compute the temperature at a particular internal grid point in the second picture; this is repeated for all the internal grid points.

2.1.3.2 Flux (Physical-Quantity)-based Solution Methodology

For the flux-based methodology, Fig. 2.5b shows computation of heat flux (at the various surfaces of the CV) as the first step, and then computation of $T_P(t + \Delta t)$ (at the centroid of the CV) as the second step. The figure shows that the approximated LAEs in the first step involves the temperatures of previous time instant (t), at the grid points which are across a surface of a CV. For example, it can be seen that the LAE for computation of the flux $q_{x+\Delta x}$ (at surface $x + \Delta x$ of a representative CV P) involves $T_E(t)$ and $T_P(t)$. The figure also shows that the fluxes computed in the first step are used to obtain net rate at which the conducted heat is entering the CV $Q_{cond}(t)$; and then to compute $T_P(t + \Delta t)$ (as a function of $T_P(t)$ and $Q_{cond}(t)$) in the second step. Note that the coefficients for the LAE in Fig. 2.5b are shown symbolically, and their exact mathematical expression with be presented in the following chapters.

The computational stencils in Fig. 2.6b shows that the two values of temperature (across all the face centers of the CVs) are picked up multiplied by their respective coefficients to compute heat flux at the various face centers in the first step; shown by a symbol which is square and inverted-triangle for computation of q_x and q_y at the various face centers, respectively. In the second step, the figure shows that the four heat fluxes (corresponding to the four face centers) in a 2D CV are picked up to compute the net heat in-flux Q_{cond} and compute the temperature $T(t + \Delta t)$ at the cell center for the second picture. Note that the figure is shown only for the explicit method, where the nature of system of LAEs are such that there is only one unknown in each LAE.

For the *explicit method*, note that the solution methodology is such that the picture (to create a movie) of temperature field for the present (or new) time instant $t + \Delta t$ is obtained using the temperature field in the picture of the previous (or old) time instant t. Since the solution of any system of LAEs is a general mathematical problem, the single-step coefficients of LAEs-based solution methodology lead to the mathematical approach of CFD development. Whereas, the two-step method involves step-by-step implementation of energy conservation law-based FVM; presented later in the Chap. 5. Thus, the physical law based FVM and the flux-based methodology leads to the *physical approach of CFD development* (Fig. 1.4).

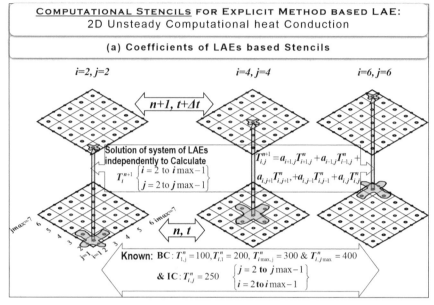

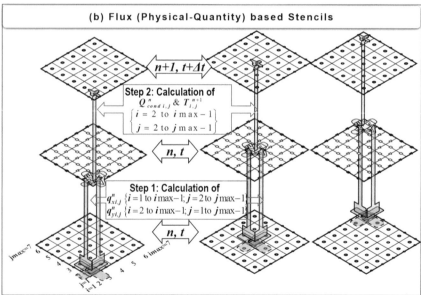

Fig. 2.6 Computational stencils for the two different approaches of solution methodology

2.1.4 Computation of Engineering Parameters: Post-Processor

After the first step on the grid generation, as pre-processor; and the second step on discretization and the third step on solution methodology, as the main-process; Fig. 2.1 shows the fourth step on computation of engineering parameters, as the post-processor. This is because the engineering parameters are obtained after computing the flow property in the first three steps (presented above). *Local value* of the parameters are computed at all the boundary grid points, using *numerical differentiation*. This involves computation of a first derivative of a flow property, using the boundary as well as nearby internal grid point values of flow properties. The local boundary grid point values are used to compute *total value* of an engineering parameter at a boundary, using *numerical integration*.

Figure 2.1 shows the engineering parameters of two types: total and local; with one value for total and spatial variation for local engineering parameters, at a boundary of the domain. The local-value corresponds to a flux term, while the total-value corresponds to the surface integral values of the flux at a boundary. Considering the 2D heat conduction problem in a square plate (Fig. 2.2), the conduction heat flux and heat transfer rate from the various surfaces of the plate are an example for the local and the total value of the parameters, respectively. The respective parameters are shear-stress and viscous-force in a fluid mechanics problem. The local conduction-fluxes are computed for various boundary grid points in Fig. 2.2, and the total heat transfer rate corresponds to one value for each surface/boundary.

It is interesting to note that the results for the spatial variation of the local engineering parameters is one dimension less than that for the flow properties. This is because the flow properties are the volumetric-information and the local engineering parameters are the surface information, computed inside and at the boundary of the computational domain, respectively. For example, considering the heat conduction problem (Fig. 2.2), the temperature distribution inside the plate $T(x, y)$ is 2D while the local heat flux distribution at a boundary is 1D; $q_x(y)$ for the east and west boundary, and $q_y(x)$ for the north and south boundary. This is also true in case of a 3D fluid dynamics problem, with 3D flow field and 2D variation of the local engineering parameters.

2.1.5 Testing

A CFD code or software developed—using the above four steps—is finally tested in the last step of CFD development, shown in Fig. 2.1. Testing is done by setting up and running the code for certain problems (called *benchmark* problems), for which accurate numerical or experimental results are available in the published literature.

Testing the accuracy of a computer simulation as compared to published analytical/numerical results is called *verification*; and the comparison with the experimental results is called *validation* of a CFD code. Verification verifies the correctness of the programming and computational implementation of a mathematical/computational model. Whereas, the validation confirms the agreement of the computer simulation with the physical reality; within certain acceptable level of inaccuracy. Verification is a mathematics and validation is a physics issue (Roache 1998).

2.2 CFD Application

CFD application involve proper computational set-up (called *pre-processor*) and running of an already developed (using the various steps in Fig. 2.1) in-house CFD code or a commercial software, for a fluid dynamics as well as heat and mass transfer problem. This is shown in Fig. 2.7 for three different problems: first, conduction heat transfer; and second and third on heat and fluid flow across a cylinder and in a plane channel, respectively. Note that the second problem is external flow problem while the third problem is an internal flow problem.

All the fluid dynamics and heat transfer problems obey the same conservation laws. However, fluid dynamic movies for each problem is different due to three main differences: first, shape and size of the flow region/domain; second, fluid properties; and third, initial and boundary conditions. Flow properties inside the domain for the first time instant/picture of a fluid dynamics movie is the initial condition;

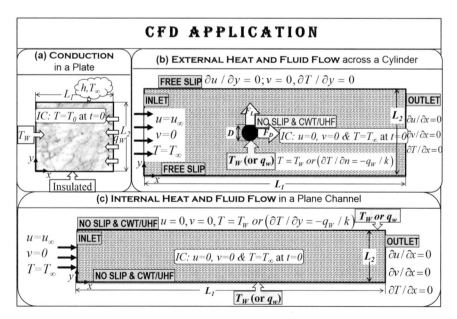

Fig. 2.7 Computational domain, consisting of initial and boundary conditions, for application of a CFD code (or software) on three representative problems

and the prescription of the properties all over the boundary of the domain for the complete time duration is the boundary condition. They are called user-inputs (for the computational setup of a problem), which act as a jury to the problem and dictate the solution. Thus, it is almost impossible to completely automate a general-purpose CFD software.

Some of the user inputs—on the shape/size of the domain and initial/boundary conditions—are shown in Fig. 2.7, for some of classical problems in fluid dynamics and heat transfer. For 2D heat conduction in a rectangular plate, Fig. 2.7a shows that the plate is taken from a furnace at a uniform temperature of T_0. It is subjected to different thermal boundary conditions: constant wall temperature on the west boundary, insulated on the south boundary, uniform heat flux on the east boundary, and convective BC on the north boundary. Similarly, for the two types of flow in fluid dynamics, the initial condition and the various types of boundary conditions are shown in Fig. 2.7b, c. The external flow across a cylinder consists of two boundaries: inner circular and outer rectangular; with the flow domain in between the two boundaries. Whereas, the internal flow in a plane channel consists of a rectangular boundary only. For both the flows, no-slip for velocity and a **C**onstant **W**all **T**emperature (CWT) (or **U**niform **H**eat **F**lux (UHF)) BCs are applied at the solid wall. Furthermore, the BC at the inlet is taken as uniform velocity and temperature, and the fully developed BC is applied at the outlet of the domain. Although the inlet conditions can also be applied to the top and bottom boundaries of the external flow, the free-slip BC is considered to be better (in terms of computational cost) at the transverse boundary.

For the BCs shown in Fig. 2.7b, c, note that all the boundaries should be sufficiently far away (from the object of interest; cylinder here) for the external flow, and the right boundary should be sufficiently far from the inlet for internal flow; to obtain an accurate numerical result. This is because an infinite flow domain in the external flow is taken as finite; and sufficient streamwise distance (downstream of the cylinder for external and length of the channel for internal flow) is needed to apply the fully developed BC at the outlet. The location of a boundary is obtained from a *domain-size independence study*. This is done to ensure that there is *almost* no change in the numerical results with further increase in the size of the computational domain. Other than the IC and BCs, there are other user inputs in the CFD application such as thermo-physical properties, governing parameters, convergence criterion, and steady-state criterion; these will be presented in a later chapter.

The overall flow chart for the application of a CFD code/software is shown in Fig. 2.8. For the explicit method-based application on the unsteady heat conduction problem (Fig. 2.2), the figure shows six steps: (i) initialization for user input along with the grid generation; (ii) the solution control for the proper input value of time step and an under-relaxation factor; (iii) monitoring the solution which corresponds to the set-up for monitoring the maximum the temperature difference between two consecutive time instant at all the grid points (Fig. 2.2); (iv) CFD calculation, corresponds to the solution of system of LAEs for temperature at the grid points; (v) check for convergence, corresponds to checking that the monitored temperature difference approaches practically zero (such as 10^{-3} or 10^{-6}; small numerical value prescribed

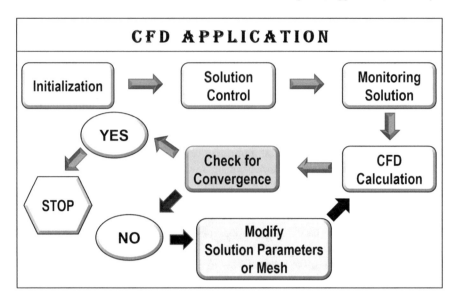

Fig. 2.8 Flow chart for the application of a CFD code (or software)

as a stopping criterion) for a steady-state problem; and (vi), if converged, stop the timewise computation otherwise reduce the time step or improve the quality of the mesh/grid and perform the CFD calculation again.

2.3   CFD Analysis

CFD analysis involves two things: first, proper collection of spatial and temporal variation of the flow properties and engineering parameters; and, second, analyzing the results after generating various figures and movies for the problem.

The analysis is presented above in Sect. 1.1.1, for a heat transfer and a fluid dynamics problems. It is demonstrated and discussed that a synchronized movie, for flow properties and engineering parameters, is needed for a unified cause-and-effect analysis of the results. For example, for vortex-shedding flow across a cylinder, a synchronized movie for the fluid flow properties (velocity vector, pressure contour, streamline, and vorticity contour) is represented as a pictures and the lift/drag force as symbols (over the line plot) at the same four discrete time instants in Fig. 1.3. Similar representation of the movie for the heat flow properties (isotherms and heatlines) and the rate of heat transfer is shown in Fig. 1.2, for conduction heat transfer in a thermal module of a heat sink. Both the figures represent a timewise synchronized change in the picture for the flow properties, and the symbols over the line plot for the engineering parameters.

2.4 Closure

Now you are firmly set and motivated to forge ahead to the remaining chapters of this book, after the eventful, thoughtful, and intuitive discussion on the first three questions on CFD: What is CFD? and Why to study CFD? in the previous chapter; and How CFD works? in this chapter.

References

Patankar, S. V. (1980). *Numerical Heat Transfer and Fluid Flow*. New York: Hemisphere Publishing Corporation.
Tonti, E. (2014). Why starting from differential equations for computational physics? *Journal of Computer Physics, 257*, 1270–1290.

Chapter 3
Essentials of Fluid Dynamics and Heat Transfer for CFD

Physical concepts and mathematical details, of fluid dynamics and heat transfer, which are considered extremely essential prerequisite for CFD are presented in this chapter. One of the essentials in fluid dynamics is a characterization, called property, which is of two types: fluid and flow. *Fluid property* is a thermophysical property of a fluid; such as kinematic viscosity and thermal diffusivity, which are physically interpreted in terms of momentum and heat penetration depth, respectively. Whereas, *flow property* is a variable which characterizes the nature of flow and are of two types: fundamental (also called as primitive in CFD) and derived (i.e., obtained from the fundamental). Velocity, pressure, and temperature are the fundamental properties and stream-function and vorticity are the derived (from velocity) properties, of a flow field.

The mind map for this chapter is presented in Fig. 3.1. The figure shows that the flow of essential ideas in fluid dynamics starts with the two types of physical laws: fundamental (or conservation) and subsidiary. This is followed by certain physical concepts on the relevant mechanisms for the transport of momentum and energy. Thereafter, a physical law-based differential formulation is presented for continuity equation from the mass conservation law; and for transport equations from the momentum and energy conservation laws (certain subsidiary laws are additionally required for the derivation). A term-by-term (unsteady, advection, diffusion, and source) derivation of the transport equations are presented in this chapter; later extended to a module-by-module development of a numerical methodology for a CFD software. After the laws and the formulations, the structure of the present chapter in Fig. 3.1 corresponds to a presentation of a generic form of the various volumetric and flux terms (encountered during the formulation), and their differential formulation. Finally, a mathematical formulation is presented first for a dimensional study and then for a non-dimensional study. It starts with the derived governing partial differential equations and is followed by initial condition and different types of boundary conditions; as the independent variables in the PDEs are temporal and spatial coor-

© The Author(s), under exclusive license to Springer Nature Switzerland AG 2022   35
A. Sharma, *Introduction to Computational Fluid Dynamics*,
https://doi.org/10.1007/978-3-030-72884-7_3

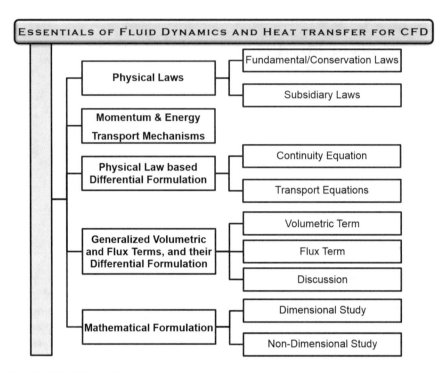

Fig. 3.1 Mind Map for Chap. 3

dinates. The formulation ends with the governing and engineering parameters for a representative external and an internal fluid dynamics problem.

3.1 Physical Laws

For fluid dynamics and heat transfer, the laws are of two types: fundamental and subsidiary. Law of conservation of mass, momentum, and energy are the fundamental laws; and Fourier's law of heat conduction and Newton's stress strain-rate relationship are the subsidiary laws.

3.1.1 Fundamental/Conservation Laws

Since the fluid dynamics movies are desired in CFD, the conservation laws (the mother of all the problems in fluid dynamics) considered are for the rate (timewise varying) process Thus, at a time instant t, Fig. 3.2 shows that the laws are stated as

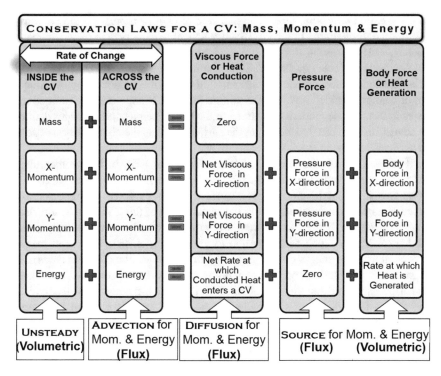

Fig. 3.2 A schematic representation of conservation laws, used in fluid dynamics as well as heat transfer; customized for CFD

the rate of change of mass is equal to zero, the rate of change of momentum is equal to force (Newton's II law of motion), and the rate of change of energy is equal to the difference between the rate at which heat is gained and work is done. These laws are shown in Fig. 3.2, for a fixed elemental control volume-based Eulerian approach in fluid dynamics; assuming that the work done is equal to zero.

These laws were originally proposed, with a Lagrangian description (following the path of individual particle) in physics. However, in fluid dynamics, it is extremely difficult to keep track of the flowing fluid particles as they continually deform and interact with adjoining particles. The interaction is also difficult to describe or model mathematically. Thus, an Eulerian description is preferred for a fluid flow where a finite control volume is defined, with a flow across the CV. Thus, instead of position vector for each particle in the Lagrangian description, the Eulerian description defines the flow properties as field variables; functions of space and time. The change from a fixed mass in the Lagrangian to a fixed volume in the Eulerian description results in a modification of the rate of change term, with two terms instead of the one term; the other terms on the RHS remains same (refer Fig. 3.2).

Figure 3.2 shows the instantaneous rate of change of mass, momentum, and energy as two separate terms: first, the rate of change for the fluid stored *inside* the CV—called *unsteady term*; and second, the rate of change due to the fluid flowing *across* the CV—called *advection term*. The *first rate of change* represents the variation of mass/momentum/energy in the CV (fixed in space) *with respect to time*; whereas, the *second change* represents the variation *with respect to spatial position* (corresponding to the various surfaces) in the CV. Thus, the first one leads to an Eulerian temporal derivative and the second one leads to a spatial derivative; presented below. Since the rate of change "inside" involves the volume of the CV, the variables defined in the unsteady term are dependent on volume of the CV. Whereas, the variables defined in the advection term are dependent on the surface area, as the change occurs "across" the surfaces enclosing the CV.

After the LHS on the rate of change, refer to the RHS of the momentum and the energy conservation law in Fig. 3.2. For the momentum conservation, the figure shows two different types of forces: surface and body force. Surface forces are those forces which are dependent on the surface area of the CV; such as the viscous force and pressure force. Body forces are those forces which are dependent on the volume of the CV; such as forces due to gravitation, electric field, and magnetic field. Similarly, for energy conservation, the figure shows two different types of heat transfer rate: conduction and heat generation, which are, respectively, dependent on the surface area and the volume of the CV. The above-discussed surface-area and the volume-dependent terms are called *flux term* and *volumetric term*, respectively. They are shown in Fig. 3.2, with the unsteady as the volumetric term, advection and diffusion as the flux term, and the source as volumetric term (heat generation) for the energy conservation and both flux (pressure) and volumetric (body force) term for the momentum conservation.

Notice the similarity in the momentum and the energy conservation law in Fig. 3.2. Both the law consist of four terms: first, unsteady; second, advection; third, diffusion (for the viscous forces and the conduction heat transfer); and fourth, source (the remaining terms).

3.1.2   Subsidiary Laws

The spatially continuous mathematical description of the flow as well as fluid properties requires a *continuum approximation*—the properties are defined for the smallest magnitude of volume of fluid before any statistical fluctuations become significant. However, in view of the particulate nature of fluid, some of the microscopic information (corresponding to a random molecular motion-driven diffusion transport mechanism) are lost by the continuum approximation; and are recovered by certain laws, called subsidiary laws. For extremely small (micro or nano)-scale fluid dynamics and heat transfer, although the conservation laws continue to remain applicable, the continuum approximation becomes weak and one should come up with a new or modified subsidiary law.

For heat transfer, the subsidiary law corresponds to Fourier's law of heat conduction; given for the conduction flux in a 2D Cartesian coordinate system as

$$\vec{q} = -k\nabla T \Longrightarrow q_x = -k\frac{\partial T}{\partial x} \text{ and } q_y = -k\frac{\partial T}{\partial y} \tag{3.1}$$

For an *incompressible*, *isotropic*, and *Newtonian* fluid flow, the subsidiary law corresponds to Newton's stress strain-rate relationship; given for a 2D Cartesian coordinate system as

$$\sigma_{xx,\text{total}} = -p + 2\mu\dot{\varepsilon}_{xx} = -p + \sigma_{xx}$$
$$\sigma_{yy,\text{total}} = -p + 2\mu\dot{\varepsilon}_{yy} == -p + \sigma_{yy}$$
$$\sigma_{xy} = \sigma_{yx} = 2\mu\dot{\varepsilon}_{xy}$$

where the total-stress $\sigma_{xx,\text{total}}$ and $\sigma_{yy,\text{total}}$ includes both pressure and viscous stress; and σ_{xx}, σ_{yy}, σ_{xy}, and σ_{yx} are the viscous stresses. The strain-rate $\dot{\varepsilon}$ in the above equation is converted in terms of velocity gradients, using flow kinematics (Cengel and Cimbala 2006). The resulting expression for viscous stress is given as

$$\sigma_{xx} = 2\mu\partial u/\partial x, \ \sigma_{yy} = 2\mu\partial v/\partial y$$
$$\sigma_{xy} = \sigma_{yx} = \mu\left(\partial v/\partial x + \partial u/\partial y\right) \tag{3.2}$$

The conduction flux (Eq. 3.1) and viscous stresses (Eq. 3.2) are involved in the diffusion term (Fig. 3.2); and are called diffusion flux. Note that the above subsidiary laws result in the diffusion flux which is directly proportional to a gradient of a flow property, with a constant of proportionality called diffusion coefficient. For the energy conservation law, the conduction flux is the diffusion flux and thermal conductivity is the diffusion coefficient; whereas, the respective terms are viscous stress and dynamic viscosity for the momentum conservation law.

3.2 Momentum and Energy Transport Mechanisms

The momentum and energy conservation laws represent a transport process— transportation of heat-energy and momentum (of a fluid) in a heat and fluid flow domain. Thus, the underlying mechanism to drive the momentum as well as energy transport process is same: diffusion and advection. An excellent discussion on the transport mechanism is given in a classical book on heat transfer by Incropera and Dewitt (1996). The diffusion occurs due to random microscopic motion of the molecules of a solid/fluid, while the advection occurs due to the bulk or macroscopic motion of a fluid. The advection refers to transport due to the collective motion of a large number of molecules, i.e., onset of flow. The molecules in the flowing fluid retain their random motion; thus, both advection as well as diffusion occurs in a flowing fluid—the cumulative transport is called convection. The diffusion mechanism-based molecular heat flux results in the conduction heat transfer rate and molecular momentum flux

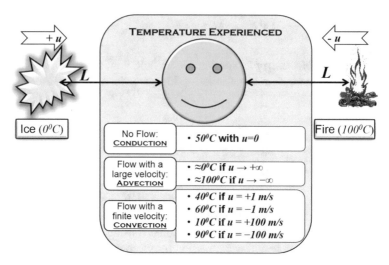

Fig. 3.3 A physical example for the *one-dimensional* heat transport mechanism: a person standing exactly in-between ice and fire, with a uniform flow from the ice or fire side

results in the surface forces (viscous as well as pressure); refer Fig. 3.2. Whereas, the advection mechanism-based bulk heat and fluid flow result in the enthalpy and momentum transport across the flow field, under the presence of temperature gradient and velocity gradient, respectively. The effect of the transport mechanism for one-dimensional flow and heat transfer (with *negligible radiation*) is presented in Fig. 3.3. The figure shows a physical example—a person standing exactly in-between ice at $0\,°C$ and fire at $100\,°C$. Furthermore, depending on the magnitude and direction of the flow, the figure shows that the various mechanisms lead to three different modes of heat transfer. *First*, no-flow results in *pure conduction/diffusion*, and the temperature experienced is $50\,°C$ (mean of $0\,°C$ and $100\,°C$ — the equal effect of ice and fire). *Second*, for an extremely large velocity of the flow results in *pure advection*, the bounding ice or fire temperature is almost experienced depending on the flow direction, i.e., *close to* 0 and $100\,°C$ for the flow from the ice and fire side, respectively. *Third*, intermediate velocity of the flow results in *convection*, and Fig. 3.3 shows that the temperature experienced is 40 and $10\,°C$ if the flow from the ice side is at a uniform velocity of 1 m/s and 100 m/s, respectively; the respective velocity of the flow from the fire side results in a temperature of 60 and $90\,°C$. The temperature experienced during convection heat transfer indicates that it depends on the magnitude as well as the direction of the flow. The extremely small and large flow velocities tend towards conduction and advection heat transfer, respectively; whereas, the intermediate range of flow velocity leads to convection heat transfer. It is interesting to note that the transportation of the information or the effect of temperature (of ice on the left and fire on the right side) to the person has a directional dependency: equal effect of both the directions for the conduction/diffusion, effect of only one direction (of the flow from the ice or fire side) for the advection, and unequal effect of both the directions (more effect from the direction of flow) for the 1D convection heat transfer.

Note that the random motion of molecules induced conduction heat transfer occurs without and with flow, resulting in pure conduction and convective heat transfer, respectively. *The convection heat transfer represents a combination of advection and diffusion* (Incropera and Dewitt 1996). Pure advection is a convection heat transfer where the magnitude of conduction—as compared to advection—is considered negligible; occurs only when the velocity of the flow is extremely high. However, such flow conditions are rarely encountered making the advection more of a hypothetical situation, and the convection a commonly encountered heat transfer phenomenon in the presence of flow.

The physical characteristics of the diffusion and advection phenomenon are presented above, with an example for heat transfer. However, they are also applicable to fluid dynamics; for the corresponding term in the momentum conservation law (Fig. 3.2). The physical understanding of the diffusion and advection phenomenon is used in the later chapters while presenting a physical (momentum and energy conservation) law-based algebraic formulation in the finite volume method. Note that the above-discussed physics of the fluid dynamics and heat transfer is followed later by the numerics in the CFD—to obtain a physically realistic computational fluid dynamics rather than a numerical artifact-based colorful fluid dynamics.

3.3 Physical Law-Based Differential Formulation

A derivation of the continuity equation from the mass conservation law and that of transport equation from the momentum/energy conservation (as well as the subsidiary) law, are presented below. The physical law-based differential formulation is presented below by the application of a conservation law on the CV, with appropriate arrows for the fluxes at the various surfaces of the CV (Figs. 3.4, 3.5, 3.6, 3.7, 3.8 and 3.9).

For a representative 2D Cartesian CV, a pictorial as well as mathematical symbol-based application of the balance conservation statements are presented in Fig. 3.4 for mass conservation, and in Figs. 3.5, 3.6, 3.7 and 3.8 for momentum and energy conservation law. The width of the 2D CV is Δx and Δy, and the depth of the CV (perpendicular to the plane of the figure) is taken as unity. Thus, for the 2D CV, the volume is $\Delta V = \Delta x \Delta y$; and the *magnitude* of surface areas are $\Delta S_{x+\Delta x} = \Delta S_x = \Delta y$ for the vertical and $\Delta S_{y+\Delta y} = \Delta S_y = \Delta x$ for the horizontal surfaces.

Figure 3.4 shows M_x/M_y as the rate at which mass enters the CV from a negative (left/bottom) surface, and M_{x+dx}/M_{y+dy} as the mass flow rate leaving from a positive (right/top) surface of the CV. Thus, at the various surfaces of the CV, the figure shows that the sense of the arrows are outward for the positive surface and inward for the negative surface; furthermore, the directions of the arrows are seen normal to the surfaces of the CV. Similar sense and direction of the arrows are seen for the

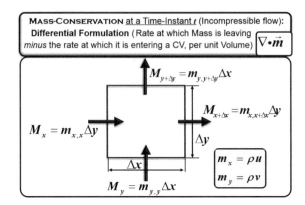

Fig. 3.4 A representative fluid CV to demonstrate a differential formulation of the balance statement for the mass flow rate across the CV, in the mass conservation law. Here, $\vec{m}$ is the mass flux and M is the mass flow rate at the left, right, bottom, and top surfaces corresponding to a constant value of $x, x + \Delta x, y,$ and $y + \Delta y$ coordinates, respectively; and $\nabla \cdot \vec{m}$ is the rate of change of mass due to the flow across the CV, per unit volume

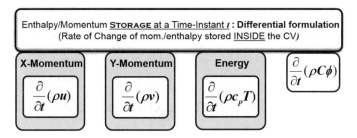

Fig. 3.5 A differential formulation of the unsteady term in the momentum/energy conservation law

advection flux (momentum flux $\vec{d}_u / \vec{d}_v$ as well as enthalpy flux $\vec{d}_T$) in Fig. 3.6; except for a stepped arrow at certain faces for the momentum flux. The normal direction corresponds to that of the normal component of the velocity/mass flux (at a surface) which drives the mass as well as advection (momentum/enthalpy) transport. For the diffusion transport, Fig. 3.7 shows the direction of the arrows are all horizontal for X−momentum and vertical for Y−momentum conservation; corresponds to the respective direction of the viscous forces in the momentum conservation. Whereas, for the diffusion transport in the energy conservation, although the sense and the direction for the conduction flux $\vec{q}$ is similar to mass flux $\vec{m}$ (Fig. 3.4), the sense of the arrows for the corresponding diffusion flux $\vec{d}_T$ is seen reversed in Fig. 3.7. This is because of the definition of the diffusion flux as $\vec{d}_T = -\vec{q}$; the negative correlation is considered here to enable a generalized differential formulation of the diffusion term (presented below).

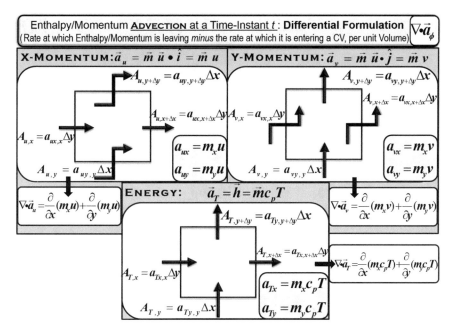

Fig. 3.6 A representative 2D CV to demonstrate a differential-formulation of the balance statement, for the momentum/enthalpy flow rate across the CV, in the momentum/energy conservation laws. Here, $A_{\phi,f}$ is the advection (momentum/enthalpy) flow rate, $\vec{a}_{\phi,f}$ is the advection flux, and $\nabla \cdot \vec{a}$ is the rate of change of advection due to the flow across the CV (per unit volume). The advected variable ϕ (used as subscript) corresponds to u, v, and T for $x-$momentum, $y-$momentum, and enthalpy transport, respectively

3.3.1 Continuity Equation

An elemental fluid CV for the application of the mass conservation law is shown in Fig. 3.4. As discussed above, using the Eulerian approach in fluid mechanics, the law of conservation of mass is stated as the rate of change of mass first due to storage inside and then due to the flow across the CV is equal to zero (Fig. 3.2). Differential formulation for the rate of change/gain of mass ($\mathcal{M} = \rho \Delta x \Delta y$) stored inside the CV M_{st}, and its volumetric term $\overline{M}_{st}$ ($\equiv M_{st}/\Delta V$), are given as

$$M_{st} = \frac{\partial \rho}{\partial t} \Delta x \Delta y \text{ and } \overline{M}_{st} = \frac{\partial \rho}{\partial t} \tag{3.3}$$

whereas, for the rate of change of mass due to the flow across the CV, a pictorial as well as mathematical symbol-based representation of the statement of mass conservation law is shown in Fig. 3.4. The figure shows the mass flow rate M and the mass flux m_x/m_y at the various faces of the CV. The rate at which mass is leaving the CV minus the rate at which it is entering the CV results in the net mass out-flux (Fig. 3.4); given as

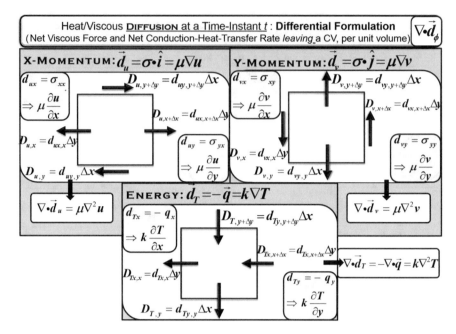

Fig. 3.7 A representative 2D CV to demonstrate a differential-formulation of the balance statement, for the viscous force/conduction heat transfer across the CV, in the momentum/energy conservation laws. Here, $\vec{d}_{\phi f}$ is the diffusion flux, $D_{\phi f}$ are the total term, and $\nabla \cdot \vec{d}_\phi$ is the net diffusion out-flux per unit volume. Note that $\nabla \cdot \vec{d}_u$ and $\nabla \cdot \vec{d}_v$ are the net viscous force acting on the CV per unit volume in the $x-$ and $y-$ direction, respectively; D_u and D_v are the viscous-force, and d_u and d_v are the viscous stress in the respective direction. Furthermore, $\nabla \cdot \vec{d}_T$ is the net conduction in-flux, D_T is the conduction heat transfer rate, and d_T is the conduction flux

$$M_{\text{net}} = (m_{x,x+\Delta x} - m_{x,x})\Delta y + (m_{y,y+\Delta y} - m_{y,y})\Delta x$$

where the first subscript for the mass flux represents its direction and the second subscript represents the location of the surface on which it is acting; x for the left, $x + \Delta x$ for the right, y for the bottom, and $y + \Delta y$ for the top surface of the CV. The above equation represents the rate of change/loss of net mass of the CV, due to the fluid flowing across the CV. Dividing the above equation by the volume $\Delta x \Delta y$ of the CV, and using the limit $\Delta x \to 0$ and $\Delta y \to 0$, the *net mass out-flux per unit volume* results in a differential formulation as

$$\overline{M}_{\text{net}} = \frac{\partial m_x}{\partial x} + \frac{\partial m_y}{\partial y} = \nabla \cdot \vec{m} \tag{3.4}$$

where $m_x = \rho u$ and $m_y = \rho v$. Using the mass conservation law per unit volume $\overline{M}_{st} + \overline{M}_{\text{net}} = 0$, and Eqs. 3.3 and 3.4, the physical law-based formulation finally results in the continuity equation as

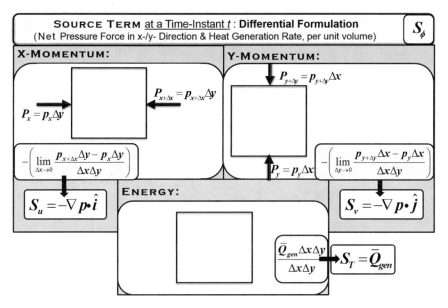

Fig. 3.8 A representative 2D CV to demonstrate a differential formulation of the balance statement for the pressure force/volumetric heat transfer, in the momentum/energy conservation laws. Here, S_u and S_v are the net pressure force per unit volume in the $x-$ and $y-$ direction, respectively; and S_T is the heat generation per unit volume

$$\frac{\partial \rho}{\partial t} + \frac{\partial (\rho u)}{\partial x} + \frac{\partial (\rho v)}{\partial y} = 0 \qquad (3.5)$$

For incompressible flow ($\rho = $ constant), the above equation is given as

$$\frac{\partial u}{\partial x} + \frac{\partial v}{\partial x} = 0 \qquad (3.6)$$

Solution methodology in CFD for the compressible as compared to incompressible flow are substantially different, mainly due to the unsteady term ($\partial \rho / \partial t$) in the continuity equation; considered for the compressible-flow (Eq. 3.5) and disappears for the incompressible flow (Eq. 3.6). A fluid dynamics movie of the density field can be created due to the presence of the unsteady term. This is done in the CFD for the compressible flow; however, the present book is on CFD for the incompressible flow where the unsteady term becomes zero (Eq. 3.6). This results in a *substantially different CFD for the incompressible as the compressible flow*, mainly due to the procedure used to compute the *pressure field*. The procedure corresponds to the pressure-density relationship (such as ideal gas law) for the compressible flow, and a detailed iterative procedure (where the continuity equation is employed as a constraint) for the incompressible flow.

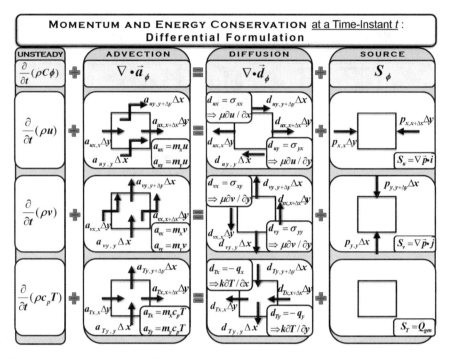

Fig. 3.9 A representative 2D CV to demonstrate a mathematical representation of the balance statement of the momentum and energy conservation laws

3.3.2 Transport Equations

For the momentum and energy conservation laws, it was discussed in the previous section that the underlying transport mechanisms are same—bulk-flow induced advection and random molecular motion-induced diffusion phenomena; however, their physical interpretation or effect are different. The advection leads to momentum flow rate and enthalpy flow rate for the momentum and energy conservation law, respectively; and the diffusion leads to the viscous force and conduction heat transfer rate for the respective law. It is important to note that the mass and energy are scalar while momentum is a vector conservation law. The change from the scalar to the vector law results in a component-by-component (in x and y direction for 2D Cartesian coordinate system) application of the momentum conservation law. This makes the physical understanding of the physical law-based differential formulation of the momentum equation much more difficult as compared to that for the mass and energy equation.

3.3.2.1 Unsteady Term

Similar to the unsteady term for the mass conservation (presented above), the unsteady term here corresponds to the rate of change/gain of momentum $\rho\Delta x\Delta y\,\vec{u}$ and enthalpy $\mathcal{H} = \rho\Delta x\Delta y c_p T$ stored inside the CV, per unit volume; and results in a general differential formulation as

$$\overline{F}_{\phi,st} = \frac{\partial(\rho C\phi)}{\partial t} \tag{3.7}$$

where $C = 1$ and $\phi = u$ and v for $x-$ and $y-$momentum equation, respectively; and $C = c_p$ and $\phi = T$ for energy equation. The general formulation is shown in Fig. 3.5.

3.3.2.2 Advection Term

For the advection term, the rate of change of momentum and enthalpy due to the flow across the CV is represented by a general variable advection (momentum/enthalpy) flow rate A_ϕ; and the corresponding flux terms as advection flux $\vec{a}_\phi$ (with $\phi = u, v$, and T for the $x-$momentum, $y-$momentum, and energy conservation, respectively). The advection flow rates are shown at the various faces of the CV in Fig. 3.6 as A_u, A_v, and A_T, for the $x-$momentum, $y-$momentum, and enthalpy flow rate, respectively. The respective advection flux variables are shown as $x-$momentum flux ($a_{ux} = m_x u$ and $a_{uy} = m_y u$), $y-$momentum flux ($a_{vx} = m_x v$ and $a_{vy} = m_y v$), and enthalpy flux ($a_{Tx} = m_x c_p T$ and $a_{Ty} = m_y c_p T$). Although $x-$momentum is expressed as the product of the mass and the $u-$velocity, $x-$momentum flow rate is considered here (for the rate process) as the product of the mass flow rate and the $u-$velocity. The corresponding $x-$momentum flux term is $m_x u$ at the vertical surfaces and $m_y u$ at the horizontal surfaces of the CV, shown in Fig. 3.6 for the $x-$momentum transport. Note in the expressions for the momentum flux, mass flux acts like an advecting/driver variable; and u, v, and T as an advected/driven variable ϕ for the $x-$momentum, $y-$momentum, and energy transport process, respectively.

For the momentum transport, the arrows for an advection flux in Fig. 3.6 consist of the horizontal as well as the vertical lines if the direction for the mass flux (driver) and u/v (driven) are orthogonal; one in the horizontal and the other in the vertical direction. For example, the direction of m_y is vertical and u is horizontal for the $x-$momentum flux $m_y u$, shown by a stepped arrow in Fig. 3.6 for the $x-$momentum transport. The rate at which momentum and enthalpy flow rate are leaving the CV minus the rate at which they are entering the CV (Fig. 3.6), dividing by the volume of the CV, and using the limit $\Delta x \to 0$ and $\Delta y \to 0$ results in the general differential formulation for the *net advection out-flux per unit volume* as

$$\overline{A}_{\phi,net} = \frac{\partial a_{\phi x}}{\partial x} + \frac{\partial a_{\phi y}}{\partial y} = \nabla \cdot \vec{a}_\phi$$

The above equation represents the rate of change/loss of net $x - /y-$momentum and enthalpy due to the fluid flowing across the CV.

3.3.2.3 Diffusion Term

For the diffusion term, the viscous force and conduction heat transfer rate are represented by a general variable diffusion D_ϕ; and the corresponding flux term as diffusion flux $\vec{d}_\phi$. They are shown at the various faces of the CV in Fig. 3.7. The figure shows D_u, D_v, and D_T as the viscous force in the $x-$direction, viscous force in the $y-$direction, and conduction heat transfer rate, respectively. The flux for the respective terms are shown as viscous stress in the $x-$direction ($d_{ux} = \sigma_{xx} = \mu \partial u/\partial x$ and $d_{uy} = \sigma_{yx} = \mu \partial u/\partial y$), viscous stress in the $y-$direction ($d_{vx} = \sigma_{xy} = \mu \partial v/\partial x$ and $d_{vy} = \sigma_{yy} = \mu \partial v/\partial y$), and conduction-flux in the normal direction ($d_{Tx} = -q_x = k \partial T/\partial x$ and $d_{Ty} = -q_y = k \partial T/\partial y$). Note the equations for the conduction flux correspond to Eq. 3.1 for the subsidiary law; whereas, the equations for viscous-stress are *simplified* form of Eq. 3.2—applicable for an *incompressible* single-fluid (not multiphase) flow.

Also note from Fig. 3.7 that all the arrows representing the viscous stress $\vec{d}_u$ and $\vec{d}_v$ are horizontal for the $x-$momentum and vertical for the $y-$momentum conservation, while the arrows for the conduction flux $\vec{d}_T(-\vec{q})$ are normal to a surface. Subtracting each of the total diffusion D_ϕ acting at the negative (left/bottom) surfaces from those at the positive (right/top) surfaces (Fig. 3.7), dividing by the volume of the CV, and using the limit $\Delta x \to 0$ and $\Delta y \to 0$ results in the general differential formulation for the *net diffusion per unit volume* as

$$\overline{D}_{\phi,\text{net}} = \frac{\partial d_{\phi x}}{\partial x} + \frac{\partial d_{\phi y}}{\partial y} = \Gamma_\phi \left(\frac{\partial^2 \phi}{\partial x^2} + \frac{\partial^2 \phi}{\partial y^2} \right)$$

where $D_{\phi,\text{net}}$ represents the net viscous force (in the positive $x - /y-$direction) or conduction in-flux. Furthermore, the 2D Cartesian components of the diffusion flux vector are given as

$$d_{\phi x} = \Gamma_\phi \frac{\partial \phi}{\partial x} \text{ and } d_{\phi y} = \Gamma_\phi \frac{\partial \phi}{\partial y} \tag{3.8}$$

with diffusion coefficient $\Gamma_\phi = \mu$ and $\phi = u/v$ for momentum equation, and $\Gamma_\phi = k$ and $\phi = T$ for energy equation—resulting in viscous-stress and conduction heat flux, respectively.

It is interesting to note from Fig. 3.7 that a diffusion flux is expressed in terms of *normal* (to the surface on which the flux is acting) gradient of the diffused variable ϕ. For example, it can be seen that the diffusion flux $d_{\phi x}$ acting on the vertical surface (normal to the $x-$direction) is expressed as $\partial \phi/\partial x$; and $d_{\phi y} = \Gamma_\phi \partial \phi/\partial y$ acts on the horizontal surface. Thus, the general expression for diffusion flux on a surface—with normal in the $\eta-$direction—is given as

$$d_{\phi\eta} = \Gamma_\phi \frac{\partial\phi}{\partial\eta}$$ (3.9)

Note that the above-generalized equation is applicable only for an incompressible and single-fluid flow; and is used in the finite volume method presented in Parts II and III of this book.

3.3.2.4 Source Term

For the differential formulation of the source term, Fig. 3.8 shows the consideration of the pressure force (which is compressive and acts normal to the surfaces of a CV) at the vertical surfaces for the x−momentum conservation, and at the horizontal surfaces for the y−momentum conservation. This is because the net pressure force acting in the positive x−direction and y−direction are used separately as the source term S_u for the x−momentum, and S_v for the y−momentum equation. The net pressure force is divided by the volume of the CV, and then using the limit $\Delta x \rightarrow 0$ and $\Delta y \rightarrow 0$ results in the differential formulation for the source term (in the momentum equation) as

$$S_u = -\frac{\partial p}{\partial x} \text{ and } S_v = -\frac{\partial p}{\partial y}$$

and the heat generation per unit volume results in the formulation for the source term in the energy equation as $S_T = \overline{Q}_{\text{gen}}$.

3.3.2.5 All Terms

Combining all the above terms, a pictorial representation of the formulation for the transport equations is shown in Fig. 3.9. The figure shows a general differential formulation as an unsteady advection-diffusion equation with a source term, given as

$$\frac{\partial(\rho C\phi)}{\partial t} + \left(\frac{\partial a_{\phi x}}{\partial x} + \frac{\partial a_{\phi y}}{\partial y}\right) = \left(\frac{\partial d_{\phi x}}{\partial x} + \frac{\partial d_{\phi y}}{\partial y}\right) + S_\phi$$

where $\phi = u/v$ and $C = 1$ for the momentum equations, and $\phi = T$ and $C = c_p$ for the energy equation. For the above equation, $a_{\phi x} = m_x C\phi$, $a_{\phi y} = m_y C\phi$, $d_{\phi x} = \Gamma_\phi \partial\phi/\partial x$, and $d_{\phi y} = \Gamma_\phi \partial\phi/\partial y$; with $\Gamma_\phi = \mu$ for the momentum and $\Gamma_\phi = k$ for the energy equation.

3.4 Generalized Volumetric and Flux Terms, and Their Differential Formulation

As mentioned above, the unsteady term is a volumetric term while the advection and diffusion terms are flux terms (Fig. 3.2). The mass flow rate is also a flux term. Thus, the rate process, corresponding to mass transport, momentum transport and energy transport, consists of two types of terms: volumetric and flux.

3.4.1 Volumetric Term

Let us define a general volumetric term as $\overline{V}_\phi = \rho C \phi$, for the rate of change inside the CV (Fig. 3.2). The term corresponds to the volumetric $x-$momentum, volumetric $y-$momentum, and volumetric enthalpy are given as $\overline{V}_u = \overline{Mu} = \rho u$, $\overline{V}_v = \overline{Mv} = \rho v$, and $\overline{V}_T = \overline{\mathcal{H}} = \overline{Mc_p T} = \rho c_p T$, respectively; here, $\mathcal{M}$ is the mass of the fluid inside the CV and the *over-bar* represents "per unit volume". Using $\overline{V}_\phi$, the rate of change of momentum/energy stored inside the CV is expressed as

$$F_{\phi,st} = \lim_{\Delta t/\Delta x/\Delta y \to 0} \frac{\left(\oiint_{\Delta V} \overline{V}_\phi \cdot dV \right)_{t+\Delta t} - \left(\oiint_{\Delta V} \overline{V}_\phi \cdot dV \right)_t}{\Delta t} = \frac{\partial \overline{V}_\phi}{\partial t} \Delta V$$

where the above equation is divided by the volume of the CV to finally derive the generalized *volumetric unsteady term* (as the rate of change per unit volume, shown in Fig. 3.5) as

$$\overline{F}_{\phi,st} = \frac{\partial \overline{V}_\phi}{\partial t} = \frac{\partial}{\partial t} (\rho C \phi) \tag{3.10}$$

where $C = 1$ and $\phi = u/v$ for $x - /y-$momentum equation; and $C = c_p$ and $\phi = T$ for energy equation (Fig. 3.7).

Other than the unsteady term presented above, the volumetric term is also encountered for the heat generation; resulting in the source term as

$$S_T = \overline{Q}_{gen} \tag{3.11}$$

3.4.2 Flux-Term

Let us define a general variable $\overrightarrow{f}$ ($\equiv f_x \hat{i} + f_y \hat{j}$) for a flux term, with $\overrightarrow{f} = \overrightarrow{m}$ for the mass flux, $\overrightarrow{f} = \overrightarrow{a}_\phi$ for the advection flux, and $\overrightarrow{f} = \overrightarrow{d}_\phi$ for the diffusion flux. The application of the generalized flux $\overrightarrow{f}$ on a 2D Cartesian CV is shown in Fig. 3.10.

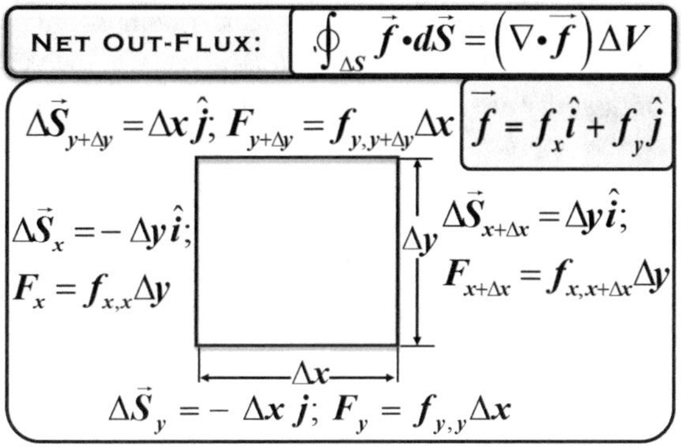

Fig. 3.10 A representative two-dimensional CV to demonstrate the balance statement of flux terms encountered in the conservation laws used in fluid dynamics and heat transfer. Here $\overrightarrow{f}$ is a flux and F is the total (flux multiplied by the surface area) term at the left, right, bottom, and top surfaces corresponding to a constant value of x, $x + \Delta x$, y, and $y + \Delta y$ coordinates, respectively

The figure also shows the surface areas of the CVs as vectors: $\overrightarrow{\Delta S}_{x+\Delta x} = -\overrightarrow{\Delta S}_x = \Delta y \hat{i}$ for the vertical surface, and $\overrightarrow{\Delta S}_{y+\Delta y} = -\overrightarrow{\Delta S}_y = \Delta x \hat{j}$ for the horizontal surface. The subscript for the $\overrightarrow{\Delta S}$ represents the location of the surface, and $\hat{i}/\hat{j}$ are the outward normal unit vector for a surface. The subscript and unit-vector correspond respectively to $x + \Delta x$ and $\hat{i}$ for the right surface, x and $-\hat{i}$ for the left surface, $y + \Delta y$ and $\hat{j}$ for the top surface, and y and $-\hat{j}$ for the bottom surface of the CV (Fig. 3.10). The figure also shows a generalized variable F, obtained by the dot product of the flux-vector $\overrightarrow{f}$ with the surface area vector $\overrightarrow{\Delta S}$. Thus, the F is a scalar quantity; such as, mass flow rate M, enthalpy flow rate H and conduction heat transfer rate Q.

However, for the momentum conservation which is a vector law, as compared to the mass and energy conservation which are scalar laws, the momentum flow rate and the force are a vector quantity and the corresponding flux term (momentum-flux $\left[\overrightarrow{m}\,\overrightarrow{u}\right]$ and viscous stress $[\sigma]$) are *second-order tensor*; given for 2D flow as

$$\overrightarrow{m}\,\overrightarrow{u} = \begin{bmatrix} m_x u & m_x v \\ m_y u & m_y v \end{bmatrix} \text{ and } \sigma = \begin{bmatrix} \sigma_{xx} & \sigma_{xy} \\ \sigma_{yx} & \sigma_{yy} \end{bmatrix} \tag{3.12}$$

where σ_{ij} is defined as the stress in the $j-$direction acting on a surface whose normal is in the $i-$direction; the first subscript i represents the direction of the plane and the second subscript j represents the direction of the stress. Similar symbolic definition is used for the momentum flux; for example, $m_y u$ is defined as the momentum flux in the $x-$direction (corresponding to u in $m_y u$) acting on a horizontal surface (whose normal is in the $y-$direction corresponding to the direction of m_y). The tensor acts on a particular surface in a particular direction; thus, the direction in which they act as well that of the surface (represented by its normal) are needed for the tensor—two directions for the tensor as compared to one direction for a vector.

Since there is a component-by-component (x and y component in a 2D Cartesian system) application of the momentum conservation for a CV, the vector components of the above tensor quantity are considered. The flux vector $\overrightarrow{f}$ is obtained by taking the dot product of the appropriate tensors with $\hat{i}$ for $x-$momentum conservation, and with $\hat{j}$ for $y-$momentum conservation. The dot product with $\hat{i}$ results in the first column, and the product with $\hat{j}$ results in the second column, of the tensors (momentum flux and stress) in Eq. 3.12. Thus, for the advection term of the momentum conservation law, the vector component $\overrightarrow{f}$ is given (refer Fig. 3.6) as

$$x - \text{Momentum flux}: \overrightarrow{d}_u = \overrightarrow{m}\,\overrightarrow{u}\,.\hat{i} = \overrightarrow{m}\,u = m_x u\,\hat{i} + m_y u\,\hat{j}$$
$$y - \text{Momentum flux}: \overrightarrow{d}_v = \overrightarrow{m}\,\overrightarrow{u}\,.\hat{j} = \overrightarrow{m}\,v = m_x v\,\hat{i} + m_y v\,\hat{j}$$

where the dot product can also be considered as $\overrightarrow{u}\,.\hat{i} = u$ leading to $\overrightarrow{m}\,\overrightarrow{u}\,.\hat{i} = \overrightarrow{m}\,u$. For the diffusion term of the momentum conservation law, the vector component $\overrightarrow{f}$ is given (refer Fig. 3.7) as

$$\text{Viscous stress in } x - \text{direction}: \overrightarrow{d}_u = \sigma\,.\hat{i} = \sigma_{xx}\,\hat{i} + \sigma_{yx}\,\hat{j}$$
$$\text{Viscous stress in } y - \text{direction}: \overrightarrow{d}_v = \sigma\,.\hat{j} = \sigma_{xy}\,\hat{i} + \sigma_{yy}\,\hat{j}$$

where $\overrightarrow{d}_u$ and $\overrightarrow{d}_v$ correspond to the above discussed first and second column of the viscous stress tensor in Eq. 3.12.

Note that the dot product of a second-order tensor with a vector results in a vector. Thus, a dot product with $\hat{i}$ reduces a tensor σ to the vector component $\sigma\,.\hat{i} = \sigma_{xx}\,\hat{i} + \sigma_{yx}\,\hat{j}$, and a vector $\overrightarrow{f}$ to a scalar component $\overrightarrow{f}\,.\hat{i} = f_x$ in the $x-$ direction. During the formulation from the momentum/*vector* conservation law, also note that the general variable F corresponds to a Cartesian component of a vector (momentum flow rate and viscous force) and the corresponding flux variable $\overrightarrow{f}$ corresponds to a vector component of second-order tensor (momentum flux and viscous stress). Whereas, the formulation from the energy/*scalar* conservation law encounters the F directly (not componentwise) as a scalar (enthalpy flow rate H and conduction heat transfer rate Q) and the $\overrightarrow{f}$ as a vector (enthalpy flux $\overrightarrow{h} = h_x\hat{i} + h_y\hat{j}$ and conduction flux $\overrightarrow{h} = h_x\hat{i} + h_y\hat{j}$).

For any flux term, a balance statement for the net flux is represented as the surface integral of the fluxes acting on the bounding surface of the CV; given as

$$\oint_{\Delta S} \overrightarrow{f}\,.d\overrightarrow{S} = \overrightarrow{f}_{x+dx}\,.\overrightarrow{\Delta S}_{x+dx} + \overrightarrow{f}_x\,.\overrightarrow{\Delta S}_x + \overrightarrow{f}_{y+dy}\,.\overrightarrow{\Delta S}_{y+dy} + \overrightarrow{f}_y\,.\overrightarrow{\Delta S}_y$$
$$= f_{x,x+dx}\Delta y - f_{x,x}\Delta y + f_{y,y+dy}\Delta x - f_{y,y}\Delta x$$
$$F_{net} \qquad = F_{x+dx} - F_x + F_{y+dy} - F_y$$

Finally, dividing by the volume of the CV, a differential formulation for the *net out-flux per unit volume* is given as

$$\overline{F}_{net} = \left(\lim_{\Delta x \to 0} \frac{\left(f_{x,x+\Delta x} - f_{x,x}\right) \Delta y}{\Delta x \Delta y} + \lim_{\Delta y \to 0} \frac{\left(f_{y,y+\Delta y} - f_{y,y}\right) \Delta x}{\Delta x \Delta y} \right)$$

$$\overline{F}_{net} = \nabla \cdot \overrightarrow{f} \tag{3.13}$$

The $F_{net} = \nabla \cdot \overrightarrow{f} \, \Delta V$ represents rate at which the mass, momentum, and enthalpy leaves a CV (F_{out}) minus the rate at which they enter the CV (F_{in}). Thus, $\nabla \cdot \overrightarrow{f}$ represents net out-flux per unit volume; due to the flow across the surfaces of the CV. This term also represents the net viscous force acting on the CV per unit volume. Note that the above formulation for $\oint_{AS} \overrightarrow{f} \cdot \overrightarrow{dS} = \left(\nabla \cdot \overrightarrow{f} \right) \Delta V$ can also be obtained directly by applying the Gauss divergence theorem. The theorem is used in a partial differential equations based algebraic formulation in finite volume method (Patankar (1980)), where the $\nabla \cdot \overrightarrow{f}$ is operated by a volume integral and converted into the surface integral $\oint_{AS} \overrightarrow{f} \cdot \overrightarrow{dS}$.

An elemental fluid CV for the advection term in Fig. 3.6, when compared with Fig. 3.10, shows the general variable $F = A$ and $\overrightarrow{f} = \overrightarrow{a}_\phi$. Similar comparison for diffusion term in Fig. 3.7 results in $F = D$ and $\overrightarrow{f} = \overrightarrow{d}_\phi$. Thus, substituting $\overrightarrow{f} = \overrightarrow{a}_\phi$ and $\overrightarrow{f} = \overrightarrow{d}_\phi$ in Eq. 3.13, the *net advection and the net diffusion per unit volume* are given as

$$\overline{A}_{\phi,net} = \nabla \cdot \overrightarrow{a}_\phi \text{ where } \overrightarrow{a}_\phi = \overrightarrow{m} \, C \phi \tag{3.14}$$

$$\overline{D}_{\phi,net} = \nabla \cdot \overrightarrow{d}_\phi \text{ where } d_\phi = \Gamma_\phi \nabla \phi \tag{3.15}$$

where $\phi = u/v$, $C = 1$ and $\Gamma_\phi = \mu$ for momentum equation; and $\phi = T$, $C = c_p$ and $\Gamma_\phi = k$ for energy equation.

An elemental fluid CV for the source term of momentum conservation in Fig. 3.8, when compared to Fig. 3.10, shows the general variable F as the total pressure force P (normal to a surface) and f as the p; note that the pressure p as a flux here is a scalar not a vector. Thus, instead of $\nabla \cdot \overrightarrow{f}$ in Eq. 3.13, we obtain the differential formulation for the pressure as a source term as

$$\overline{F}_{net} = \nabla p = \frac{\partial p}{\partial x} + \frac{\partial p}{\partial y}$$

where the *negative* value of the first and second term on the RHS represents the *net pressure force per unit volume* in the *x−* and *y−direction*, respectively; given as

$$S_u = -\nabla p \cdot \hat{i} \text{ for x-momentum equation} \tag{3.16}$$
$$S_v = -\nabla p \cdot \hat{j} \text{ for y-momentum equation}$$

3.4.3   Discussion

The mass, momentum, and energy conservation law are represented as a general unsteady advection-diffusion equation with a source term; given as

$$\overline{F}_{\phi,st} + \overline{F}_{\phi,adv} = \overline{F}_{\phi,diff} + S_\phi$$

$$\frac{\partial \overline{\mathcal{V}}_\phi}{\partial t} + \nabla \cdot \overrightarrow{a}_\phi = \nabla \cdot \overrightarrow{d}_\phi + S_\phi$$

where the generic variables $\overline{\mathcal{V}}$, F, $\overrightarrow{f}$, and F_{net} for the various conservation laws are shown in Table 3.1.

It is also interesting to note that both enthalpy and momentum are encountered here in the unsteady and the advection term, but not in the diffusion and the source term. However, there is a temporal variation of the enthalpy and momentum for the unsteady term, and a spatial variation of the enthalpy flow rate and momentum flow rate for the advection term. Also, note that both enthalpy $\mathcal{H}(\mathcal{M}c_pT)$ and momentum $\mathcal{M}\overrightarrow{u}$ involves mass $\mathcal{M}(\rho\Delta V)$ of fluid inside a CV, and the enthalpy flow rate $H(Mc_pT)$ and the momentum flow rate $M\overrightarrow{u}$ involves mass flow rate $M(\overrightarrow{m}\cdot\Delta\overrightarrow{S})$ at the various surfaces of a CV. Thus, both enthalpy and momentum are dependent on volume ΔV, and the enthalpy flow rate and momentum flow rate are dependent on the surface area $\Delta\overrightarrow{S}$ of a CV; the former variables are volumetric and latter are flux terms, with enthalpy flux as a vector $\overrightarrow{h} = \overrightarrow{m}c_pT$ and momentum flux as a second-order tensor $\overrightarrow{m}\overrightarrow{u}$.

Table 3.1 Specific variables corresponding to the various generic variables (volumetric $\overline{\mathcal{V}}$, total surface term F, flux $\overrightarrow{f}$, and net out-flux/balance F_{net}) encountered in the conservation laws used in fluid dynamics and heat transfer. A_ϕ is the flow rate, $\overrightarrow{a}_\phi$ is the flux, and $\nabla\cdot\overrightarrow{a}_\phi$ is the net out-flux for momentum/enthalpy

Laws→			Mass	Momentum		Energy
Terms ↓				$x-$	$y-$	
Volumetric (Unsteady)		$\overline{\mathcal{V}}$	$\overline{\mathcal{M}}=\rho$	$\overline{\mathcal{M}\sqcap}=\rho u$	$\overline{\mathcal{M}\sqsubset}=\rho v$	$\overline{\mathcal{H}}=\rho c_p T$
Flux	Adv.	F	M	A_u	A_v	A_T
		$\overrightarrow{f}$	$\overrightarrow{m}$	$\overrightarrow{a}_u = \overrightarrow{m}\overrightarrow{u}\cdot\hat{\imath}$	$\overrightarrow{a}_v = \overrightarrow{m}\overrightarrow{u}\cdot\hat{\jmath}$	$\overrightarrow{a}_T = \overrightarrow{h} = \overrightarrow{m}c_pT$
		$\overline{F}_{net}$	$\nabla\cdot\overrightarrow{m}$	$\nabla\cdot\overrightarrow{a}_u = \nabla\cdot\overrightarrow{m}u$	$\nabla\cdot\overrightarrow{a}_v = \nabla\cdot\overrightarrow{m}v$	$\nabla\cdot\overrightarrow{a}_T = \nabla\cdot\overrightarrow{m}c_pT$
	Diff.	F	0	D_u	D_v	D_T
		$\overrightarrow{f}$	0	$\overrightarrow{d}_u = \sigma\cdot\hat{\imath}$	$\overrightarrow{d}_v = \sigma\cdot\hat{\jmath}$	$\overrightarrow{d}_T = -\overrightarrow{q}$
		$\overline{F}_{net}$	0	$\nabla\cdot\overrightarrow{d}_u = \mu\nabla^2 u$	$\nabla\cdot\overrightarrow{d}_v = \mu\nabla^2 v$	$\nabla\cdot\overrightarrow{d}_T = k\nabla^2 T$
Source		S_ϕ	0	$S_u = \nabla p\cdot\hat{\imath}$	$S_v = \nabla p\cdot\hat{\jmath}$	$S_T = \overline{Q}_{gen}$

3.5 Mathematical Formulation

The system of governing PDEs derived above, from the conservation and subsidiary laws, are called Navier-Stokes (NS) equations—the mathematical model for fluid dynamics and heat transfer. The final mathematical formulation in a fluid dynamics and heat transfer study consists of the NS equations, initial and boundary conditions, and governing and engineering parameters. The formulation is presented below first for a dimensional study, and then for a non-dimensional study.

3.5.1 Dimensional Study

3.5.1.1 Navier-Stokes Equations

From the formulation in the previous section, the general form of the NS equations are given as

$$\text{Continuity: } \frac{\partial \rho}{\partial t} + \nabla \cdot \vec{m} = 0$$

$$\text{Momentum/Energy: } \frac{\partial \overline{\mathcal{V}}_\phi}{\partial t} + \nabla \cdot \vec{a}_\phi = \nabla \cdot \vec{d}_\phi + S_\phi$$

For *incompressible Newtonian fluid flow with constant thermophysical properties*, substituting the various terms in the above equation from Table 3.1, the NS equations are given as

$$\text{Continuity: } \nabla \cdot \vec{u} = 0$$

$$\text{Momentum/Energy: } \rho C \frac{\partial \phi}{\partial t} + \vec{m} C \cdot \nabla \phi = \Gamma_\phi \nabla^2 \phi + S_\phi$$

where, using the appropriate ϕ, C, and Γ_ϕ, the final expanded form of the Navier-Stokes equations are given for a 2D Cartesian coordinate system as

$$\text{Continuity: } \frac{\partial u}{\partial x} + \frac{\partial v}{\partial y} = 0 \tag{3.17}$$

$$x\text{-Momentum: } \rho \frac{\partial u}{\partial t} + \rho u \frac{\partial u}{\partial x} + \rho v \frac{\partial u}{\partial y} = \mu \left(\frac{\partial^2 u}{\partial x^2} + \frac{\partial^2 u}{\partial y^2} \right) - \frac{\partial p}{\partial x}$$

$$y\text{-Momentum: } \rho \frac{\partial v}{\partial t} + \rho u \frac{\partial v}{\partial x} + \rho v \frac{\partial v}{\partial y} = \mu \left(\frac{\partial^2 v}{\partial x^2} + \frac{\partial^2 v}{\partial y^2} \right) - \frac{\partial p}{\partial y}$$

$$\text{Energy: } \rho c_p \frac{\partial T}{\partial t} + \rho c_p u \frac{\partial T}{\partial x} + \rho c_p v \frac{\partial T}{\partial y} = k \left(\frac{\partial^2 T}{\partial x^2} + \frac{\partial^2 T}{\partial y^2} \right) + \overline{Q}_{\text{gen}}$$

Fig. 3.11 Conduction heat transfer in a rectangular plate, subjected to initial and different types of thermal boundary conditions

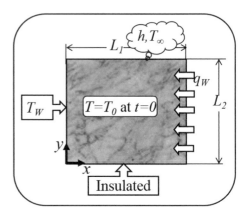

It is important to notice the similarity in the above momentum and energy equation, as an unsteady advection-diffusion equation with a source term. This generality is used to realize the advection as one and diffusion as the other component for the development of a CFD software—a component corresponds to a *function/subroutine* in the computer program. Thus, the advection and diffusion components are developed as separate generalized function/subroutine, which are used by momentum as well as energy equations for the computation of the appropriate terms; discussed in the later chapters of this book.

3.5.1.2   Initial and Boundary Conditions

All the fluid dynamics and heat transfer problems obey the conservation laws; thus, the governing partial differential equations (Eq. 3.17) are applicable to almost all the problems of incompressible Newtonian fluid flow with constant thermophysical properties. Initial condition is a prescription of flow properties inside the domain at a time instant, corresponding to the first picture of a fluid dynamics movie. Whereas, boundary condition is a condition of the flow properties all over the boundary of the domain, for the complete time duration of the movie.

Since the independent variables in the governing PDEs are time and spatial coordinates, conditions are needed at a certain time and spatial coordinates. The highest derivative of the independent variables in the Eq. 3.17 is one for the time, and two for the $x-$coordinate as well as $y-$coordinate. Time is a one-way coordinate system with the events of the future depending on the history of the event; but not vice versa. Thus, for the one initial time instant, the value of the dependent variable $(u/v/T)$ is taken at the initial time $t = 0$—called the initial condition. The values for the two bounding spatial location, in each of the coordinate direction, is taken at the boundary $x = 0$ and L, and $y = 0$ and H (for a rectangular domain of size $L \times H$)—called the boundary condition. The initial and boundary conditions are essential for a CFD solution of a fluid dynamics problem; however, they should be physically realistic for an accurate solution. They act as a jury to the problem and dictate the solution.

Let us consider an example of 2D unsteady-state heat conduction in a rectangular plate shown in Fig. 3.11. The figure shows that the plate is taken from a furnace at a uniform temperature of T_0 and subjected to the different thermal boundary conditions. Mathematically, the boundary conditions are classified into three types: *Dirichlet*, if the value of a dependent variable $(T / \vec{u} / p)$ is prescribed; *Neumann*, if the gradient of the variable is prescribed; and *Robin or mixed*, if a linear combination of the variable and its gradient is prescribed at the boundary. For the example problem in Fig. 3.11, the different types of BCs are as follows:

1. *Dirichlet BC*: $T = T_W$ at $x = 0$, for the **C**onstant **W**all **T**emperature (CWT) BC at the left wall
2. *Neumann BC*: $\partial T / \partial y = 0$ at $y = 0$, for the *insulated* BC at the bottom wall; and $\partial T / \partial x = -q_W / k$ at $x = L$ for the **U**niform **H**eat **F**lux (UHF) BC at the right wall.
3. *Robin or mixed BC*: $a\,T + b\,\partial T / \partial y = c$ at $y = H$, for the *convective* BC at the top wall. Here, $a = h$, $b = k$, and $c = hT_\infty$, with h as the heat transfer coefficient and T_∞ as the ambient temperature.

For the two types of flow (internal and external) in fluid dynamics, the initial and the boundary conditions are already presented in the previous chapter (Fig. 2.7b, c); these BCs also correspond to the three types presented above.

3.5.1.3   Governing and Engineering Parameters

Parameters in fluid dynamics and heat transfer are of two types: governing and engineering; the former is the input and the latter is the output parameter.

Governing-parameters are those parameters which govern/affect the results in a fluid dynamics and heat transfer problem. The *governing parameters* are of three types: first, the *thermo-physical properties* of the fluid (or solid in conduction); second, *geometrical parameters*; and third, *BCs related heat/fluid flow parameters*. For example, considering the heat conduction problem in Fig. 3.11, the governing thermophysical properties are the thermal conductivity for the steady state results; also density and specific heat for the unsteady results. Furthermore, the geometrical governing parameters are the dimensions of the plate, and the BCs related governing parameters are the wall temperature T_W, the heat flux q_W, the convection coefficient h, and the ambient temperature T_∞. For most of the fluid dynamics problems, the steady-state results are independent of the initial condition; however, the IC is also a governing parameter for the transient/unsteady results.

Engineering parameters are those parameters which are of interest to engineers, as these parameters lead to an increase in the cost or/and affects the efficient operation of an engineering system. Mostly, these parameters are calculated at the surface or boundary of a solid in an external/internal heat and fluid flow problem. Thus, the calculation involves surface integral of the spatial variation of various fluxes (such as heat flux, viscous stress, and pressure) on the solid surface. For example, for the heat conduction problem in Fig. 3.11, the parameter is rate of heat transfer from the plate. Furthermore, for the external flow in Fig. 2.7b, the relevant parameters are the lift and drag force acting on the cylinder, and the rate of heat transfer from the cylinder. Finally, for the internal flow in Fig. 2.7c, they correspond to shear force acting on the walls of the channel and pressure gradient in the axial direction; and the rate of heat transfer from the walls for the CWT BC and the temperature of the walls for the UHF BC.

The general expression for the drag force, lift force, and rate of heat transfer are given (Muralidhar and Biswas 2005) as

$$F_D = \oint (\hat{n} \cdot \sigma_{\text{total}}) \cdot \hat{i} dS, \; F_L = \oint (\hat{n} \cdot \sigma_{\text{total}}) \cdot \hat{j} dS \text{ and } Q = \oint \hat{n} \cdot \vec{q} \, dS$$

where $\hat{n}$ is the unit vector normal to a surface (varies from point-to-point in a curved surface). Furthermore, σ_{total} is the total (viscous as well as pressure) stress and $\vec{q}$ is the heat flux at the surface. Since σ is second-order tensor with two direction, $\hat{n} \cdot \sigma_{total}$ represents a vector component of the stress tensor in the normal direction. Further dot product with $\hat{i}$ and $\hat{j}$ results in the $x-$ and $y-$ components of the vector ($\hat{n} \cdot \sigma_{\text{total}}$), whose surface integral represents drag and lift force, respectively. Furthermore, $\hat{n} \cdot \vec{q}$ represents a scalar component of the heat flux vector in the normal direction, whose surface integral represents the rate of heat transfer.

For example, let us consider a 2D flow over an inclined plate whose normal unit vector is given as $\hat{n} = a\hat{i} + b\hat{j}$. Then, the integral parameters are given as

$$
\begin{aligned}
\text{Lift Force:} \quad & F_D = \oint \left(a\sigma_{\text{total},xx} + b\sigma_{yx} \right) dS \\
\text{Drag Force:} \quad & F_L = \oint \left(a\sigma_{xy} + b\sigma_{\text{total},yy} \right) dS \\
\text{Heat Transfer Rate:} \quad & Q = \oint \left(aq_x + bq_y \right) dS
\end{aligned}
\tag{3.18}
$$

where, the total normal stress and the viscous shear stress are given as

$$
\begin{aligned}
& \sigma_{\text{total},xx} = -p + 2\mu \partial u/\partial x, \; \sigma_{\text{total},yy} = -p + 2\mu \partial v/\partial y \\
& \sigma_{xy} = \sigma_{yx} = \mu \left(\partial v/\partial x + \partial u/\partial y \right)
\end{aligned}
\tag{3.19}
$$

3.5.2 Non-Dimensional Study

The fluid dynamics and heat transfer results are mostly reported in non-dimensional form; this makes the results more generic. The non-dimensional results are applicable to various types of fluids and large range of length and velocity scale, corresponding to the same problem; provided that the non-dimensional parameters *governing* the problem remains unchanged. For example, consider a free-stream flow across a cylinder with the free stream velocity u_∞ as the velocity scale, and the diameter of the cylinder D as the length scale. For a constant value of the non-dimensional governing parameter, Reynolds number $Re \equiv u_\infty D/v$, the same non-dimensional results are obtained for the various combinations of u_∞, D, and ν—which results in the same Re.

Experimental work are performed only in dimensional form; thereafter, the governing parameters as well as results are calculated and presented in non-dimensional form. However, in a numerical work, it is possible to directly perform non-dimensional CFD simulations, and obtain the non-dimensional results for the flow properties and engineering parameters. This is done by considering the non-dimensional governing equations, consisting of non-dimensional governing parameters. The non-dimensional form of governing equations, flow properties, governing parameters, and engineering parameters are presented below.

3.5.2.1 Navier-Stokes Equations

The non-dimensional NS equations are obtained from Eq. 3.17 as

$$\text{Continuity: } \frac{\partial U}{\partial X} + \frac{\partial V}{\partial Y} = 0 \tag{3.20}$$

$$\text{X-Momentum: } \frac{\partial U}{\partial \tau} + U\frac{\partial U}{\partial X} + V\frac{\partial V}{\partial Y} = \frac{1}{Re}\left(\frac{\partial^2 U}{\partial X^2} + \frac{\partial^2 U}{\partial Y^2}\right) - \frac{\partial P}{\partial X}$$

$$\text{Y-Momentum: } \frac{\partial V}{\partial \tau} + U\frac{\partial V}{\partial X} + V\frac{\partial V}{\partial Y} = \frac{1}{Re}\left(\frac{\partial^2 V}{\partial X^2} + \frac{\partial^2 V}{\partial Y^2}\right) - \frac{\partial P}{\partial Y}$$

$$\text{Energy: } \frac{\partial \theta}{\partial \tau} + U\frac{\partial \theta}{\partial X} + V\frac{\partial \theta}{\partial Y} = \frac{1}{Re\,Pr}\left(\frac{\partial^2 \theta}{\partial X^2} + \frac{\partial^2 \theta}{\partial Y^2}\right) + \overline{Q}^*_{gen}$$

where the above non-dimensional spatial as well as temporal coordinates are defined as

$$X = \frac{x}{l_c}, \; Y = \frac{y}{l_c}, \; \tau = \frac{t}{l_c/u_c} \tag{3.21}$$

and the non-dimensional flow properties are defined as

$$U = \frac{u}{u_c}, \; V = \frac{v}{u_c}, \; P = \frac{p}{\rho u_c^2}, \; \theta = \frac{T - T_\infty}{T_W - T_\infty} \left(\text{or} = \frac{T - T_\infty}{q_W l_c/k}\right) \tag{3.22}$$

Note the appearance of two non-dimensional governing parameters in Eq. 3.20. They are Reynolds number (Re) and Prandtl number (Pr), defined as

$$Re = \frac{u_c l_c}{\nu}, \; Pr = \frac{\nu}{\alpha} \tag{3.23}$$

For the above definitions, l_c is characteristic length scale and u_c is the velocity scale; T_W is the wall temperature for CWT BC and q_W is the wall heat flux for UHF BC;

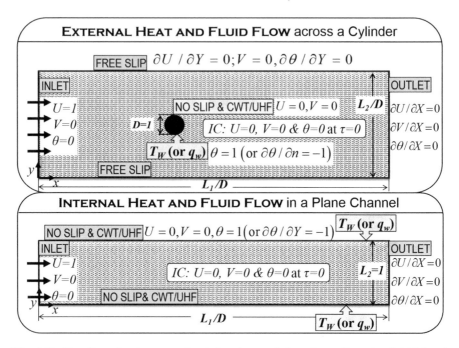

Fig. 3.12 Non-dimensional computational domain, consisting of non-dimensional initial and boundary conditions, for an external and an internal flow problem

T_∞ is the free-stream temperature; and $v \, (= \mu/\rho)$ is the kinematic viscosity and α $(= k/\rho c_p)$ is the thermal diffusivity. Finally, the non-dimensional volumetric heat generation is given as

$$\overline{Q}_{\text{gen}}^* = \frac{\overline{Q}_{\text{gen}}}{\rho c_p u_c (T_W - T_\infty)/l_c} \quad \text{for CWT BC}$$

$$= \frac{\overline{Q}_{\text{gen}}}{\rho c_p u_c q_W/k} \quad \text{for UHF BC}$$

3.5.2.2 Initial and Boundary Conditions

For the external flow across a cylinder, and the internal flow in a channel, the IC and the BCs for a dimensional study are already shown in Fig. 2.7b, c; similar figure for a non-dimensional study is presented here in Fig. 3.12. Note that the figure shows the IC and the BCs, considering the non-dimensional form of the spatial and temporal coordinates (Eq. 3.21), and the flow properties (Eq. 3.22).

The non-dimensionalization considers the non-dimensional length scale as $l_c = D$ for the external flow, and $l_c = L_2$ for the internal flow problem; and the velocity scale as $u_c = u_\infty$ for both the problems. Using the definition of the non-dimensional temperature (Eq. 3.22) at the cylinder/channel wall, note from the figure that that the constant wall temperature BC results in $\theta = 1$, and uniform heat flux BC results in $\partial\theta/\partial\eta = -1$ (where η is the coordinate normal to the solid surface) at the walls.

3.5.2.3 Governing and Engineering Parameters

The non-dimensionalization results in an appropriate combination of the three types of the dimensional governing properties (presented in Sect. 3.5.1.3), defined as the non-dimensional governing parameters. For example, for the flow across a cylinder, a dimensional study results in the thermo-physical property, diameter of the cylinder (geometrical) and inlet velocity (BC related), as the three types of dimensional governing parameter. Their appropriate combination results in the non-dimensional governing parameter as Reynolds number $Re = u_\infty D/\nu$ and the Prandtl number $Pr = \nu/\alpha$. Note that the non-dimensionalization results in a substantial reduction of the governing/input parameters; thereby, making the CFD analysis easier. Furthermore, for certain problems, smaller or larger magnitude of certain non-dimensional governing parameters indicate the importance of certain terms (such as advection or diffusion) in the flow; thereby, neglecting certain terms in a fluid dynamics study.

For the flow across a cylinder and channel, although the thermo-physical properties, the geometrical dimensions, and the BCs related parameters disappears as the governing properties in the non-dimensional study, they can appear as the non-dimensional governing parameters for certain problems. For example, the nondimensional thermo-physical property is governing parameter, for a multiphase flow; the nondimensional width of the cylinder is a geometrical governing parameter, if the shape of the cylinder is rectangular instead of circular; and the amplitude as well as frequency of oscillation of the cylinder (if the cylinder is oscillating periodically like a pendulum) are the BCs related governing parameter. Also note that the above (dimensional as well as non-dimensional) differential formulation are presented for a *single phase forced convection heat transfer*. However, for the other heat and fluid flow situations, there are additional non-dimensional governing equations; such as, Grashoff number for the natural convection, Richardson number for the mixed convection, and Weber number and Froude number for the multi-phase flow. The respective numbers are given as

$$Gr = \frac{g\beta(T_W - T_\infty)l_c^3}{\nu^2}, \ Ri = \frac{Gr}{re^2}, \ We = \frac{\rho u_c^2 l_c}{\sigma} \ \text{and} \ Fr = \frac{u_c}{\sqrt{gl_c}}$$

Engineering parameters such as lift/drag force and rate of heat transfer are discussed above (Sect. 3.5.1.3) in dimensional form. The non-dimensional form of the forces are called the lift and drag coefficient, given as

$$C_L = \frac{F_L}{\rho u_c^2 l_c/2} \ \text{and} \ C_D = \frac{F_D}{\rho u_c^2 l_c/2}$$

The above non-dimensional engineering parameters are for an external flow; whereas, for the internal flow, the parameters are skin friction coefficient and friction factor; given as

$$c_f = \frac{\tau_w}{\rho u_c^2} \text{ and } f = \frac{l_c}{\rho u_c^2} \frac{\partial p}{\partial x}$$

The respective parameters are non-dimensional representation of wall shear stress and pressure gradient (in the axial direction). The non-dimensional engineering parameter in heat transfer is the Nusselt number; given (Sharma and Eswaran 2004) as

$$Nu = -\frac{\partial \theta}{\partial \eta} \text{ for CWT BC, and } Nu = \frac{1}{\theta_W} \text{ for UHF BC}$$

where η is the coordinate normal to a solid surface.

3.6   Closure

In this chapter, the physical laws are customized for CFD, the transport mechanisms are discussed with a physical example, the physical law based differential formulations are demonstrated, the flux and volumetric terms (encountered later during a physical law-based algebraic formulation in CFD) are highlighted, boundary conditions are presented in a generalized form, and the governing as well as engineering parameters commonly encountered in CFD are presented in dimensional as well as non-dimensional form.

References

Cengel, Y. A., & Cimbala, J. M. (2006). *Fluid mechanics: fundamentals and applications.* New Delhi: Tata McGraw Hill.

Incropera, F. P., & Dewitt, D. P. (1996). *Fundamentals of heat and mass transfer* (4th ed.). New York: Wiley.

Muralidhar, K., & Biswas, G. (2005). *Advanced fluid mechanics* (2nd ed.). New Delhi: Narosa Publishing House.

Patankar, S. V. (1980). *Numerical heat transfer and fluid flow.* New York: Hemisphere Publishing Corporation.

Sharma, A., & Eswaran, V. (2004). Heat and fluid flow across a square cylinder in the two-dimensional laminar flow regime. *Numer. Heat Transfer A, 45*(3), 247–269.

Chapter 4
Essentials of Numerical-Methods for CFD

Numerical method is an approximate method for solving mathematical problems, taking into account the extent of possible errors. The method is needed as many mathematical problems do not lead to an exact solution; for0 example, solving a general polynomial of degree five or higher and most of the non-linear differential equations. The origin of numerical methods was much before the invention of computers. However, the development of computer hardware as well as programming language led to an enormous development in the field of numerical methods.

Computer results in a *machine error*, during a numerical solution. This is due to a certain fixed number of digits used to represent a number in a digital computer called precision. The digital form of continuous variable results in an approximate or discrete representation—with a single or double precision. The machine errors are of three types: first, round-off (or truncation) errors; second, overflow; and third, underflow. *Round-off error* corresponds to the difference between the computed approximation of a number and its exact mathematical value. *Overflow* occurs when a computer attempts to handle a number which is close to infinity—too large for the space designated to hold it; vice versa for *underflow*—number too close to zero. For example, if your computer supports eight decimal places of precision and a calculation produces the number 0.000000008 (with nine decimal places), an underflow condition occurs.

Numerical method results in a *discretization error*, during a discretization procedure (FDM or FVM) to convert a differential equation to a system of linear algebraic equations. The discretization error corresponds to a difference between the exact solution of the differential equation and the exact solution of the LAEs. An algorithm is *numerically stable* if an error, once it is generated, does not grow too much during the computation. Art of a numerical method is to find a stable (or well-conditioned) algorithm for solving a mathematical problem. The essential numerical methods for CFD considered in this chapter corresponds to the various steps of the CFD development, presented in the Chap. 2 (Fig. 2.1). The mind map for this chapter

© The Author(s), under exclusive license to Springer Nature Switzerland AG 2022
A. Sharma, *Introduction to Computational Fluid Dynamics*,
https://doi.org/10.1007/978-3-030-72884-7_4

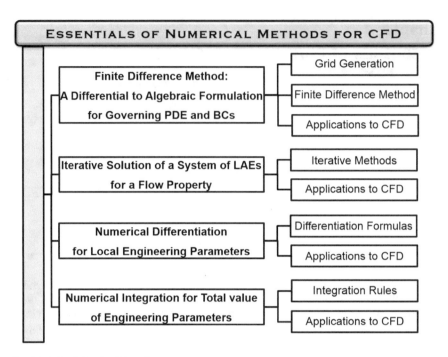

Fig. 4.1 Mind Map for Chap. 4

is presented in Fig. 4.1. The figure shows that this chapter presents four numerical methods: first, finite difference method; second, iterative solution of the system of LAEs; third, numerical differentiation and fourth, numerical integration. The first method is a discretization method, the second method is used as a part of the solution methodology, and the third as well as fourth methods are used for the computation of engineering parameters. The respective methods correspond to the second step, third step, and fourth step of the CFD development; refer Fig. 2.1. The first step in the figure, on grid generation, is also presented here in the first section of the finite difference method (Fig. 4.1); and the last step, on testing is presented later in an example problem. Figure 4.1 shows that each numerical method is introduced here with a formulation and followed by its application to CFD.

Although the FVM is recent method and FDM is an old method in CFD, FDM is gaining importance nowadays for a large class of CFD problems; such as moving boundary problems using Cartesian grid-based immersed boundary method (Mittal and Iaccarino 2005). Note that the second step in Fig. 2.1 shows application of FDM to boundary grid points and FVM to internal grid points; however, the FDM is presented in this chapter for both internal and boundary grid points. Since the main focus of this book is on FVM, FDM-based numerical methodology in this chapter is limited to a representative 2D steady-state heat conduction problem. For

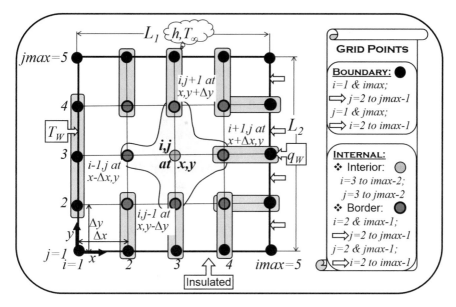

Fig. 4.2 A uniform body-fitted Cartesian grid generation, for a finite difference method. The figure also shows the various types of grid points, computational stencils (for an internal grid point and various boundary grid points), and the BCs for a 2D heat conduction problem

the later chapters of this book, FVM is used to obtain the LAEs from governing PDE/conservation laws; and the FDM is indeed used for the discretization of BCs and computation of local engineering parameters by numerical differentiation. Note that the numerical methodology is presented in this chapter for FDM, using a steady-state formulation; and will be presented in the later chapters for FVM, using an unsteady-state formulation.

4.1 Finite Difference Method: A Differential to Algebraic Formulation for Governing PDE and BCs

CFD results in the conversion of the continuous problem to a discrete problem, first by *domain discretization,* and then the continuous *equation discretization.* The former discretization of spatial region is called grid generation and the latter discretization of the continuous PDEs and BCs is called a discretization method. FDM is a discretization method to convert the *continuous* differential formulation to a *discrete* algebraic formulation for a flow property. Both the domain and equation discretization, grid generation and FDM, are presented in separate subsection below.

4.1.1   Grid Generation

Grid generation is a method for the *spatial discretization* of the inside and boundary region of the domain called internal and boundary grid points, respectively. For a simple geometry problem in a Cartesian Coordinate system, Fig. 4.2 shows the simplest form of grid generation obtained by drawing equi-spaced horizontal and vertical lines—called grid lines. It is ensured that the boundary of the domain is represented by one of the grid lines, called as *body-fitted grid*. Similar grid generation procedure is already introduced in Fig. 2.2 for the FVM. However, the circles in both the figures shows that the grid points are defined differently in the FVM and the FDM − at the *vertex* of the CVs for the *FDM* (Fig. 4.2) and at the *centroid* of the CVs for the FVM (Fig. 2.2).

Figure 4.2 shows that the length of the domain is L_1 in $x−$direction and L_2 in $y−$direction. It can be seen that the spacing between the consecutive horizontal line is Δx and the vertical line is Δy—called *grid spacing*. Since the grid lines/points are equi-spaced, the distribution of grid is called *uniform grid*. The grid points are classified here into two types: boundary and internal, shown in the figure by black and gray color filled circles, respectively. The classification is done due to the fact that the grid points for the application of FDM to PDE and BCs are different—FDM is applied to the PDE at the internal grid points, and to the BCs at the boundary grid points. For a better explanation of the numerical methodology below, the internal grid points are further classified as interior and border—differentiated in Fig. 4.2 by a thicker boundary of the filled circle for the border as compared to the interior points. The figure shows the grid point at (3,3) is the only interior grid point, and the other *internal* grid points are border grid points. The *border grid point is an internal grid point* which has at least one of the adjoining (east or west or north or south) neighbor as the boundary grid point.

Figure 4.2 also shows i and j as the *running indices*—consisting of integer values—which act as a tag (or address) for a grid point. The respective running indices corresponds to the rows and columns of a matrix − a computationally efficient data structure to store the values of a flow property at the grid points shown in the figure. Although (i, j) for a matrix represent an element by i^{th} row and j^{th} column, note that the running indices (i, j) for a grid point here (Fig. 4.2) corresponds to i^{th} column and j^{th} row; the difference is due to a variation in the definition of the origin: west-south corner for the *Cartesian* domain (Fig. 4.2), and north-west corner for the definition of a matrix. Also note that the *neighboring information is inbuilt in a matrix*. For example, for a representative element (i, j) of a matrix, Fig. 4.2 shows that $(i + 1, j)$, $(i − 1, j)$, $(i, j + 1)$, and $(i, j − 1)$ correspond to the adjoining east, west, north, and south neighboring element, respectively. The figure also shows that the respective neighboring grid points are located at $(x + \Delta x, y)$, $(x − \Delta x, y)$, $(x, y + \Delta y)$, and $(x, y − \Delta y)$.

The in-built neighboring information in the matrix—as the data structure—makes the computer programming in CFD much easier. Using a second-order FDM on the governing PDE for a 2D steady-state conduction, it is presented below that the temperature at *any* internal grid point is represented in terms of *four* (east, west, north, and south; not all) adjoining neighboring grid points. This functional relationship is presented schematically in Fig. 4.2, for *one* of the representative internal grid point, and is called as *computational stencil*. Note that the functional dependence remains same for all the other internal grid points. The stencil for the Dirichlet BC at the west boundary, and non-Dirichlet BCs at the other boundaries, are also shown in Fig. 4.2. Using first-order FDM for the non-Dirichlet BCs, it is presented below that the temperature at a boundary grid point is a function of one adjoining border grid point. Thus, the figure shows that the functional dependence for the boundary grid points and the computational stencil is same for any type of non-Dirichlet BC; Neumann BC at the south and east boundary, and mixed BC at the north boundary. For the grid point-by-point recursive computation, a range of values of the running indices (*i* and *j*) is shown in Fig. 4.2—separately for the internal and boundary grid points, since their computational stencils are different. The recursive computation corresponds to scrolling through all the internal grid points for an iterative solution of the corresponding LAE, with the help of a "for" or "do" loop in a computer program.

4.1.2 Finite Difference Method

FDM is a discretization method for algebraic representation of the governing PDE and BCs, at each of the internal and boundary grid points, respectively. Since the continuous equations for the PDE and the BCs are quite different, the FDM results in a different type of LAEs for the internal grid points as compared to that for the boundary grid points; presented below.

The four essential numerical methods are presented in this chapter, for a 2D steady-state heat conduction problem with a volumetric heat generation. The governing equation (a Poisson equation) and the various types of thermal boundary conditions considered (Fig. 4.2) for the 2D conduction problem, are given as

$$\textbf{G.E}: \frac{\partial^2 T}{\partial x^2} + \frac{\partial^2 T}{\partial y^2} + \frac{\overline{Q}_{gen}}{k} = 0 \tag{4.1}$$

$$\textbf{BCs:} \; At\, x = 0 \Rightarrow T = T_W$$

$$At\, x = L_1 \Rightarrow -k \left(\frac{dT}{dx} \right)_{x=L_1} = -q_W$$

$$At\, y = 0 \Rightarrow \left(\frac{dT}{dy} \right)_{y=0} = 0$$

$$At\, y = L_2 \Rightarrow -k \left(\frac{dT}{dy} \right)_{y=L_2} = h(T_{y=L_2} - T_\infty) \tag{4.2}$$

where the negative value of q_W is due to the heat flux in the negative x−direction on the east boundary, shown in Fig. 4.2. Note that the steady-state form of the above GE for heat conduction, results in a steady-state formulation, for the FDM-based CFD development, in this chapter. However, in the later chapters, FVM-based CFD development is presented for an unsteady-state formulation; considering the unsteady form of the conservation laws.

Using the above GEs and BCs, FDM is used to obtain an approximate algebraic formulation for each of the differential term. Note from that the GE (Eq. 4.1) consist of second derivative, and the non-Dirichlet BCs (Eq. 4.2) consist of only first derivative in space. Thus, presentation of FDM below is limited to the first and second derivatives in space; mostly encountered not only in heat transfer but also in fluid dynamics (Eq. 3.17).

The presentation of the FDM below considers a general symbol ϕ corresponding to a flow property—$\phi = T$ for heat conduction. Using a second-order FDM (presented below), for the first and second derivatives in the GEs and BC, the resulting algebraic equation consisting of discrete values of ϕ at the point under consideration (i, j) and the first adjoining neighboring grid point on the either side; east $(i + 1, j)$ and west $(i − 1, j)$ for the derivatives in x−direction, and north $(i, j + 1)$ and south $(i, j − 1)$ for the derivative in y−direction. A *symbolic* representation of a FDM-based approximated algebraic equation, for the derivatives in the x and y direction, are given as

$$\left(\frac{\partial\phi}{\partial x}\right)_{i,j} \Delta x, \ \left(\frac{\partial^2\phi}{\partial x^2}\right)_{i,j} \Delta x^2 \approx f\phi_{i+1,j} + c\phi_{i,j} + b\phi_{i-1,j}$$

$$\left(\frac{\partial\phi}{\partial y}\right)_{i,j} \Delta y, \text{ and } \left(\frac{\partial^2\phi}{\partial y^2}\right)_{i,j} \Delta y^2 \approx f\phi_{i,j+1} + c\phi_{i,j} + b\phi_{i,j-1} \qquad (4.3)$$

where $\phi_{i+1,j}$, $\phi_{i,j}$, and $\phi_{i-1,j}$ are the values at a grid point *forward* at $(x + \Delta x, y)$, *under-consideration* at (x, y), and *backward* at $(x − \Delta x, y)$, respectively; the respective grid points in the y−direction are $\phi_{i,j+1}$, $\phi_{i,j}$, and $\phi_{i,j-1}$. Furthermore, the coefficient for the respective $\phi's$ are presented in the above equation as f, c, and b—for both first and second derivatives. However, note that the value for each coefficient $(f, c,$ and $b)$ is different for the first and second derivatives; shown later. The coefficients are some constant values, presented below after a detailed formulation.

4.1.2.1 Taylor Series

A Taylor series is an *infinite series* representing any function $\phi_{x+\Delta x}$ (ϕ at a location $x + \Delta x$) as an infinite sum of terms, consisting of the value and the derivatives of the function ϕ at a *single point* x; given as

$$\phi(x + \Delta x) = \phi(x) + \left(\frac{d\phi}{dx}\right)_x \Delta x + \left(\frac{d^2\phi}{dx^2}\right)_x \frac{\Delta x^2}{2!} + \left(\frac{d^3\phi}{dx^3}\right)_x \frac{\Delta x^3}{3!} \cdots \left(\frac{d^n\phi}{dx^2}\right)_x \frac{\Delta x^n}{n!}$$
$$\qquad (4.4)$$

where Δx is any positive or negative spacing from the single representative location x. Considering upto first, second, third, and fourth term, and truncating the remaining terms, an approximate solution of $\phi(x + \Delta x)$ is obtained as ϕ_1, ϕ_2, ϕ_3, and ϕ_4, respectively; given as

$$\phi_1(x + \Delta x) \approx \phi(x)$$
$$\phi_2(x + \Delta x) \approx \phi(x) + \left(\frac{d\phi}{dx}\right)_x \Delta x$$
$$\phi_3(x + \Delta x) \approx \phi(x) + \left(\frac{d\phi}{dx}\right)_x \Delta x + \left(\frac{d^2\phi}{dx^2}\right)_x \frac{\Delta x^2}{2!} \tag{4.5}$$
$$\phi_4(x + \Delta x) \approx \phi(x) + \left(\frac{d\phi}{dx}\right)_x \Delta x + \left(\frac{d^2\phi}{dx^2}\right)_x \frac{\Delta x^2}{2!} + \left(\frac{d^3\phi}{dx^3}\right)_x \frac{\Delta x^3}{3!}$$

The accuracy of the above approximate representation of $\phi(x + \Delta x)$, by the Taylor series expansion, depends on both the number of terms considered and the value of the spacing Δx (Eq. 4.5); demonstrated below with the help of an example problem.

Example 4.1: Taylor Series Expansion.
Consider a continuous and differentiable function: $\phi(x) = e^{-x}$. Considering upto fourth term of the Taylor series expansion (Eq. 4.5), obtain the various approximate value of the function $\phi(x)$ at $x = 0$, using (a) $\Delta x = 1$ and (b) $\Delta x = 0.5$. Also determine the %error of the approximate as compared to the exact solution. Show the approximate value of ϕ in a figure, and discuss the error for the various approximate representation as compared to exact solution.

Solution:

(a) Substituting $\phi(x) = e^{-x}$, $x = 0$, and $\Delta x = 1$ in Eq. 4.4, the Taylor series expansion is given as

$$\phi(1) = e^{-1} = 1 - 1 + \frac{1}{2} - \frac{1}{6} \cdots (-1)^n \frac{1}{n!} \tag{4.6}$$

Considering upto first, second, third, and fourth term, and truncating the remaining terms, an approximate solution of $\phi(1)$ is obtained as ϕ_1, ϕ_2, ϕ_3, and ϕ_4, respectively; similar to Eq. 4.5 above. The approximate and the exact solution are given as

$$
\begin{array}{llll}
\phi_1(1) & \approx 1 & = 1 & (171.81\%) \\
\phi_2(1) & \approx 1 - 1 & = 0 & (100.00\%) \\
\phi_3(1) & \approx 1 - 1 + \frac{1}{2} & = 0.5 & (35.91\%) \\
\phi_4(1) & \approx 1 - 1 + \frac{1}{2} - \frac{1}{6} & = 0.3334 & (09.38\%) \\
\phi_{exact}(1) & = e^{-1} & = 0.3679 &
\end{array} \tag{4.7}
$$

where the term within the bracket represents the % error, given as

$$\% \text{ error} = \frac{|\phi_{m=1,4} - \phi_{exact}|}{\phi_{exact}} \times 100$$

(b) Similarly, the Taylor series for $\phi(x) = e^{-x}$ at $x = 0$ and $\Delta x = 0.5$ is obtained from Eq. 4.4 as

$$e^{-0.5} = 1 - 1 \times 0.5 + \frac{1}{2} \times (0.5)^2 - \frac{1}{6} \times (0.5)^3 \cdots (-1)^n \frac{\Delta x^n}{n!} \quad (4.8)$$

Then, the solutions and the corresponding errors are given as

$$
\begin{array}{lll}
\phi_1(0.5) & \approx 1 & = 1 \quad (64.88\%) \\
\phi_2(0.5) & \approx 1 - 0.5 & = 0.5 \quad (17.56\%) \\
\phi_3(0.5) & \approx 1 - 1 \times 0.5 + \frac{1}{2} \times (0.5)^2 & = 0.625 \quad (03.05\%) \\
\phi_4(0.5) & \approx 1 - 1 \times 0.5 + \frac{1}{2} \times (0.5)^2 - \frac{1}{6} \times (0.5)^3 & = 0.6042 \quad (00.38\%) \\
\phi_{exact}(0.5) & = e^{-0.5} & = 0.6065
\end{array}
$$

$$(4.9)$$

Discussion:

Figure 4.3 shows a graphical comparison of the Taylor series-based approximate solutions and the exact solution, for $\Delta x = 0.5$ and 1. The above equations (Eq. 4.7 and 4.9) and the figure clearly shows that the error reduces substantially with an increase in the number of terms (of the Taylor series expansion) considered in the approximate solution. Furthermore, a more accurate Taylor series-based solution of $\phi(x + \Delta x)$ is obtained with $\Delta x = 0.5$ as compared to 1. Thus, the error reduces by considering larger number of terms as well as a smaller Δx. Also note from Eq. 4.5 and the figure that $\partial\phi/\partial x$ in ϕ_2 incorporates the slope and $\partial^2\phi/\partial x^2$ in ϕ_3 incorporates the curvature at $x = 0$.

4.1.2.2   Finite Difference Equation

Using a FDM, a derivative term is approximated as a linear algebraic equation—the LAE is called **F**inite **D**ifference **E**quation (FDE). However, the same FDE can be derived either by a Taylor series expansion or a local polynomial profile assumption; presented below separately.

Taylor Series-based Algebraic Formulation

Although the Taylor series is introduced in the previous subsection for a function $\phi(x)$ of one independent variable x, it is also applicable for a function of more than

Fig. 4.3 Illustration of a representative function $\phi(x) = e^{-x}$, and approximate solutions from a Taylor series expansion at $x = 0$. The function is shown by the solid line, and the solutions by the symbols; filled by black color for $\Delta x = 1$ and gray color for $\Delta x = 0.5$. The solution ϕ_1, ϕ_2, ϕ_3, and ϕ_4 considers upto first, second, third, and fourth term of the Taylor series (Eq. 4.5)

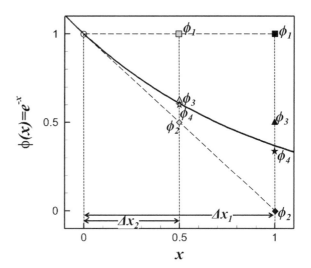

one independent variable; with a partial instead of total derivative. For a 2D problem with x and y as the independent variables, the Taylor series expansion is given as

$$\phi(x + \Delta x, y) = \phi(x, y) + \left(\frac{\partial \phi}{\partial x}\right)_{x,y} \Delta x + \left(\frac{\partial^2 \phi}{\partial x^2}\right)_{x,y} \frac{\Delta x^2}{2!} + \cdots \left(\frac{\partial^n \phi}{\partial x^n}\right)_{x,y} \frac{\Delta x^n}{n!}$$
(4.10)

$$\phi(x - \Delta x, y) = \phi(x, y) - \left(\frac{\partial \phi}{\partial x}\right)_{x,y} \Delta x + \left(\frac{\partial^2 \phi}{\partial x^2}\right)_{x,y} \frac{\Delta x^2}{2!} \cdots (-1)^n \left(\frac{\partial^n \phi}{\partial x^n}\right)_{x,y} \frac{\Delta x^n}{n!}$$
(4.11)

where the LHS of the above equations consist of ϕ at a spacing of Δx and $-\Delta x$ from x, and same y. From *each* of the above equation, the first derivative is given as

$$\left(\frac{\partial \phi}{\partial x}\right)_{x,y} = \frac{\phi(x + \Delta x, y) - \phi(x, y)}{\Delta x} - \left[\left(\frac{\partial^2 \phi}{\partial x^2}\right)_{x,y} \frac{\Delta x}{2!} + \cdots\right]$$
(4.12)

$$\left(\frac{\partial \phi}{\partial x}\right)_{x,y} = \frac{\phi(x, y) - \phi(x - \Delta x, y)}{\Delta x} + \left[\left(\frac{\partial^2 \phi}{\partial x^2}\right)_{x,y} \frac{\Delta x^2}{2!} + \cdots\right]$$
(4.13)

Using *both* the equations, subtracting Eq. 4.11 from Eq. 4.10 results in a FDE for the first derivative, and adding both the equations results in a FDE for the second derivative; given as

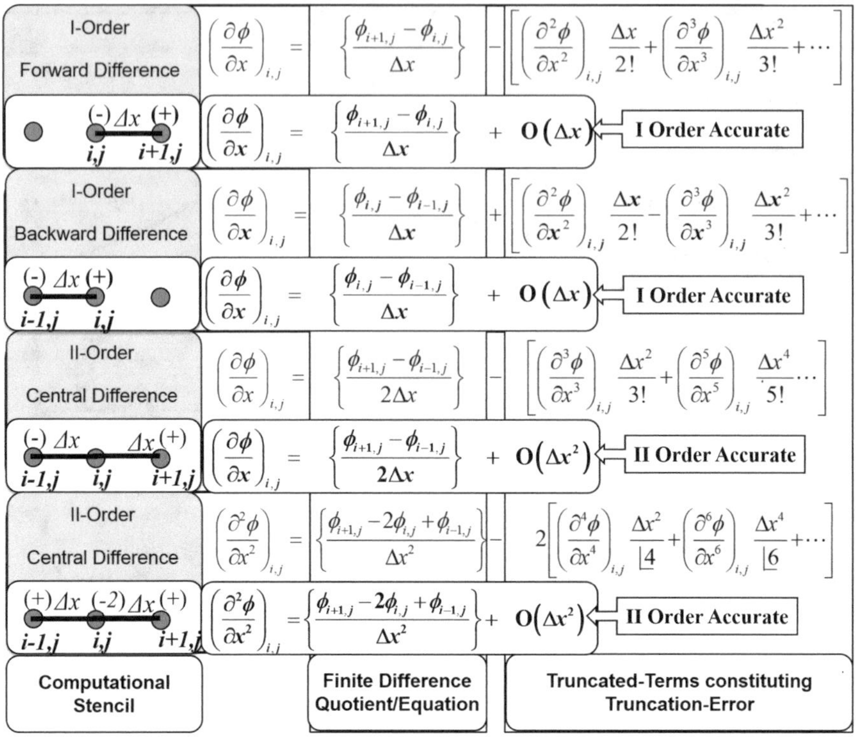

Computational Stencil	Finite Difference Quotient/Equation	Truncated-Terms constituting Truncation-Error

Fig. 4.4 Illustration of the various types of FDM-based algebraic representation of the first and second derivatives, using Taylor series expansion

$$\left(\frac{\partial \phi}{\partial x}\right)_{x,y} = \frac{\phi(x + \Delta x, y) - \phi(x - \Delta x, y)}{2\Delta x} - \left[\left(\frac{\partial^3 \phi}{\partial x^3}\right)_{x,y} \frac{\Delta x^2}{3!} + \cdots\right] \quad (4.14)$$

$$\left(\frac{\partial^2 \phi}{\partial x^2}\right)_{x,y} = \frac{\phi(x + \Delta x, y) - 2\phi(x, y) + \phi(x - \Delta x, y)}{\Delta x^2} + 2\left[\left(\frac{\partial^4 \phi}{\partial x^4}\right)_{x,y} \frac{\Delta x^2}{4!} + \cdots\right] \quad (4.15)$$

Substituting the spatial coordinates in the above equations as running indices: (x, y) as (i, j), $(x + \Delta x, y)$ as $(i + 1, j)$, and $(x - \Delta x, y)$ as $(i - 1, j)$), a FDM-based representation of the first and second derivative are shown in Fig. 4.4. Division of a dividend with a divisor results in a quotient and a remainder; similarly, for the RHS of the above Eqs. 4.12–4.15, the first term is called *finite difference quotient* (Anderson 1995) and the second term (within a square bracket) is called truncated term—results in a *truncation error*. The finite difference quotient and the truncated terms are shown more clearly in Fig. 4.4. The figure shows the various finite difference quotient as an algebraic term. A quotient which involves one-point forward, one-point backward, and both forward and backward points is called forward, backward, and central difference method, respectively. The points involved in the quotient are also shown schematically (joined by straight lines) in the figure called *computational stencil*. Furthermore, the truncated terms in the figure shows higher order derivatives multiplied by a certain power of Δx. The power of the Δx in the lowest order derivative of the truncated terms decides the *order of accuracy* of a finite difference representation; seen in the figure as first order for forward and backward difference method, and second order for central difference method.

Finally, the finite difference quotient or equation (corresponding to FDM-based *approximate* algebraic formulation of the derivatives) are given as

$$
\begin{array}{lll}
I - \text{Order Forward Difference:} & \left(\dfrac{\partial \phi}{\partial x}\right)_{i,j} \approx \dfrac{\phi_{i+1,j} - \phi_{i,j}}{\Delta x} & (4.16) \\[3ex]
I - \text{Order Backward Difference:} & \left(\dfrac{\partial \phi}{\partial x}\right)_{i,j} \approx \dfrac{\phi_{i,j} - \phi_{i-1,j}}{\Delta x} &
\end{array}
$$

$$
\begin{array}{lll}
II - \text{Order Central Difference:} & \left(\dfrac{\partial \phi}{\partial x}\right)_{i,j} \approx \dfrac{\phi_{i+1,j} - \phi_{i-1,j}}{2\Delta x} & (4.17) \\[3ex]
II - \text{Order Central Difference:} & \left(\dfrac{\partial^2 \phi}{\partial x^2}\right)_{i,j} \approx \dfrac{\phi_{i+1,j} - 2\phi_{i,j} + \phi_{i-1,j}}{\Delta x^2} &
\end{array}
$$

Further *higher order* one-sided (forward/backward) or both-sided (central) FDEs— for the first and second derivatives — can also be derived using the above procedure. Note that the procedure corresponds to *algebraic elimination* from the Taylor series for ϕ at neighboring grid points. For example, Eq. 4.10 for $\phi_{i+1,j}$, and Eq. 4.11 for $\phi_{i-1,j}$, are used independently for forward/backward difference method; whereas, for central difference method, both the equations are used together—subtracted for first and added for second derivative. The subtraction led to Eq. 4.14, with the elimination of the second derivative term; and the addition led to Eq. 4.15, with the elimination of first as well as third derivative term. Taylor series equation for each neighboring grid-point result in one equation, and the number of neighboring grid point or equations are one for the first-order forward and backward difference method, and two for the second-order central difference method.

The *order of accuracy* is equal to the number of the *neighboring* grid points in the finite difference equation. This is shown in Fig. 4.4, and also presented below in the example problems for the higher order FDE.

Example 4.2: Second-order *one-sided* forward and backward difference method for the first derivative.

Present a Taylor series-based algebraic formulation (finite difference equation) for $\partial\phi/\partial x$.

Solution:

As discussed above, the order of accuracy is equal to number of neighboring grid points in the FDE. Thus, second-order forward and backward difference will require two one-sided neighboring grid points in the FDE: ϕ_{i+2} and ϕ_{i+1} for the forward difference method, and ϕ_{i-1} and ϕ_{i-2} for the backward difference method; one more neighbor as compared to the first-order forward and backward difference method (Eq. 4.16). Thus, for second-order forward difference method, the formulation will start with the Taylor series expansion for the two neighboring grid points as

$$\phi_{i+1,j} = \phi_{i,j} + \left(\frac{\partial\phi}{\partial x}\right)_{i,j}\Delta x + \left(\frac{\partial^2\phi}{\partial x^2}\right)_{i,j}\frac{\Delta x^2}{2!} + \left(\frac{\partial^3\phi}{\partial x^3}\right)_{i,j}\frac{\Delta x^3}{3!} + \cdots$$

$$(4.18)$$

$$\phi_{i+2,j} = \phi_{i,j} + \left(\frac{\partial\phi}{dx}\right)_{i,j}2\Delta x + \left(\frac{\partial^2\phi}{\partial x^2}\right)_{i,j}\frac{4\Delta x^2}{2!} + \left(\frac{\partial^3\phi}{\partial x^3}\right)_{i,j}\frac{8\Delta x^3}{3!} + \cdots$$

$$(4.19)$$

Thereafter, the derivation involves *algebraic elimination* using the above Taylor series expansions, i.e., an algebraic operation to eliminate the *II* derivative term. Thus, $(4\times\text{Eqs. }4.18-4.19)$ is given as

$$\left(\frac{\partial\phi}{\partial x}\right)_{i,j} = \frac{-3\phi_{i,j} + 4\phi_{i+1,j} - \phi_{i+2,j}}{2\Delta x} - \left[\left(\frac{\partial^3\phi}{\partial x^3}\right)_{i,j}\frac{2\Delta x^2}{3!} + \cdots\right] \quad (4.20)$$

Similarly, for second-order backward difference method, similar algebraic elimination from the Taylor series expansion for $\phi_{i-1,j}$ and $\phi_{i-2,j}$ is given as

$$\left(\frac{\partial\phi}{\partial x}\right)_{i,j} = \frac{3\phi_{i,j} - 4\phi_{i-1,j} + \phi_{i-2,j}}{2\Delta x} + \left[\left(\frac{\partial^3\phi}{\partial x^3}\right)_{i,j}\frac{2\Delta x^2}{3!} + \cdots\right] \quad (4.21)$$

Thus, the above equations results in the FDEs as

$$
\begin{aligned}
II-\text{Order Forward Difference:} \left(\frac{\partial\phi}{\partial x}\right)_{i,j} &\approx \frac{-3\phi_{i,j} + 4\phi_{i+1,j} - \phi_{i+2,j}}{2\Delta x} \\
II-\text{Order Backward Difference:} \left(\frac{\partial\phi}{\partial x}\right)_{i,j} &\approx \frac{3\phi_{i,j} - 4\phi_{i-1,j} + \phi_{i-2,j}}{2\Delta x}
\end{aligned}
$$

(4.22)

Example 4.3: Fourth-order *both-sided* central difference method for the first as well as second derivative.

Present a Taylor series-based algebraic formulation (finite difference equation) for (a) $\partial\phi/\partial x$ and (b) $\partial^2\phi/\partial x^2$.

Solution:

For the second derivative, the fourth-order as compared to second-order both-sided central difference method (Eq. 4.22) will require Taylor series expansion for two additional neighboring grid point on either side: $\phi_{i+2,j}$ and $\phi_{i,j-2}$. Thus, the derivation will start with the Taylor series expansion for the four neighboring grid points as

$$
\phi_{i+1,j} = \phi_{i,j} + \left(\frac{\partial\phi}{\partial x}\right)_{i,j}\Delta x + \left(\frac{\partial^2\phi}{\partial x^2}\right)_{i,j}\frac{\Delta x^2}{2!} + \left(\frac{\partial^3\phi}{\partial x^3}\right)_{i,j}\frac{\Delta x^3}{3!} + \left(\frac{\partial^4\phi}{\partial x^4}\right)_{i,j}\frac{\Delta x^4}{4!}\cdots
$$

(4.23)

$$
\phi_{i-1,j} = \phi_{i,j} - \left(\frac{\partial\phi}{\partial x}\right)_{i,j}\Delta x + \left(\frac{\partial^2\phi}{\partial x^2}\right)_{i,j}\frac{\Delta x^2}{2!} - \left(\frac{\partial^3\phi}{\partial x^3}\right)_{i,j}\frac{\Delta x^3}{3!} + \left(\frac{\partial^4\phi}{\partial x^4}\right)_{i,j}\frac{\Delta x^4}{4!}\cdots
$$

(4.24)

$$
\phi_{i+2,j} = \phi_{i,j} + \left(\frac{\partial\phi}{\partial x}\right)_{i,j}2\Delta x + \left(\frac{\partial^2\phi}{\partial x^2}\right)_{i,j}\frac{4\Delta x^2}{2!} + \left(\frac{\partial^3\phi}{\partial x^3}\right)_{i,j}\frac{8\Delta x^3}{3!} + \left(\frac{\partial^4\phi}{\partial x^4}\right)_{i,j}\frac{16\Delta x^4}{3!}\cdots
$$

(4.25)

$$
\phi_{i-2,j} = \phi_{i,j} - \left(\frac{\partial\phi}{\partial x}\right)_{i,j}2\Delta x + \left(\frac{\partial^2\phi}{\partial x^2}\right)_{i,j}\frac{4\Delta x^2}{2!} - \left(\frac{\partial^3\phi}{\partial x^3}\right)_{i,j}\frac{8\Delta x^3}{3!} + \left(\frac{\partial^4\phi}{\partial x^4}\right)_{i,j}\frac{16\Delta x^4}{3!}\cdots
$$

(4.26)

Thereafter, *for* $\partial\phi/\partial x$, the derivation involves an algebraic eliminations from the above equations, to eliminate the II, III and IV derivative terms in the above equations. In the first step, the II and IV derivative terms are eliminated by the algebraic operations as (Eqs. 4.23–4.24) and (Eqs. 4.25–4.26), respectively; given as

$$\phi_{i+1,j} - \phi_{i-1,j} = \left(\frac{\partial\phi}{\partial x}\right)_{i,j} 2\Delta x + \left(\frac{\partial^3\phi}{\partial x^3}\right)_{i,j} \frac{2\Delta x^3}{3!} + \left(\frac{\partial^5\phi}{\partial x^5}\right)_{i,j} \frac{2\Delta x^5}{5!} \cdots$$

(4.27)

$$\phi_{i+2,j} - \phi_{i-2,j} = \left(\frac{\partial\phi}{\partial x}\right)_{i,j} 4\Delta x + \left(\frac{\partial^3\phi}{\partial x^3}\right)_{i,j} \frac{16\Delta x^3}{3!} + \left(\frac{\partial^5\phi}{\partial x^5}\right)_{i,j} \frac{64\Delta x^5}{5!} \cdots$$

(4.28)

Finally, III derivative term is eliminated from the above equations by another algebraic operation as ($8\times$Eqs. 4.27–4.28); given as

$$\left(\frac{\partial\phi}{\partial x}\right)_{i,j} = \frac{-\phi_{i+2,j} + 8\phi_{i+1,j} - 8\phi_{i-1,j} + \phi_{i-2,j}}{12\Delta x} - \left[\left(\frac{\partial^5\phi}{\partial x^5}\right)_{i,j} \frac{4\Delta x^4}{5!} + \cdots\right]$$

Similarly, for fourth-order central difference representation of $\partial\phi^2/\partial x^2$, algebraic eliminations are done to eliminate I as well as III, IV and V derivative term in Eqs. 4.23–4.26. The I, III, and V derivative term are eliminated by the algebraic operations as (Eqs. 4.23+4.24) and (Eqs. 4.25+4.26). The resulting equations are subjected to another algebraic elimination of the IV derivative term; given as

$$\left(\frac{\partial^2\phi}{\partial x^2}\right)_{i,j} = \frac{-\phi_{i+2,j} + 16\phi_{i+1,j} - 30\phi_{i,j} + 16\phi_{i-1,j} - \phi_{i-2,j}}{12\Delta x^2}$$

$$+ \left[\left(\frac{\partial^6\phi}{\partial x^6}\right)_{i,j} \frac{8\Delta x^4}{6!} + \cdots\right]$$

Thus, the FDEs are derived as

Fourth-Order Central Difference:

$$\left(\frac{\partial\phi}{\partial x}\right)_{i,j} \approx \frac{-\phi_{i+2,j} + 8\phi_{i+1,j} - 8\phi_{i-1,j} + \phi_{i-2,j}}{12\Delta x}$$

(4.29)

$$\left(\frac{\partial^2\phi}{\partial x^2}\right)_{i,j} \approx \frac{-\phi_{i+2,j} + 16\phi_{i+1,j} - 30\phi_{i,j} + 16\phi_{i-1,j} - \phi_{i-2,j}}{12\Delta x^2}$$

Local/Piecewise Polynomial Profile-based Algebraic-Formulation

The above one-sided forward/backward and two-sided central difference equations can also be derived by assuming a polynomial variation *locally* in between the associated grid points. This is shown in Fig. 4.5, with the *local profile assumption* as

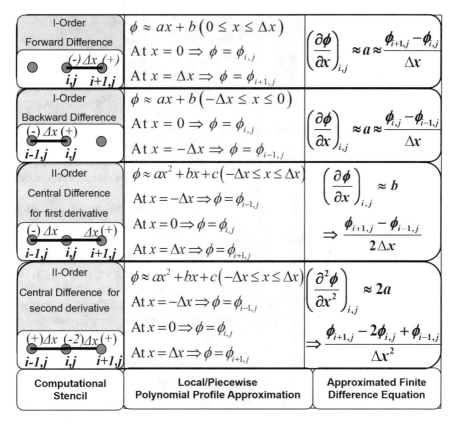

Fig. 4.5 Illustration of the various types of FDM-based algebraic representation of the first and second derivatives, using local/piecewise profile assumption as linear for first-order FDM and quadratic for second-order FDM

linear within $x_{i,j} \leq x \leq x_{i+1,j}$ and $x_{i-1,j} \leq x \leq x_{i,j}$ for the first-*order* forward and backward difference method, respectively; and *quadratic* within $x_{i-1,j} \leq x \leq x_{i+1,j}$ for the second-*order* central difference method. Since the assumption is local, the x-coordinate at the point under consideration is taken as $x_i = 0$, with $x_{i+1,j} = \Delta x$ and $x_{i-1,j} = -\Delta x$. The figure shows the equations for the local profile assumption along with the associated grid-point values of the local x-coordinate and ϕ, which are used in the derivation of the finite difference equation.

The figure shows that the first and second derivatives at the point under consideration ($x = 0$) is obtained in terms of the coefficients: a in the linear and b in the quadratic approximation for $(\partial\phi/\partial x)_{x=0}$; and $2a$ in the quadratic approximation for $(\partial^2\phi/\partial x^2)_{x=0}$. The derivation is presented in detail for the quadratic profile assumption below.

$$\phi \approx ax^2 + bx + c \Rightarrow \left(\frac{\partial \phi}{\partial x}\right)_{x=0} \approx b \;\&\; \left(\frac{\partial^2 \phi}{\partial x^2}\right)_{x=0} \approx 2a$$

$$\text{At } x = 0, \; \phi = \phi_{i,j} \Rightarrow c = \phi_{i,j}$$

$$\text{At } x = \Delta x, \; \phi = \phi_{i+1,j} \Rightarrow \phi_{i+1,j} - \phi_{i,j} = a\Delta x^2 + b\Delta x$$

$$\text{At } x = -\Delta x, \; \phi = \phi_{i-1,j} \Rightarrow \phi_{i-1,j} - \phi_{i,j} = a\Delta x^2 - b\Delta x$$

Solving the above algebraic equations, the coefficients a and b are obtained in terms of $\phi_{i,j}$, $\phi_{i+1,j}$, and $\phi_{i-1,j}$; shown in Fig. 4.5. The figure shows that the first-order forward and backward difference, and the second-order central difference equation (derived by the local polynomial approximation), are exactly same as that derived by the Taylor series approximation (Fig. 4.4).

Moreover, Eq. 4.22, for the second- *order* forward and backward difference method can also be derived by assuming a local *quadratic* variation; within $x_{i,j} \leq x \leq x_{i+2,j}$ for the forward difference and $x_{i-2,j} \leq x \leq x_{i,j}$ for the backward difference method. Furthermore, Eq. 4.29, for the fourth-*order* central difference method, can be derived by assuming a local *quartic* variation within $x_{i-2,j} \leq x \leq x_{i+2,j}$. For the local piecewise polynomial profile-based FDE, notice from the Fig. 4.5 that the *order of accuracy* of a FDE is equal to the *order of the polynomial*; and the *number of neighboring grid points* involved in the FDE are *one less* than the number of coefficients in the polynomial.

Coefficients for the LAEs in the Finite Difference Equations

The above FDEs (Eqs. 4.16, 4.17, 4.22, and 4.29) are also applicable for the derivatives in the $y-$direction, with certain modifications: $\phi_{i+2,j}$ to $\phi_{i,j+2}$, $\phi_{i+1,j}$ to $\phi_{i,j+1}$, $\phi_{i-1,j}$ to $\phi_{i,j-1}$, $\phi_{i-2,j}$ to $\phi_{i,j-2}$, Δx to Δy, and Δx^2 to Δy^2. Note that the grid point considered in the FDE for $\partial \phi/\partial x$ and $\partial \phi/\partial y$ are on a horizontal and vertical line, with a constant $y-$coordinate and $x-$coordinate, respectively. Thus, the *local polynomial profile assumption is one-dimensional*: $\phi = f(x)$ to derive the FDEs for $\partial \phi/\partial x$, and $\phi = f(y)$ in the derivation for $\partial \phi/\partial y$.

The generic finite difference equation, presented above as Eq. 4.3, for the first-order forward (and backward) and second-order central difference method can be extended to second-order forward (and backward) and fourth-order central difference method; given as

$$\left(\frac{\partial \phi}{\partial x}\right)_{i,j} \Delta x, \; \left(\frac{\partial^2 \phi}{\partial x^2}\right)_{i,j} \Delta x^2 \approx ff\phi_{i+2,j} + f\phi_{i+1,j} + c\phi_{i,j} + b\phi_{i-1,j} + bb\phi_{i-2,j}$$

$$\left(\frac{\partial \phi}{\partial y}\right)_{i,j} \Delta y, \; \left(\frac{\partial^2 \phi}{\partial y^2}\right)_{i,j} \Delta y^2 \approx ff\phi_{i,j+2} + f\phi_{i,j+1} + c\phi_{i,j} + b\phi_{i,j-1} + bb\phi_{i,j-2}$$

$$(4.30)$$

Comparing the generic FDE above with a specific FDE (Eqs. 4.16, 4.17, 4.22, and 4.29), the coefficients of ϕ for the various FDMs are shown in Table 4.1. Note from the table that the coefficients are some constant values which is same for a derivative

Table 4.1 The coefficients in the general FDE for the first and the second derivatives in x/y direction (Eq. 4.30)—on a uniform Cartesian grid

Derivatives	FDM	Order of Accuracy	bb	b	c	f	ff
First	Forward	I			-1	1	
		II			$-3/2$	2	$-1/2$
	Back-ward	I		-1	1		
		II	$1/2$	-2	$3/2$		
	Central	II		$-1/2$	0	$1/2$	
		IV	$1/12$	$-2/3$	0	$2/3$	$-1/12$
Second	Central	II		1	-2	1	
		IV	$-1/12$	$4/3$	$-5/2$	$4/3$	$-1/12$

in the x and y direction, but different for the first as compared to the second derivative. Also note from the table that the coefficient f is equal to b, and ff is equal to bb; however, the equality is with a *change in sign* of the coefficients for the $\partial\phi/\partial x$ but not for the $\partial^2\phi/\partial x^2$. Furthermore, the sum of the coefficients is equal to zero—the approximate representation ensures that the derivative is zero for a uniform flow property, *i.e.*, $\phi_{i+2,j} = \phi_{i+1,j} = \phi_{i,j} = \phi_{i-1,j} = \phi_{i,j+1} = \phi_{i,j-1} = \phi_{i,j-2}$.

The table also shows that the FDE with larger order of accuracy involves more neighboring grid points, *i.e..*, longer expression. Thus, the *computational cost* per grid point is *more* for the larger as compared to the smaller order FDE. However, their is a *reduction* in the *grid size* ($imax \times jmax$; Fig. 4.2) required by the higher order FDE to achieve results of almost *same* accuracy as that obtained by the lower order FDE. *Quality* of a numerical solution is judged by the *accuracy*, and the associated *price* corresponds to the *computational cost*. Thus, with the increase in the order of accuracy of a FDE at a constant grid size, both quality and price increases—*pay more to achieve a better quality*. However, to achieve an almost same accuracy of results, if the computational cost required by the higher as compared to the lower order accurate FDE is smaller, then the higher order FDE is preferred; otherwise, lower order FDE is preferred. Thus, a higher order scheme is not always preferred, and *a second-order accurate FDM has been widely accepted in the CFD development* (Anderson 1995); this is also true for *FVM*, presented later.

4.1.3 Applications to CFD

For the 2D steady-state heat conduction with volumetric heat generation. FDEs for the governing equation (Eq. 4.1) and the various types of thermal boundary conditions (Eq. 4.2) are given as

GE and FDE:

$$\frac{\partial^2 T}{\partial x^2} + \frac{\partial^2 T}{\partial y^2} + \frac{\overline{Q}_{gen}}{k} = 0 \Rightarrow \frac{T_{i+1,j} - 2T_{i,j} + T_{i-1,j}}{\Delta x^2} \tag{4.31}$$
$$+ \frac{T_{i,j+1} - 2T_{i,j} + T_{i,j-1}}{\Delta y^2} + \frac{\overline{Q}_{gen,i,j}}{k} = 0$$

BCs and FDEs:

$$T_{x=0} = T_W \qquad\qquad \Rightarrow T_{1,j} = T_W \equiv T_{wb}$$

$$-k\left(\frac{\partial T}{\partial x}\right)_{x=L_1} = -q_w \qquad \Rightarrow T_{imax,j} = T_{imax-1,j} + \frac{q_w \Delta x}{k} \equiv T_{eb}$$

$$\left(\frac{\partial T}{\partial y}\right)_{y=0} = 0 \qquad\qquad \Rightarrow T_{i,1} = T_{i,2} \equiv T_{sb}$$

$$-k\left(\frac{\partial T}{\partial y}\right)_{y=L2} = h(T_{y=L_2} - T_\infty) \Rightarrow T_{i,jmax} = \frac{kT_{i,jmax-1} + h\Delta y T_\infty}{(k + h\Delta y)} \equiv T_{nb}$$

$$\tag{4.32}$$

where second-order *central difference method* (Eq. 4.16) is used for the second derivative in the *GE* and first-order *forward (or backward) difference method* (Eq. 4.17) is used for the first derivative in the *BCs*. The application of the forward difference at the south boundary, and backward difference at the east and north boundaries, is decided by the one-sided forward or backward location of the border grid points adjoining the corresponding boundary grid points (Fig. 4.2).

For the *non-Dirichlet BCs*, a schematic representation of the grid points involved in the above FDE—called computational stencil—is shown in Fig. 4.2. Note from the stencil and the FDE above (Eq. 4.32) that the temperature at the east, south, and north *boundary* grid points (T_{eb}, T_{sb}, and T_{nb}) are expressed in terms of its neighboring *border* grid-point value; and it evolves with iteration for the steady problem here and with time for an unsteady problem (presented in the next chapter).

Example 4.4: Higher order FDE for the governing equation and boundary conditions, for 2D steady-state heat conduction with volumetric heat generation.

Present the discretization as (a) fourth-order for the second derivative in the GE and (b) second-order for the first derivative in the BCs.

Solution:

Using the Eq. 4.29 for the GE (Eq. 4.1) and Eq. 4.22 for the various types of thermal BCs (Eq. 4.2), the higher order FDEs-based discretization is given as

$$\textbf{G.E}: \frac{-T_{i+2,j} + 16T_{i+1,j} - 30T_{i,j} + 16T_{i-1,j} - T_{i-2,j}}{12\Delta x^2} \qquad (4.33)$$

$$+ \frac{-T_{i,j+2} + 16T_{i,j+1} - 30T_{i,j} + 16T_{i,j-1} - T_{i,j-2}}{12\Delta y^2} + \frac{\overline{Q}_{gen,i,j}}{k} = 0$$

BCs: At $x = 0$ **FDEs** : $T_{1,j} = T_W \equiv T_{wb}$

At $x = L_1$ $T_{imax,j} = \dfrac{4T_{imax-1,j} - T_{imax-2,j}}{3} + \dfrac{2q_W\Delta x}{3k} \equiv T_{eb}$

At $y = 0$ $T_{i,1} = \dfrac{4T_{i,2} - T_{i,3}}{3} \equiv T_{sb}$

At $y = L_2$ $T_{i,jmax} = \dfrac{k\left(4T_{i,jmax-1} - T_{i,jmax-2}\right) + 2h\Delta yT_\infty}{(3k + 2h\Delta y)} \equiv T_{nb}$

$$(4.34)$$

4.2 Iterative Solution of System of LAEs for a Flow Property

A general form of system of LAEs, for ϕ as a flow property with n equations and n unknowns, is given as

$$a_{11}\phi_1 + a_{12}\phi_2 + a_{13}\phi_3 + \cdots + a_{1n}\phi_n = b_1$$
$$a_{21}\phi_1 + a_{22}\phi_2 + a_{23}\phi_3 + \cdots + a_{2n}\phi_n = b_2$$
$$\vdots \qquad \vdots \qquad \vdots \qquad \vdots \quad = \vdots$$
$$a_{n1}\phi_1 + a_{n2}\phi_2 + a_{n3}\phi_3 + \cdots + a_{nn}\phi_n = b_n \qquad (4.35)$$

It can be also be expressed in a matrix form as $\mathbf{A}\Phi = \mathbf{B}$, where $\mathbf{A}$ is called *coefficient matrix*, Φ as *unknown vector* and $\mathbf{B}$ as *non-homogenous vector*; given as

$$\mathbf{A} = \begin{bmatrix} a_{11} & a_{12} & \cdots & a_{1n} \\ a_{21} & a_{22} & \cdots & a_{2n} \\ \vdots & \vdots & \ddots & \vdots \\ a_{n1} & a_{n2} & \cdots & a_{nn} \end{bmatrix}, \quad \Phi = \begin{bmatrix} \phi_1 \\ \phi_2 \\ \vdots \\ \phi_n \end{bmatrix} \quad \text{and} \quad \mathbf{B} = \begin{bmatrix} b_1 \\ b_2 \\ \vdots \\ b_n \end{bmatrix}$$

There are two broad approaches for the solution of the system of LAEs: direct and indirect. *Direct-method* is based on *algebraic eliminations* for obtaining solution in a fixed number of operations; for example, matrix inversion and Gauss elimination. *Indirect-method* is used to obtain the solution asymptotically by an *iterative* procedure; for example, Jacobi and Gauss-Seidel method. Where ever needed, the iterative method is commonly used in CFD; presented below.

4.2.1 Iterative Methods

An iterative method is initiated by assuming an *initial guess* and obtain an *improved or approximated solution* (called first iterative value), which is checked for a condition called *convergence*. If the condition is not obeyed, the first iterative value is used to obtain the second iterative value, and this procedure is repeated iteration by iteration till convergence. *Convergence* of an iterative method is defined by an error approaching towards practically zero − say 10^{-3} or 10^{-5}, as a *convergence-tolerance* (ϵ) for a computer.

Considering $N + 1$ as the *present* and N as the *previous* iteration number, and ϕ_i^{N+1} and ϕ_i^N as the respective iterative value, the actual *error* in the solution ($\phi_i^{N+1} - \phi_i^{\text{exact}}$) is approximated as $\Delta\phi_i \equiv \phi_i^{N+1} - \phi_i^N$ (Hoffman 1993). This error is computed for all the n values of ϕ and should be reduced *simultaneously* for all of them, to ensure the minimum accuracy desired (as per the value of ϵ) for the system of equations. One of the most common convergence criterion is given as

$$\left|\Delta\phi_{i=1,\,2\cdots n}\right|_{max} \leq \epsilon \tag{4.36}$$

where the LHS represents absolute value of the maximum from all the $\Delta\phi_i$. Ideally, with increase in the iteration number, a monotonic decay in the error is expected; however, an iterative procedure should to terminated if there is almost no change in the error.

Iterative or indirect methods are generally used under *four conditions* (Hoffman 1993): *first*, the number of equation is large (100 or more); *second*, most of the coefficient a_{ij} are zero (*sparse matrix*); *third*, the system of equations is *diagonally dominant* (if the *absolute* value of each term on the major diagonal of its *coefficient matrix* is *equal to, or larger than*, the sum of the absolute value of all the other terms in that row of the matrix); and *fourth*, the system of equations is *not ill-conditioned*. These conditions are commonly encountered in CFD: the first condition due to large grid size, the second condition due to the discretization method (FDM or FVM) resulting in a LAE with sparse coefficient matrix (presented below in Example 4.6), the third condition is ensured during the discretization, and the fourth condition is avoided during the discretization. Thus, it is generally more efficient in terms of computational time and storage requirements to solve the system of LAEs in CFD by the iterative than the direct methods.

Jacobi and Gauss-Seidel method are two of the simplest iterative methods; and will be presented here. In these methods, each of the LAE is converted into an *explicit* equation for one of the element of the unknown vector, and the values of the other elements are taken from the *previous iteration* for *Jacobi* and the *improved (or latest) value* for *Gauss-Seidel* method. They are represented mathematically as

$$\text{Jacobi: } \phi_i^{N+1} = \frac{\sum_{i=1, i \neq j}^{n} (a_{ij}\phi_i^N - b_i)}{a_{ii}} \tag{4.37}$$

$$\text{Gauss-Seidel: } \phi_i^{N+1} = \frac{\sum_{i=1}^{j<i} (a_{ij}\phi_i^{N+1} - b_i) + \sum_{j>i}^{n} (a_{ij}\phi_i^N - b_i)}{a_{ii}} \tag{4.38}$$

where the $n-$algebraic equations (Eq. 4.35) are solved in a *sequence* $i = 1, 2 \cdots n$. The above equations show that an iterative solution for the LAEs are *simultaneous* in Jacobi and *successive* in Gauss-Seidel method—the sequence of the solution of the LAEs is relevant for the Gauss-Seidel but not for the Jacobi method. The diagonal dominance is a *sufficient* condition for the convergence of the Jacobi and Gauss-Seidel methods; for any initial guess. If a system of LAEs is not diagonally dominant, the sequence of the LAEs can be rearranged to make them diagonally dominant. This is presented below with the help of an example problem below.

Example 4.5: Jacobi and Gauss-Seidel Method.
   Using a convergence tolerance of 10^{-4}, solve a system of algebraic equation:

$$9\phi_1 + \phi_2 - \phi_3 = 9$$
$$\phi_1 + 2\phi_2 + 9\phi_3 = 12$$
$$\phi_1 - 8\phi_2 + 3\phi_3 = -4$$

Solution:

The first step in the iterative solution is to make the above system of LAEs diagonally dominant. As discussed earlier, this is done by a rearrangement of the above equations—by row interchanges. Thus, considering the first, third, and second LAE as an explicit equation for the solution of ϕ_1, ϕ_2, and ϕ_3, respectively; the system of LAEs is given as

$$\begin{bmatrix} 9 & 1 & -1 \\ 1 & -8 & 3 \\ 1 & 2 & 9 \end{bmatrix} \begin{bmatrix} \phi_1 \\ \phi_2 \\ \phi_3 \end{bmatrix} = \begin{bmatrix} 9 \\ -4 \\ 12 \end{bmatrix}$$

The above equation is solved by a Jacobi method (Eq. 4.37) as

$$\phi_1 = f(\phi_{2,old}, \phi_{3,old}) = (9 - \phi_{2,old} + \phi_{3,old})/9$$
$$\phi_2 = f(\phi_{1,old}, \phi_{3,old}) = (4 + \phi_{1,old} + 3\phi_{3,old})/8$$
$$\phi_3 = f(\phi_{1,old}, \phi_{2,old}) = (12 - \phi_{1,old} - 2\phi_{2,old})/9$$

and by Gauss-Seidel method (Eq. 4.38) as

$$
\begin{aligned}
\phi_1 &= f(\phi_{2,old}, \phi_{3,old}) = \left(9 - \phi_{2,old} + \phi_{3,old}\right)/9 \\
\phi_2 &= f(\phi_1, \phi_{3,old}) \quad\;\; = \left(4 + \phi_1 + 3\phi_{3,old}\right)/8 \\
\phi_3 &= f(\phi_1, \phi_2) \quad\quad\;\; = \left(12 - \phi_1 - 2\phi_2\right)/9
\end{aligned}
$$

where the subscript *old* corresponds to the previous iterative value. In these methods, using an initial guess $\phi_{1,old} = \phi_{2,old} = \phi_{3,old} = 0$ in first equation above, first iterative (or improved) value of ϕ_1 is calculated. Then, for calculating the first iteration value of ϕ_2, the value of ϕ_1 is the guessed value for the Jacobi method and first iteration (or the latest) value for the Gauss-Seidel method; and the value of ϕ_3 is the guessed value for both the methods. Similarly, for the calculation of the first iteration value of ϕ_3, the values of ϕ_1 and ϕ_2 are the guessed ones for the Jacobi and the first-iterative values for the Gauss-Seidel method. This completes the first iteration and the convergence is checked by the convergence criterion; Eq. 4.36 with $\left|\Delta\phi_{i=1,2,3}\right|_{max} = max\left(\left|\phi_1 - \phi_{1,old}\right|, \left|\phi_2 - \phi_{2,old}\right|, \left|\phi_3 - \phi_{3,old}\right|\right)$ and $\epsilon = 10^{-4}$ (a user-input value). If the condition is not obeyed, then the iteration is continued with $\phi_{n,old} = \phi_n$ ($n = 1$ to 3) and the first iteration values are used to obtain the second iteration values (using the above equations); and this process is repeated till convergence. Note that each LAE in the above system of equation are solved in a sequence; relevant for the Gauss-Seidel but not for the Jacobi method.

Using the solution methodology discussed above, a computer program for the Jacobi and Gauss-Seidel method is presented in Program 4.1.

Listing 4.1   Scilab code for Jacobi and Gauss-Seidel Method.

```
1    p(1)=0; p(2)=0; p(3)=0; // Initial Guess
2    epsilon=0.0001; //convergence Criterion
3    N=0; // Counter for iteration number
4    Error=1; // some number greater than epsilon to start the first iteration
            in the while loop below
5    pm=p'; //Matrix to store different iterative values
6    while Error>=epsilon //start of while loop
7    p_old=p; // present iterative value stored as old one
8    N=N+1 // increase in the iteration counter
9    //Jacobi Method
10   p(1)=(9−p_old(2)+p_old(3))/9;
11   p(2)=(4+p_old(1)+3*p_old(3))/8;
12   p(3)=(12−p_old(1)−2*p_old(2))/9;
13   Error=max(abs(p−p_old));// parameter to check convergence; LHS of Eq. 4.36.
14   pm=[pm;p']
15   end //end of while loop

//Gauss–Seidel Method: Replace Line No. 9–11 as follows
p(1)=(9−p(2)+p(3))/9;
p(2)=(4+p(1)+3*p(3))/8; p(3)=(12−p(1)−2*p(2))/9;
```

Table 4.2 Iterative solution and convergence by the Jacobi (J) and Gauss-Seidel (GS) method

$N+1$	ϕ_1		ϕ_2		ϕ_3		$\left|\phi_i^{N+1} - \phi_i^N\right|_{max}$	
	J	GS	J	GS	J	GS	J	GS
0	0	0	0	0	0	0	–	–
1	1	1	0.5	0.625	1.3333	1.0833	1.3333	1.0833
2	1.0926	1.0509	1.125	1.0376	1.1111	0.986	0.625	0.4126
3	0.9985	0.9943	1.0532	0.994	0.9619	1.002	0.1492	0.0567
4	0.9899	1.0009	0.9855	1.0008	0.9883	0.9997	0.0677	0.0068
5	1.0003	0.9999	0.9944	0.9999	1.0043	1	0.016	0.001
6	1.0011	1	1.0017	1	1.0012	1	0.0073	0.0001
⋮	⋮	⋮	⋮	⋮	⋮	⋮	⋮	
10	1		1		1		0.0001	

Using the computer programs, iterative solution for the two methods are shown in Table 4.2.

Discussion:

The table shows that the solution is obtained as $\phi_1 = \phi_2 = \phi_3 = 1$, with the number of iterations required for convergence as 10 for Jacobi and 6 for Gauss-Seidel method; using a convergence tolerance of $\epsilon = 10^{-4}$. Convergence is generally faster for the Gauss-Seidel as compared to the Jacobi method.

The number of iterations required to achieve convergence depends on *four conditions* (Hoffman 1993): *first*, the dominance (in terms of magnitude) of the diagonal coefficients; *second*, the initial guess for the unknowns; *third*, the iterative method used; and *fourth*, the prescribed value of convergence criterion. The iteration-by-iteration marching of solution in an iterative method can be used to create a movie. However, note that the movie will be with respect to iteration (not time) and will represent a numerical artifact—route followed by improved (or iterative) solution from initial guess to the numerically approximated converged solution.

4.2.2 Applications to CFD

For the 2D steady-state heat conduction with volumetric heat generation, a representative problem and its grid generations is already introduced in Fig. 4.2. Furthermore, its finite difference equation is presented in Eq. 4.31 for the GE, and Eq. 4.32 for the

BCs. The respective equations corresponds to internal and boundary grid points (Fig. 4.2); and are presented here as LAE for the uniform grid (with $\Delta x = \Delta y$) as

$$-T_{i,j-1} - T_{i-1,j} + 4T_{i,j} - T_{i+1,j} - T_{i,j+1} = S_T \left.\begin{array}{l} \\ \end{array}\right\} \begin{array}{l} \text{for } i = 2 \text{ to } imax - 1 \\ \text{for } j = 2 \text{ to } jmax - 1 \end{array}$$
$$\text{where } S_T = \overline{Q}_{gen,i,j} \Delta x^2 / k$$

$$\left.\begin{array}{l} T_{1,j} = T_{wb} \text{ (west boundary)} \\ T_{imax,j} = T_{eb} \text{ (east boundary)} \end{array}\right\} \text{ for } j = 2 \text{ to } jmax - 1$$

$$\left.\begin{array}{l} T_{i,1} = T_{sb} \text{ (south boundary)} \\ T_{i,jmax} = T_{nb} \text{ (north boundary)} \end{array}\right\} \text{ for } i = 2 \text{ to } imax - 1$$

$$(4.39)$$

The boundary temperature value above is a constant for a Dirichlet BC, and gets updated (after every iteration, using Eq. 4.32) for a non-Dirichlet BC. Since the temperature at a particular grid point depends on the temperature at adjoining *four not all* the grid points in the domain, the system of LAEs for Eq. 4.39 results in a *penta diagonal* coefficient matrix. The matrix consists of five non-zero diagonal vectors: one main ($aP_{i,j}$), two lower ($aS_{i,j}$ and $aW_{i,j}$), and two upper ($aE_{i,j}$ and $aN_{i,j}$) diagonal. The non-zero value corresponds to 4 for the main diagonal, and -1 for both the upper and lower diagonals. This is presented below with the help of an example problem.

Example 4.6: Matrix form of FDM-based system of LAEs, for a 2D steady state heat conduction problem with volumetric heat generation.

Consider the conduction in a 2D plate subjected to a BCs, shown earlier in Fig. 4.2. This is shown below in Fig. 4.6, with the temperature at the various boundary: $T_{wb} = T_W$, $T_{eb} = f(k, q_W, T_{imax-1,j})$, $T_{sb} = T_{i,2}$, and $T_{nb} = f(h, T_\infty, T_{i,jmax-1})$; refer Eq. 4.32. Figure 4.6 also shows the various types of grid points, generated for a finite difference method-based solution of the problem. Consider the internal grid point-by-point application of the second-order central difference-based finite difference method, as per the sequence shown in Fig. 4.6 (starting from $i = j = 2$ and ending at $i = j = 4$), and determine the matrix form of the resulting system of LAEs. Thereafter, discuss the nature of the coefficient matrix and the resulting constant vector.

Solution:

The the second-order FDE (Eq. 4.39) is given for the internal grid points (Fig. 4.6) as

$$-T_{i,j-1} - T_{i-1,j} + 4T_{i,j} - T_{i+1,j} - T_{i,j+1} = S_T \left.\begin{array}{l} \\ \end{array}\right\} \begin{array}{l} \text{for } i = 2 \text{ to } 4 \\ \text{for } j = 2 \text{ to } 4 \end{array} \quad (4.40)$$
$$\text{where } S_T = \overline{Q}_{gen,i,j} \Delta x^2 / k$$

The above equation involves 9 internal (1 interior and 8 border) and 3 boundary grid points in each of the four boundaries, seen in Fig. 4.6; note that the

boundary grid points at the corner of the domain are not involved. The application of the FDE (Eq. 4.40) for a border (an internal) grid point involves boundary
grid points. The corresponding computational stencil for the border grid-point
temperature $T_{2,2}$ is shown in Fig. 4.6, with T_{wb} and T_{sb} as the boundary temperature involved in the FDE; similarly, T_{sb} occurs in the FDE for $T_{3,2}$. Thus, the
FDE for a border grid point involves one or two boundary grid points; the two
for those border grid points which are at the corner of the domain ($T_{2,2}, T_{4,2}$,
$T_{2,4}$, and $T_{4,4}$).

Applying Eq. 4.40 results in a sequence of internal grid point-by-point LAE
(the sequence shown in Fig. 4.6) as

$$
\begin{aligned}
i = 2,\ i = 2: \quad & 4T_{2,2} - T_{3,2} - T_{2,3} & = T_{wb} + T_{sb} + S_T \\
i = 3,\ j = 2: \quad & -T_{2,2} + 4T_{3,2} - T_{4,2} - T_{3,3} & = T_{sb} + S_T \\
i = 4,\ j = 2: \quad & -T_{3,2} + 4T_{4,2} - T_{4,3} & = T_{sb} + S_T + T_{eb} \\
i = 2,\ j = 3: \quad & -T_{2,2} + 4T_{2,3} - T_{3,3} - T_{2,4} & = T_{wb} + S_T \\
i = 3,\ j = 3: \quad & -T_{3,2} - T_{2,3} + 4T_{3,3} - T_{4,3} - T_{3,4} & = S_T \\
i = 4,\ j = 3: \quad & -T_{4,2} - T_{3,3} + 4T_{4,3} - T_{4,4} & = S_T + T_{eb} \\
i = 2,\ j = 4: \quad & -T_{2,3} + 4T_{2,4} - T_{3,4} & = T_{wb} + S_T + T_{nb} \\
i = 3,\ j = 4: \quad & -T_{3,3} - T_{2,4} + 4T_{3,4} - T_{4,4} & = S_T + T_{nb} \\
i = 4,\ j = 4: \quad & -T_{4,3} - T_{3,4} + 4T_{4,4} & = S_T + T_{eb} + T_{nb}
\end{aligned}
$$

where the coefficient of the LAE is -1 for the temperature at the neighboring
grid points, and $+4$ for the grid point under consideration; and the source term S_T
appears in the non-homogeneous term. This term also consists of the temperature
at the boundary grid points, in the above LAE for the border grid points; this is
because the coefficient and the temperature for certain neighboring grid points,
which are at the boundary, moves from the LHS to RHS of the above LAEs and
appears with S_T as the non-homogeneous term.

The above systems of LAEs is represented in a matrix form as

$$
\begin{bmatrix}
4 & -1 & 0 & -1 & 0 & 0 & 0 & 0 & 0 \\
-1 & 4 & -1 & 0 & -1 & 0 & 0 & 0 & 0 \\
0 & -1 & 4 & 0 & 0 & -1 & 0 & 0 & 0 \\
-1 & 0 & 0 & 4 & -1 & 0 & -1 & 0 & 0 \\
0 & -1 & 0 & -1 & 4 & -1 & 0 & -1 & 0 \\
0 & 0 & -1 & 0 & -1 & 4 & 0 & 0 & -1 \\
0 & 0 & 0 & -1 & 0 & 0 & 4 & -1 & 0 \\
0 & 0 & 0 & 0 & -1 & 0 & -1 & 4 & -1 \\
0 & 0 & 0 & 0 & 0 & -1 & 0 & -1 & 4
\end{bmatrix}
\begin{bmatrix}
T_{2,2} \\ T_{3,2} \\ T_{4,2} \\ T_{2,3} \\ T_{3,3} \\ T_{4,3} \\ T_{2,4} \\ T_{3,4} \\ T_{4,4}
\end{bmatrix}
=
\begin{bmatrix}
T_{wb} + T_{sb} + S_T \\
T_{sb} + S_T \\
T_{sb} + S_T + T_{eb} \\
T_{wb} + S_T \\
S_T \\
S_T + T_{eb} \\
T_{wb} + S_T + T_{nb} \\
S_T + T_{nb} \\
S_T + T_{eb} + T_{nb}
\end{bmatrix}
$$

$$(4.41)$$

Discussion:

The coefficient matrix above is a penta-diagonal matrix. It can be clearly seen
that it is a *sparse matrix* with more number of zero as compared to the non-
zero values; the difference increase substantially due to a much larger grid
size commonly used in CFD. Furthermore, the coefficient-matrix (Eq. 4.41) is
diagonal-dominant—the major diagonal value (equal to 4) for a row as compared
to the sum of the absolute value of all the other terms in that row is equal for
the interior grid point, and greater for the border grid points. Thus, for CFD, the
system of LAEs is ideally suited for the iterative solution method.

Considering a general form of the LAEs as $aS_{i,j}T_{i,j-1} + aW_{i,j}T_{i-1,j} +$
$aP_{i,j}T_{i,j} + aE_{i,j}T_{i+1,j} + aN_{i,j}T_{i,j+1} = b_{i,j}$ ($j = 2$ to 4 and $i = 2$ to 4),   the
above coefficient matrix consists of *five-diagonals*: main $aP_{i,j}$, adjoining-lower
$aW_{i,j}$, further-lower $aS_{i,j}$, adjoining-upper $aE_{i,j}$, and further-upper $aN_{i,j}$.
Note that an element of the upper-diagonal $aE_{i,j}$ is zero for the LAE where
T_{eb} appears in the constant vector $b_{i,j}$; similarly, the lower-diagonal $aW_{i,j}$ is
zero due to the appearance of T_{wb} in the $b_{i,j}$. It can be seen that the number of
non-zero elements in any of the lower or upper diagonals is 6 out of the total of
9 elements; the three elements with zero values are due to three boundary grid
points at the east, west, north, and south boundaries (shown encircled for each
of the boundary in Fig. 4.6) which are involved in the FDE for the border grid
points. Thus, the diagonal, unknown, and constant vectors are given as

$$
\begin{bmatrix} aS_{i,j} \end{bmatrix}
\begin{bmatrix} aW_{i,j} \end{bmatrix}
\begin{bmatrix} aP_{i,j} \end{bmatrix}
\begin{bmatrix} aE_{i,j} \end{bmatrix}
\begin{bmatrix} aN_{i,j} \end{bmatrix}
\begin{bmatrix} T_{i,j} \end{bmatrix}
\begin{bmatrix} b_{i,j} \end{bmatrix}
$$

$$
\begin{bmatrix} 0 \\ 0 \\ 0 \\ -1 \\ -1 \\ -1 \\ -1 \\ -1 \\ -1 \end{bmatrix}
\begin{bmatrix} 0 \\ -1 \\ -1 \\ 0 \\ -1 \\ -1 \\ 0 \\ -1 \\ -1 \end{bmatrix}
\begin{bmatrix} 4 \\ 4 \\ 4 \\ 4 \\ 4 \\ 4 \\ 4 \\ 4 \\ 4 \end{bmatrix}
\begin{bmatrix} -1 \\ -1 \\ 0 \\ -1 \\ -1 \\ 0 \\ -1 \\ -1 \\ 0 \end{bmatrix}
\begin{bmatrix} -1 \\ -1 \\ -1 \\ -1 \\ -1 \\ -1 \\ 0 \\ 0 \\ 0 \end{bmatrix}
\begin{bmatrix} T_{2,2} \\ T_{3,2} \\ T_{4,2} \\ T_{2,3} \\ T_{3,3} \\ T_{4,3} \\ T_{2,4} \\ T_{3,4} \\ T_{4,4} \end{bmatrix}
\begin{bmatrix} T_{wb} + T_{sb} + S_T \\ T_{sb} + S_T \\ T_{sb} + S_T + T_{eb} \\ T_{wb} + S_T \\ S_T \\ S_T + T_{eb} \\ T_{wb} + S_T + T_{nb} \\ S_T + T_{nb} \\ S_T + T_{eb} + T_{nb} \end{bmatrix}
$$

**Example 4.7: Testing of FDM-based code, developed using the steady-state
formulation and Gauss-Seidel method, for the 2D heat conduction problem.**

Consider a *long* stainless-steel ($k = 16.7 \ W/m.K$) plate subjected to constant
temperature at the various boundaries (Fig. 4.6): $T_{wb} = T_{sb} = T_{eb} = T_\infty$, and
$T_{nb} = T_W$; and no heat generation, $\overline{Q}_{gen} = 0$.

1. Develop a computer program to obtain the steady-state temperature distribution in the plate, using the central difference-based FDE for the governing

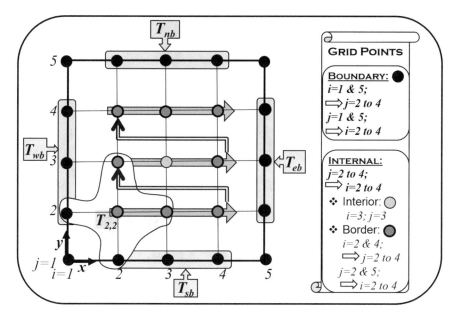

Fig. 4.6 Various types of grid points and the BCs for a 2D heat conduction problem, along with the sequence (shown by arrows inside the domain) in the point-by-point solution of LAEs for internal grid points

equation (Eq. 4.39). Run the code for $L_1 = L_2 = 1\,m$, $imax = jmax = 5$, $\theta_{wb} = \theta_{sb} = \theta_{eb} = 0$, $\theta_{nb} = 1$, and a convergence tolerance of $\epsilon = 10^{-4}$.

2. For the BCs considered here, there is an analytical solution (Incropera and Dewitt 1996):

$$\theta(x, y) \left(\equiv \frac{T - T_\infty}{T_W - T_\infty} \right) = \frac{2}{\pi} \sum_{n=1}^{\infty} \frac{(-1)^{n+1} + 1}{n} \sin \frac{n\pi x}{L_1} \frac{\sinh\left(n\pi y/L_1\right)}{\sinh\left(n\pi L_2/L_1\right)}$$

(4.42)

Present the numerical as well as analytical results in a tabular form.

Solution:

Using the FDEs (Eq. 4.39) and the Gauss-Seidel method discussed above, a computer program for the present problem is presented in Program 4.2. Furthermore, Program 4.3 presents a computer program for the analytical solution.

Listing 4.2 Scilab code for the *numerical* solution of the 2D steady-state conduction problem, by the Gauss-Seidel Method.

```
1   imax=5;jmax=5; L1=1;L2=1;Dx=L1/(imax−1);Dy=Dx;
2   Th=0.5*ones(imax,jmax);  // Initial Guess
3   Th(:,1)=0; Th(:,jmax)=1;  // BC for bottom & top boundaries
```

```
4   Th(1,:)=0; Th(imax,:)=0; // BC for left & right boundaries
5   epsilon=0.0001; N=0; Error=1;
6   while Error>=epsilon
7     Th_old=Th; // present iterative value stored as old one
8     N=N+1 // increase in the iteration counter
9     for j=2:jmax−1
10    for i=2:imax−1
11    Th(i,j)=Th(i+1,j)+Th(i−1,j)+Th(i,j+1)+Th(i,j−1); Th(i,j)=Th(i,j)/4;
12    end end
13    Error=max(abs(Th−Th_old)) // convergence parameter
14    end
15    Th=round(Th*10000)/10000; //Round−off: 4 decimal places
```

Listing 4.3 Scilab code for *analytical* solution of the 2D steady-state conduction problem, by the method of separation of variables.

```
1   imax=5;jmax=5; L1=1;Dx=L1/(imax−1);L2=1;Dy=L2/(jmax−1);
2   [Y,X]=meshgrid(0:Dy:1,0:Dx:1);
3   epsilon=0.0001; theta(1:imax,jmax)=1;theta(1,1:jmax)=0;
4   theta(1:imax,1)=0;theta(imax,1:jmax)=0;
5   for j=2:jmax−1
6   for i=2:imax−1
7     Sum=0; Error=1;n=1;
8     while Error>=epsilon
9       Sum_old=Sum;
10      t1=(−1)^(n+1)+1;t1=t1/n;t2=sin(n*%pi*X(i,j)/L1);
11      t3=sinh(n*%pi*Y(i,j)/L1);t4=sinh(n*%pi*L2/L1);
12      Sum=Sum+(t1*t2*t3)/t4; Error=abs(Sum−Sum_old);
13      n=n+2; //Avoid summation, as the next term is zero
14    end
15    theta(i,j)=2*Sum/%pi;
16  end end
17  theta=round(theta*10000)/10000;// 4 decimal places
```

Using these programs, the numerical and analytical results are shown in Table 4.3.

Discussion:

Table 4.3 shows a very good agreement between the numerical and analytical results, with a maximum error close to 5%. This leads to the verification of the present FDM-based code for the 2D steady-state conduction; called verification-study. The non-dimensional temperature at the centroid is obtained as 0.2501 by the numerical and 0.25 by the analytical result (Table 4.3). Note that the temperature is equal to the mean of the temperature at the four walls of the plate. This is due to an equal thermal effect of the walls located at the same normal distance from the centroid of the 2D plate—characteristics of any diffusion phenomenon (Fig. 3.3).

Table 4.3 Converged numerical solution obtained by Gauss-Seidel method, and the analytical solution obtained by the method of separation of variables

$\theta_{i,5}$		1	1	1	1	1	1	
		Numerical			Analytical			
$\theta_{i,4}$	0	0.4286	0.5268	0.4286	0.432	0.5405	0.432	0
$\theta_{i,3}$	0	0.1876	0.2501	0.1875	0.182	0.25	0.182	0
$\theta_{i,2}$	0	0.0715	0.0983	0.0715	0.068	0.0954	0.068	0
$\theta_{i,1}$	0	0	0	0	0	0	0	0
	$\theta_{1,j}$	$\theta_{2,j}$	$\theta_{3,j}$	$\theta_{4,j}$	$\theta_{2,j}$	$\theta_{3,j}$	$\theta_{4,j}$	$\theta_{5,j}$

4.3 Numerical Differentiation for Local Engineering parameters

For the 2D steady-state heat conduction, the governing equations in CFD, FDM-based discretization procedure, and then solution by an iterative method are presented in the previous section. It is shown that the flow property (temperature in the previous section) is computed at discrete grid points in the computational domain. Thereafter, for the computation of the local engineering parameters, it is necessary to use numerical differentiation for evaluation of first derivatives at the various boundaries of the domain. The computational stencil for the implementation of the numerical-differentiation is shown as arrows in Fig. 4.7, corresponding to the computation of $\partial T/\partial x$ (at the east and west boundaries) and $\partial T/\partial y$ (at the north and south boundary). The stencil is shown for the first-order forward and backward differences at the west/south and east/north boundaries, respectively; for the computation of local conduction flux at the boundary grid points. Although the figure is shown for heat transfer, similar procedure is used in fluid dynamics; first for the computation of velocity gradients (from the velocity field), and then for the determination of the wall shear stress at the boundary of the domain. Thus, after obtaining the heat or/and fluid flow field, the numerical differentiation is used to compute the discrete (boundary grid point-by-point) variation of local engineering parameters at the boundary of the domain.

4.3.1 Differentiation Formulas

Differentiation formulas are derived from the Taylor-series or the local polynomial-profile assumption, using the finite difference method; presented above. Since the formulas are applied at the boundary, the adjoining border grid points are available only on one (not both) side of the boundary. Thus, one-sided finite difference equation acts as the differentiation formulas; given in a general form (for upto third-order accurate FDE) as

Fig. 4.7 Computational stencil for boundary grid point-by-point numerical differentiation by first-order forward and backward difference method; and for numerical integration by considering all the boundary grid-point values. The stencil is shown as arrows which are *one-sided* for the differentiation and *two-sided* for the integration. The figure also shows the various types of grid points and the BCs for a 2D heat conduction problem

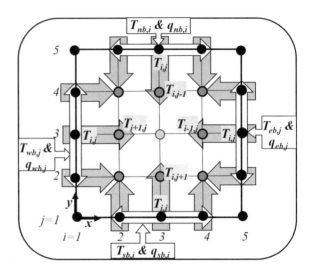

Table 4.4 The coefficients in the general FDE for the first and the second derivatives in x/y direction (Eq. 4.43)—on a uniform Cartesian grid

FDM	Order of Accuracy	bbb	bb	b	c	f	ff	fff
Forward	I				-1	1		
	II				$-3/2$	2	$-1/2$	
	III				$-11/6$	3	$-3/2$	1/3
Backward	I			-1	1			
	II		$1/2$	-2	$3/2$			
	III	$-1/3$	$3/2$	-3	$11/6$			

$$\left.\frac{\partial \phi}{\partial x}\right)_{i,j} \Delta x \approx fff\phi_{i+3,j} + ff\phi_{i+2,j} + f\phi_{i+1,j} + c\phi_{i,j} +$$
$$b\phi_{i-1,j} + bb\phi_{i-2,j} + bbb\phi_{i-3,j}$$
$$\left.\frac{\partial \phi}{\partial y}\right)_{i,j} \Delta y \approx fff\phi_{i,j+3} + ff\phi_{i,j+2} + f\phi_{i,j+1} + c\phi_{i,j} +$$
$$b\phi_{i,j-1} + bb\phi_{i,j-2} + bbb\phi_{i,j-3}$$

$$(4.43)$$

where the coefficients of ϕ for the various order of FDM are shown in Table 4.4. Note from the table that the coefficient f, ff, and fff for the forward difference formula is equal to the *negative* value of b, bb, and bbb for the backward difference formula, respectively.

4.3.2   Applications to CFD

An application of the numerical differentiation is presented below, with the help of an example problem—for computation of local conduction flux in the 2D steady-state heat conduction problem introduced above.

Example 4.8: Numerical Differentiation for the 2D steady-state heat conduction problem.

Consider the FDM method-based numerical solution of the previous example problem, with $T_\infty = 30\,^\circ C$ and $T_W = 10\,^\circ C$ and the *long* stainless-steel ($k = 16.7\,W/m.K$) plate of dimension $1m \times 1m$. Using the Scilab code presented in Program 4.2, the steady-state non-dimensional temperature field ($\theta = (T - T_\infty)/(T_W - T_\infty)$) is computed on a uniform grid size of 11×11. Near the north boundary of the plate, the temperature field is obtained as

$\theta_{i,11}$	1	1	1	1	1	1	1	1	1	
$\theta_{i,10}$	0	0.489	0.6749	0.7546	0.7892	0.799	0.7891	0.7545	0.6749	0.489	0
$\theta_{i,9}$	0	0.2811	0.4562	0.5541	0.603	0.6179	0.603	0.554	0.4561	0.281	0
$\theta_{i,8}$	0	0.1791	0.3145	0.4026	0.451	0.4663	0.4509	0.4025	0.3144	0.179	0
$\theta_{i,7}$	0	0.1207	0.2202	0.2907	0.3318	0.3452	0.3317	0.2906	0.2201	0.1205	0
	$\vdots$	$\vdots$	$\vdots$	$\vdots$	$\vdots$	$\vdots$	$\vdots$	$\vdots$	$\vdots$	$\vdots$	$\vdots$
	$\theta_{1,j}$	$\theta_{2,j}$	$\theta_{3,j}$	$\theta_{4,j}$	$\theta_{5,j}$	$\theta_{6,j}$	$\theta_{7,j}$	$\theta_{8,j}$	$\theta_{9,j}$	$\theta_{10,j}$	$\theta_{11,j}$

Since the above solution is symmetric about $\theta_{6,j}$, corresponding to the vertical centerline of the plate, calculate local conduction flux $q_{nb,2} - q_{nb,6}$ at the north boundary (for $0 < x \leq 0.5m$ and $y = L_2$) by three different differentiation formulas:

1. First-order backward difference
2. Second-order backward difference
3. Third-order backward difference

Solution:

The formulas given in Eq. 4.43 and Table 4.4 are used here, with backward difference for the top wall. The calculation of the local conduction flux is presented below for a representative grid point at $x = 0.5$; however, a computer program is given in Program 4.4 for the computation of the flux at all the grid points on the top wall.

Listing 4.4 Scilab code for numerical differentiation, using the steady-state solution of 2D conduction (Program 4.2).

```
1   k=16.7;T_W=10;T_inf=30;temp=-k*(T_W-T_inf);
2   for i=2:imax-1
3   // 1-Oder Backward Difference
```

```
4    dTh_dy1=(Th(i ,jmax)–Th(i ,jmax–1))/Dy; q_nb1(i)=temp*dTh_dy1;
5    // Second–Order Backward Difference
6    dTh_dy2=(3*Th(i ,jmax)–4*Th(i ,jmax–1)+Th(i ,jmax–2))/(2*Dy);
7    q_nb2(i)=temp*dTh_dy2;
8    // Third–order Backward Difference
9    dTh_dy3=(11*Th(i ,jmax)–18*Th(i ,jmax–1)+9*Th(i ,jmax–2)–2*Th(i ,jmax–3))/(6*Dy
     ); q_nb3(i)=temp*dTh_dy3;
10   end
```

Using the results $\theta_{6,j}$ at $x = 0.5$ (shown shaded in the nondimensional temperature-field θ given above), $k = 16.7 \ W/m.K$, $T_W = 10\,^\circ C$, $T_\infty = 30\,^\circ C$, and $\Delta y = 0.1$, the flux at the north boundary is calculated as follows:

$$(q_{nb})_{x=0.5} = -k \left(\frac{\partial T}{\partial y} \right)_{x=0.5, \ y=1}$$

$$= -k(T_W - T_\infty) \left(\frac{\partial \theta}{\partial y} \right)_{i=6, \ j=11} \qquad \left\{ \theta \equiv \frac{T - T_\infty}{T_W - T_\infty} \right\}$$

(a) First-order numerical differentiation

$$\left(\frac{\partial \theta}{\partial y} \right)_{(6,11)} = \frac{\theta_{6,11} - \theta_{6,10}}{\Delta y} = \frac{(1 - 0.799)}{0.1} = 2.01 \ (/m)$$

$$q_{nb,6} = -k(T_W - T_\infty) \left(\frac{\partial \theta}{\partial y} \right)_{(6,11)} = 671.34 \ W/m^2$$

(b) Second-order numerical differentiation

$$\left(\frac{\partial \theta}{\partial y} \right)_{(6,11)} = \frac{3\theta_{6,11} - 4\theta_{6,10} + \theta_{6,9}}{2\Delta y}$$

$$= \frac{3 \times 1 - 4 \times 0.799 + 0.6179}{0.2} = 2.1095 \ (/m)$$

$$q_{nb,6} = -k(T_W - T_\infty) \left(\frac{\partial \theta}{\partial y} \right)_{(6,11)} = 704.573 \ W/m^2$$

(c) Third-order numerical differentiation

$$\left(\frac{\partial \theta}{\partial y} \right)_{(6,11)} = \frac{11\theta_{6,11} - 18\theta_{6,10} + 9\theta_{6,9} - 2\theta_{6,8}}{6\Delta y}$$

$$= \frac{11 \times 1 - 18 \times 0.799 + 9 \times 0.6179 - 2 \times 0.4663}{0.6} = 2.0775 \ (/m)$$

Fig. 4.8 Boundary grid point-by-point variation of local conduction flux at the north boundary, obtained by the application of the various numerical differentiation formulas on the 2D steady-state temperature field; on a grid size of 11×11

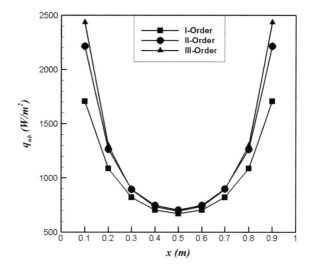

$$q_{nb,6} = -k(T_W - T_\infty) \left(\frac{\partial \theta}{\partial y} \right)_{(6,11)} = 693.885 \, \text{W}/m^2$$

The results obtained above are shown shaded in Table 4.5. The table also shows the results obtained for the other grid points at the north boundary. The results are obtained by running a computer program given in Program 4.4. The variation of local heat flux on the north boundary is shown in Fig. 4.8.

Discussion:

Figure 4.8 shows that the difference between the results reduces with an increase in the order of differentiation formulas. However, the figure shows a much larger reduction in the difference of the results between the first and second orders, as compared to the difference between the second- and third-order differentiation. For example, at $x = 0.1$, the respective difference is 22.87% and 9%; whereas, as compared to the result from the third-order numerical differentiation, the difference is 29.83% for the first-order and 9% for the second-order differentiation. The second-order accurate representation is considered sufficient for most of the CFD solutions; as mentioned above.

Table 4.5 Local conduction heat fluxes at the north boundary grid points, obtained by the application of first-, second-, and third-order numerical differentiation formulas

Flux →	$q_{nb,2}/q_{nb,10}$	$q_{nb,3}/q_{nb,9}$	$q_{nb,4}/q_{nb,8}$	$q_{nb,5}/q_{nb,7}$	$q_{nb,6}$
First-Order	1706.74	1085.834	819.636	704.072	671.34
Second-Order	2212.917	1263.522	894.619	745.154	704.573
Third-Order	2432.4663	1296.254	890.0543	734.466	693.885
x (m) →	0.1 & 0.9	0.2 & 0.8	0.3 & 0.7	0.4 & 0.6	0.5

4.4 Numerical Integration for the Total Value of Engineering Parameters

After the computation of a local engineering parameter at the boundary of the domain, it is necessary to use numerical-integration for the evaluation of total value of the engineering parameter. For the computation of total heat transfer rate at the various boundary ($Q_{eb}/Q_{wb}/Q_{nb}/Q_{sb}$), Fig. 4.7 shows the computational stencil as the *two-sided arrows* at the domain boundary, which involves grid-point values of $q_{eb,j}$ at the east boundary, $q_{wb,j}$ at the west boundary, and $q_{nb,i}$ at the north boundary, and $q_{sb,i}$ at the south boundary.

Although the figure is shown for heat transfer, similar procedure is used in fluid dynamics—for the computation of shear force (using the wall shear stress) at the various boundary of the domain. Thus, after obtaining the discrete (boundary grid point-by-point) variation of a local engineering parameter by numerical differentiation, the numerical integration is used to compute a total/single value of the engineering parameter—at the various boundary of the domain. The integration involves surface integral of the flux terms; line integral for 2D problems.

4.4.1 Integration Rules

The input for a numerical integration of a function $f(x)$ (may be the conduction flux $q_y(x)$ at a horizontal boundary of the domain) is a discrete/tabular information for f_i at certain x_i ($1 \leq i \leq imax$), shown in Fig. 4.9. The figure shows that the domain of integration L_1 is subdivided into ($imax - 1$) number of *panels* (of width Δx), and ($imax - 1$)/n number of *sub-intervals*; n is the number of panels in each sub-interval. The figure shows $n = 1$ for the linear and $n = 2$ for the quadratic polynomial-approximation, leading to the first- and second-order integration rules, respectively.

Numerical integration of a function is a summation of piecewise integrals (I_m, where m is the numbering of the sub-intervals, shown in Fig. 4.9), with the integrand (within each piece/sub-interval) approximated as a polynomial $p(x)$; given as

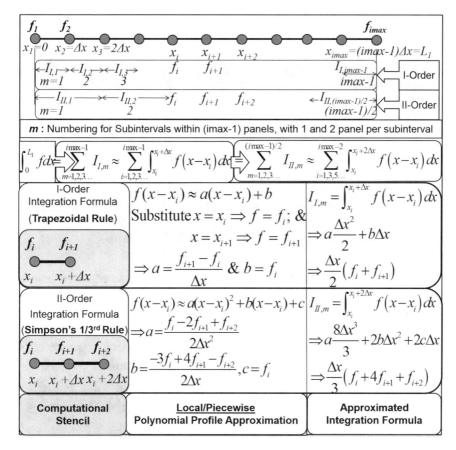

Fig. 4.9 Illustration of local/piecewise profile assumption-based algebraic representation of the surface integral of local engineering parameters $f(x)$ (may be the local conduction flux at the north boundary $q_{nb}(x)$ and at the south boundary $q_{sb}(x)$, for the conduction problem in Fig. 4.7). The assumption is linear for first-order and quadratic for second-order integration rules

$$\int_0^{L_1} f(x)dx = \sum_{m=1,2,3\ldots}^{(imax-1)/n} I_m \approx \sum_{i=1,1+n,1+2n\ldots}^{imax-n} \int_{x_i}^{x_i+n\Delta x} p(x)dx \qquad (4.44)$$

Note that the function is approximated here as a piecewise polynomial. A local/-piecewise polynomial profile assumption-based derivation of first- and second-order integration rules are shown in Fig. 4.9. It can be seen that the piecewise profile is linear and quadratic between two and three consecutive grid points, respectively. Thus, the number of grid points or panels involved in a sub-interval increases with the increase in the order of the polynomial. The figure shows first the determination of the constants a, b, and c of the polynomial by substituting $x = x_i \Rightarrow f = f_i$, $x = x_{i+1} \Rightarrow f = f_{i+1}$, and $x = x_{i+2} \Rightarrow f = f_{i+2}$ in the polynomials. Then, the constants are substituted in the expression obtained after the piecewise integration to obtain the integration rule for a sub-interval. Summing up for all the sub-intervals, the final expression for the first- and second-order integration rules are given as

TrapizoidalRule :

$$\int_0^{L_1} f(x)dx \approx \sum_{i=1,2,3\cdots}^{imax-1} \frac{\Delta x}{2}(f_i + f_{i+1}) = \frac{\Delta x}{2}(f_1 + 2\sum_{i=2,3\cdots}^{imax-1} f_i + f_{imax}) \tag{4.45}$$

Simpson's$1/3^{rd}$ Rule :

$$\int_0^{L_1} f(x)dx \approx \sum_{i=1,3\cdots}^{imax-2} \frac{\Delta x}{3}(f_i + 4f_{i+1} + f_{i+2}) \tag{4.46}$$

$$\Rightarrow \frac{\Delta x}{3}(f_1 + 4\sum_{i=2,4\cdots}^{imax-1} f_i + 2\sum_{i=3,5\cdots}^{imax-2} f_i + f_{imax})$$

Although the formulas in equations above and Fig. 4.9 are shown for the integration in the $x-$direction (north and south boundaries in a 2D Cartesian domain), they are also applicable for the integration in $y-$direction by replacing i with j, and Δx with Δy.

Example 4.9: Simpson's 3/8 th rule for Numerical Integration.

Present a local/piecewise profile-based derivation of the 3/8th rule. Consider the profile as a third-order polynomial—within $x_i \le x \le x_i + 3\Delta x$ for $0 \le x \le L_1$, and maximum number of grid points in $x-$direction as $imax$.

Solution:

Substituting $n = 3$ and $p(x) = a(x - x_i)^3 + b(x - x_i)^2 + c(x - x_i) + d$ in Eq. 4.44, the integration rule is given as

$$\int_0^{L_1} f(x)dx$$

$$\approx \sum_{i=1,4,7\cdots}^{imax-3} \int_{x_i}^{x_i+3\Delta x} \left(a(x - x_i)^3 + b(x - x_i)^2 + c(x - x_i) + d\right) dx \tag{4.47}$$

$$= \sum_{i=1,4,7\cdots}^{imax-3} \left(a \times \frac{81\Delta x^3}{4} + b \times 9\Delta x^2 + c \times 9\frac{\Delta x}{2} + 3d\right) \Delta x$$

Now, obtain the coefficient a, b, c, and d from the cubic polynomial by using the four discrete values of the functions as

$$f(x) = a(x - x_i)^3 + b(x - x_i)^2 + c(x - x_i) + d$$
$$\text{At } x = x_i \Rightarrow f = f_i, \; x = x_{i+1} \Rightarrow f = f_{i+1}$$
$$\text{At } x = x_{i+2} \Rightarrow f = f_{i+2}, \; x = x_{i+3} \Rightarrow f = f_{i+3}$$

The solution of above equations results in the coefficients as

$$a = \frac{-f_i + 3f_{i+1} - 3f_{i+2} + f_{i+3}}{6\Delta x^3}, \; b = \frac{2f_i - 5f_{i+1} + 4f_{i+2} - f_{i+3}}{2\Delta x^2}$$
$$c = \frac{-11f_i + 18f_{i+1} - 9f_{i+2} + 2f_{i+3}}{6\Delta x} \text{ and } d = f_i$$

Substituting the above coefficients in Eq. 4.47, we get the integration rule as

Simpson's 3/8th Rule :

$$\int_0^{L_1} f(x)dx \approx \sum_{i=1,4,7\ldots}^{imax-3} \frac{3\Delta x}{8}(f_i + 3f_{i+1} + 3f_{i+2} + f_{i+3})$$

4.4.2 Applications to CFD

An application of the numerical integration is presented below, with the help of an example problem—for computation of total rate of heat transfer in the 2D steady-state heat conduction problem introduced above.

Example 4.10: Numerical Integration for the 2D steady-state heat conduction problem.

Consider the local conduction flux at the north boundary computed in previous example problem with $\Delta x = 0.1$, presented in Table 4.5. Calculate the rate of heat transfer from the north boundary using the three different differentiation formulas

1. Trapezoidal rule
2. Simpson's $1/3rd$ rule

Solution:

Table 4.5 shows the local conduction heat flux at the north boundary at nine grid points ($i = 2$ to 10), obtained by I, II, and III order numerical differentiation; reproduced in Table 4.6. Since the temperature distribution for the 2D steady-state heat conduction problem is symmetric about the vertical centerline of the plate, the flux values are given as

$$q_{nb,2} = q_{nb,10}, \ q_{nb,3} = q_{nb,9}, \ q_{nb,4} = q_{nb,8}, \ q_{nb,5} = q_{nb,7} \qquad (4.48)$$

1. Trapezoidal rule: Using Eq. 4.45 with $i = 2$ to 10, the rate of heat transfer at the north boundary is given as

$$Q_{nb,trap} = \frac{\Delta x}{2}(q_2 + \sum_{i=3,4\cdots9} 2q_i + q_{10})$$

Using Eq. 4.48 and the results for first-order numerical differentiation (shown shaded in Table 4.6), the above equation is given as

$$Q_{nb,trap} = \frac{\Delta x}{2}[2(q_2 + q_6) + 4(q_3 + q_4 + q_5)]$$

$$\Rightarrow \frac{0.1}{2}[(2 \times (1706.74 + 671.34)) +$$
$$(4 \times (1085.834 + 819.636 + 704.072))] = 759.7832$$

2. Simpson's $1/3rd$ rule: Using Eq. 4.46 with $i = 2$ to 10, the rate of heat transfer is given as

$$Q_{nb,simp1/3} = \frac{\Delta x}{3}(q_2 + \sum_{i=3,5,7,9} 4q_i + \sum_{i=4,6,8} 2q_i + q_{10})$$

Using Eq. 4.48 and the results for first-order numerical differentiation (shown shaded in Table 4.6), the numerical integration by the Simpson's $1/3rd$ rule is given as

$$Q_{nb,simp1/3} = \frac{\Delta x}{3}[2(q_2 + q_6) + 8(q_3 + q_5) + 4q_4]$$

$$\Rightarrow \frac{0.1}{3}[(2 \times (1706.74 + 671.34)) +$$
$$(8 \times (1085.834 + 704.072)) + 4 \times 819.636] = 745.1985$$

Table 4.6 Local conduction heat fluxes at the north boundary grid points (obtained by the application of first-, second-, and third-order numerical differentiation formulas), which is used for trapezoidal and Simpson's $1/3rd$ rule-based numerical integration—to calculate conduction heat transfer rate at the north boundary

Results from Num. Differentiation for $q_{nb,i}$ ↓					Num. Integration	
$i \rightarrow$	2&10	3&9	4&8	5&7	6	$Q_{nb,trap}$	$Q_{nb,simp1/3}$
I	1706.74	1085.834	819.636	704.072	671.34	759.7832	745.1985
II	2212.917	1263.522	894.619	745.154	704.573	872.4999	849.5234
III	2432.4663	1296.254	890.0543	734.466	693.885	896.9041	868.747

The calculation for numerical integration is given above, for results obtained from first-order numerical differentiation. Whereas, for the other results (*II* and *III* order numerical differentiation), similar calculations can be done and the results are given in Table 4.6. The results are obtained by a program given in Program 4.5.

Listing 4.5 Scilab code for numerical integration, using the results obtained by numerical differentiation (Program 4.5).

```
1  //q_nb1, q_nb2, & q_nb3 are the results for
        first-, second-, & third-order num.
        diffferentiations, respectively.
2  //Trapizoidal Rule
3  Q_nb1_trap=(Dx/2)*(q_nb1(2)+2*sum(q_nb1(3:9))+
        q_nb1(10));
4  Q_nb2_trap=(Dx/2)*(q_nb2(2)+2*sum(q_nb2(3:9))+
        q_nb2(10));
5  Q_nb3_trap=(Dx/2)*(q_nb3(2)+2*sum(q_nb3(3:9))+
        q_nb3(10));
6  //Simpson's 1/3rd Rule
7  Q_nb1_sim1_3=(Dx/3)*(q_nb1(2)+4*sum(q_nb1(3:2:9)
        )+2*sum(q_nb1(4:2:8))+q_nb1(10));
8  Q_nb2_sim1_3=(Dx/3)*(q_nb2(2)+4*sum(q_nb2(3:2:9)
        )+2*sum(q_nb2(4:2:8))+q_nb2(10));
9  Q_nb3_sim1_3=(Dx/3)*(q_nb3(2)+4*sum(q_nb3(3:2:9)
        )+2*sum(q_nb3(4:2:8))+q_nb3(10));
```

4.5 Closure

All the elements of CFD development (Fig. 2.1) in presented in this chapter, considering the four essential numerical-methods for CFD. *Finite difference method*-based CFD development is demonstrated, for the 2D steady-state heat conduction—using the *steady-state* formulation. Using the *unsteady-state* form of the conservation laws, all the later chapters are on *finite volume method*-based CFD development, application, and analysis. This will be presented first for heat transfer (conduction, advection, and convection) and then for fluid dynamics, in the Cartesian-geometry (Part-II) and complex geometry (Part-III).

Problems

4.1 Using the local/piecewise polynomial profile-based algebraic formulation, presented in Sect. 4.1.2.2, derive the second-order forward and backward difference finite difference equation (Eq. 4.22).

4.2 Consider 2D steady-state heat conduction in a *long* stainless-steel plate ($k = 16.2\,W/m.K$), with a square cross-section as $L_1 = L_2 = 1\,m$, shown in Fig. 4.2. The figure shows the computational domain and the various types of thermal BCs; constant temperature $T_{wb} = 100\,°C$ at the west boundary, insulated at the south boundary, constant incident heat flux of $10\,kW/m^2$ at the east boundary, and $h = 100\,W/m^2.K$ and $T_\infty = 30\,°C$ at the north boundary. Modify the non-dimensional code (presented for *Example 4.7*) to make it dimensional; and implement the discretized form of the non-Dirichlet BCs (Fig. 4.2), using second-order forward/backward difference method (Eq. 4.34). Run the code for a uniform grid size $imax = jmax = 11$, volumetric heat generation of 0 and $50\,kW/m^3$, and a convergence tolerance of $\epsilon = 10^{-4}$. Present the results as steady-state isotherms (temperature contours), and discuss the effect of heat generation on the temperature distribution in the plate.

4.3 Using the steady-state temperature field, obtained from the code in the previous problem, incorporate the implementation of the second-order forward/backward differentiation formulas (Sect. 4.3.1) in the code. This is done to compute the local heat flux at all the boundary grid points (Fig. 4.2). Plot as well as discuss the steady-state heat flux profile at the east and west boundary grid points (q_x versus $y-$coordinate), and along the north and south boundary ($x-$coordinate versus q_y), of the plate.

4.4 Using the steady-state heat flux profile in the previous problem, incorporate the implementation of the trapezoidal rule (Sect. 4.4.1) in the code. This is done for the computation of the total rate of heat transfer from the east, south, west and north boundaries of the plate (Fig. 4.2). Considering the steady-state heat transfer rates, present an application of the energy conservation law for the complete domain as the CV.

References

1. Anderson, J. D. (1995). *Computational Fluid Dynamics: The basics with applications*. New York: McGraw Hill.
2. Hoffman, J. D. (1993). *Numerical Methods for Engineers and Scientists*. New York: McGraw Hill.
3. Incropera, F. P. & Dewitt, D. P. (1996). *Fundamentals of Heat and Mass Transfer* 4th ed. New York: Wiley.
4. Mittal, R., & Iaccarino, G. (2005). Immersed boundary methods. *Annual Review of Fluid Mechanics, 37,* 239–261.

Part II
CFD for a Cartesian-Geometry

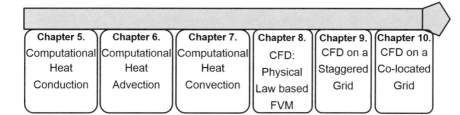

After building the foundation for a course in CFD in the Part I, the CFD is introduced on a simpler (Cartesian) geometry. This part consist of the first three chapters on computational heat transfer (CHT), and the next three chapters on computational fluid dynamics (CFD) consisting of combined fluid-dynamics and heat-transfer. The CHT is presented first separately for the molecular-motion driven diffusion and bulk-motion driven advection heat transfer; and then their combination as convection. The presentation of CHT is modular while CFD is presented together for advection-diffusion with a source term. The flow field, needed for the advection and convection, is considered known for CHT and is determined in the CFD. CFD is presented for both staggered and colocated grid. The complete development methodology is presented in each of the three chapters on CHT; whereas, for CFD, the finite volume method and solution methodology are presented in the separate chapters for a co-located grid system; and together for a staggered grid. CHT as well as CFD development methodology for a Cartesian geometry is presented on uniform as well as non-uniform Cartesian grid, and for explicit as well as implicit method.

Chapter 5
Computational Heat Conduction

Conduction is a transport mechanism due to a random motion of the molecule. For energy transport due to the conduction, the numerical methodology for the computation of temperature distribution (in space and time) is presented in this chapter. The mind map for this chapter is shown in Fig. 5.1. The chapter starts with an algebraic formulation by the physical law based finite volume method. The formulation starts with energy conservation law for a CV, followed by algebraic formulation, approximations, and approximated algebraic formulation, and ends with a discussion.

The resulting LAEs are used to compute the temperature at the internal grid points. Whereas, for the boundary grid points, the algebraic formulation for the temperature is presented by the finite difference method.

Solution methodology is presented by two different approaches: flux-based methodology and coefficient of LAE-based methodology. Since the former one follows a physical quantity-based approach, the solution methodology is introduced by the flux-based methodology; for an explicit method, on a uniform grid. Thereafter, a coefficient of LAE-based methodology is presented, for explicit as well as implicit methods, on a non-uniform grid. Both the types of methodologies are presented first for 1D and then for 2D conduction heat transfers.

5.1 Physical Law-based Finite Volume Method

System of LAEs, consisting of one algebraic equation for each of the grid points in the domain, are the governing equations in CFD. Finite volume method is an approximate method of algebraic formulation, for the internal grid points in a computational domain. Although the FVM presented in almost all the books on CFD (such as Patankar (1980)) starts from a differential formulation (PDE-based FVM), a physical law-based FVM is presented in this book which starts from an application of an appropriate form of physical law on a control volume—a *time interval-based*

© The Author(s), under exclusive license to Springer Nature Switzerland AG 2022
A. Sharma, *Introduction to Computational Fluid Dynamics*,
https://doi.org/10.1007/978-3-030-72884-7_5

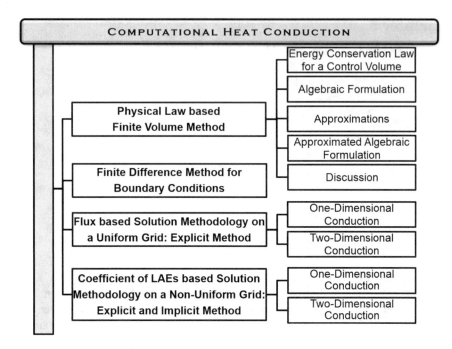

Fig. 5.1 Mind map for Chap. 5

form of energy conservation law for conduction here. The application results in an algebraic formulations which require two approximations to obtain a suitable form of approximated algebraic formulation (in terms of temperature at the grid points); presented below.

5.1.1 Energy Conservation Law for a Control Volume

For a representative CV, the law may be stated in *two different forms* (Incropera and Dewitt 1996). For the heat conduction, they are shown in Fig. 5.2 and presented here as

1. At an instant t (*instantaneous*)

The *rate* at which the conducted thermal energy enters a CV Q_{in}, plus the

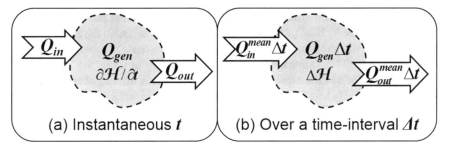

Fig. 5.2 Two different forms of conservation of energy for a CV, subjected to heat conduction and heat generation. (**a**) leads to a differential and (**b**) leads to an algebraic formulation

---

rate at which the thermal energy is generated within the CV Q_{gen}, minus the *rate* at which the conducted thermal energy leaves the CV Q_{out} must be equal to the *rate* of increase of enthalpy $\mathcal{H}$ stored within the CV.

$$Q_{in} + Q_{gen} - Q_{out} = \frac{\partial \mathcal{H}}{\partial t} \tag{5.1}$$

---

2. Over a *time interval* Δt

The *amount* of conducted thermal energy that enters a CV $Q_{in}^{mean} \Delta t$, plus the *amount* of thermal energy that is generated within the CV $Q_{gen} \Delta t$, minus the *amount* of conducted thermal energy that leaves the CV $Q_{out}^{mean} \Delta t$ must be equal to the *amount* of increase of enthalpy $\Delta \mathcal{H}$ stored within the CV.

$$\left(Q_{in}^{mean} + Q_{gen} - Q_{out}^{mean}\right) \Delta t = \Delta \mathcal{H} \left[= \mathcal{H}^{t+\Delta \sqcup} - \mathcal{H}^{t}\right]$$
$$\text{where } Q^{mean} = \frac{1}{\Delta t} \int_{t}^{t+\Delta t} Q dt \text{ and } \Delta \mathcal{H} = \int_{t}^{t+\Delta t} \frac{\partial \mathcal{H}}{\partial t} dt \tag{5.2}$$

---

where the superscript *mean* represents the time-averaged value over the time interval Δt. Note that the above-conducted heat, over the time interval Δt, is given as $\int_{t}^{t+\Delta t} Q dt$; and is equal to $Q^{mean} \Delta t$. Also the heat generation term, which is associated with conversion from some other form (chemical, electrical, electromagnetic, or nuclear) to thermal energy, is considered to remain constant within the time interval $Q_{gen}^{mean} = Q_{gen}$; encountered in most of the engineering applications. The above conservation statements are for the solid which is not experiencing a change in phase—latent energy is not pertinent.

The instantaneous law is used for a differential formulation in a heat transfer course (Incropera and Dewitt 1996), and the formulation is used to obtain the exact solution. Whereas, since the temperature T^t (at a certain time instant t) is used to determine the temperature $T^{t+\Delta t}$ (after a certain time interval Δt) in CFD, the law over the time interval results in the algebraic formulation which is used to obtain numerical solution; presented below. Thus, the differential and algebraic formulations are based on infinitesimal and finite (reasonably small) time interval, respectively. However, note that both the forms of the energy conservation law-based equations (Eqs. 5.1 and 5.2) consist of all the variables on rate basis; conduction heat transfer rate Q, heat generation rate Q_{gen}, and rate of change of enthalpy ($\partial \mathcal{H}/\partial t$ and $\Delta \mathcal{H}/\Delta t$).

5.1.2 Algebraic Formulation

The application of time interval Δt-based energy conservation law is shown in Fig. 5.2b, for an arbitrary-shaped CV. Whereas, for a 2D Cartesian CV, Fig. 5.3 shows the conservation of energy for one of the representative internal CV of the computational domain. The figure shows a subdivision of a 2D Cartesian domain into contiguous non-overlapping non-uniform CVs—also called as *cells* in an FVM. The CVs are generated by a non-uniform Cartesian grid generation method which leads to the non-uniformly spaced horizontal and vertical lines. The circles in the figure are the symbolic representation of the centroids of the CVs, which are also called as *cell centers* or *internal grid points*. They are also represented by P for a representative 2D Cartesian CV; and by E, W, N, and S for the east, west, north, and south neighboring cell centers, respectively. The width of the representative CV is shown as Δx_P in the x-direction, and Δy_P in the y-direction. For all the CVs in the domain, the Δx_P and Δy_P are same for the uniform grid and different for a non-uniform grid. Along with the centroid of the CVs, the figures also shows the centroids of the various faces, called as *face centers*; represented by e, w, n, and s for the east, west, north, and south surfaces, respectively. They are differentiated by symbols, as a square for the e and w face centers at the vertical surfaces, and triangle for the n and s face centers at the horizontal surfaces. The pentagon symbol in the figure represents the corners of the representative CV and are indicated by ne, nw, se, and sw. Here, note that the capital letters (E, W, N, S) correspond to the cell centers/grid points, and small letters correspond to face centers (e, w, n, s) as well as corners (ne, nw, se, sw) of a CV. The grid points, face centers, and vertices are shown in the figure by both alphabetical and geometric shape-based representations. The representation results in a better understanding and effective implementation of the algebraic formulations in a solution methodology; presented later.

Figures 5.2b and 5.3 show that the application of the time interval-based conservation of energy for a CV involves three variables: enthalpy $\mathcal{H}$, conduction heat transfer rate Q and heat generation rate Q_{gen}. There is a point-to-point variation of $\mathcal{H}$ and Q_{gen} within the volume ΔV of the CV, and that of Q at the various surfaces ΔS of the CV. Thus, the enthalpy storage and the heat generation are volumetric phenomena,

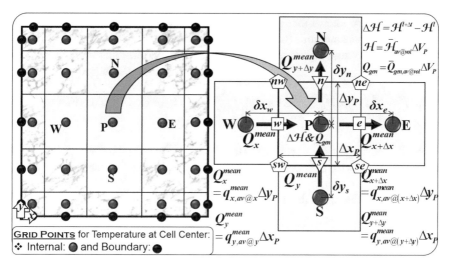

Fig. 5.3 Application of the time interval-based law of conservation of energy (for heat conduction) to a representative internal CV in a 2D Cartesian computational domain, with a non-uniform grid

and the inflow and outflow of conduction heat transfer are surface phenomena. The respective phenomenon results in two types of terms: volumetric term $\overline{\mathcal{H}}$ ($\equiv \mathcal{H}/\Delta V$) and $\overline{Q}_{gen} \equiv Q_{gen}/\Delta V$, and flux term $q \equiv Q/\Delta S$. Note that the classification of a term (as volumetric and flux) is presented here for heat transfer and will be presented later for fluid dynamics. Similar to attention drawn (in thermodynamics) to a thermodynamic property as extensive or intensive; here, for CFD, the attention is needed on whether a flow property is a volumetric term or flux term.

Accounting the spatial variation of the $\mathcal{H}$ and Q_{gen} as the volumetric phenomenon and Q as the surface phenomenon, Eq. 5.2 is given for a 2D Cartesian CV (Fig. 5.3) as

$$
\frac{\int_{\Delta V_P} \overline{\mathcal{H}}^{t+\Delta t} dV - \int_{\Delta V_P} \overline{\mathcal{H}}^{t} dV}{\Delta t} = Q_{cond}^{mean} + \int_{\Delta V_P} \overline{Q}_{gen} dV
$$

$$
\text{where } Q_{cond}^{mean} = \left(\int_{y_{sw}}^{y_{nw}} q_{x,x}^{mean} dy - \int_{y_{se}}^{y_{ne}} q_{x,x+\Delta x}^{mean} dy \right) \quad (5.3)
$$

$$
+ \left(\int_{x_{sw}}^{x_{se}} q_{y,y}^{mean} dx - \int_{x_{nw}}^{x_{ne}} q_{y,y+\Delta y}^{mean} dx \right)
$$

where the volumetric term and flux term are represented by volume and surface integral, respectively; and Q_{cond}^{mean} is the mean rate at which the net (entering minus leaving) conducted thermal energy enters a CV. Also ΔV_P is the volume of the CV; $\Delta x_P \Delta y_P$ for the 2D CV in Fig. 5.3. The limit of the above surface integral corresponds to the appropriate coordinates of the vertices seen in the figure. Furthermore, for conduction flux q, the first subscript represents its 2D Cartesian direction (x or y) and the second subscript x, y, $x + \Delta x$ and $y + \Delta y$ correspond to the *west*, *south*, *east* and *north* surface of the CV, respectively; refer Fig. 5.3.

Figure 5.3 also shows that an *average value* based representation of the above volume and surface integrals (Eq. 5.3) as $av@vol$ and $av@face$, respectively. Thus, the one-point average representation for the volumetric term and the flux term are given as

$$\int_{\Delta V_P} \overline{\mathcal{H}} dV = \overline{\mathcal{H}}_{av@vol} \Delta V_P \text{ and } \int_{\Delta V_P} \overline{\mathcal{Q}}_{gen} dV = \overline{\mathcal{Q}}_{gen,av@vol} \Delta V_P$$
$$\int_{\Delta S} q^{mean}_{x/y,face} dS = q^{mean}_{x/y,av@face} \Delta S \tag{5.4}$$

where the subscript *face* corresponds to the various faces of the CV and ΔS is their surface-area; Δx_P for the *north* (at $y + \Delta y$) and *south* (at y) surfaces, and Δy_P for the *east* (at $x + \Delta x$) and *west* (at x) surfaces in Fig. 5.3.

Substituting the integrals in Eqs. 5.4 –5.3, an algebraic formulation is given as

$$\frac{\left(\overline{\mathcal{H}}^{t+\Delta t}_{av@vol} - \overline{\mathcal{H}}^{t}_{av@vol}\right) \Delta x_P \Delta y_P}{\Delta t} = Q^{mean}_{cond} + \overline{\mathcal{Q}}_{gen,av@vol} \Delta x_P \Delta y_P$$
$$\text{where } Q^{mean}_{cond} = \left(q^{mean}_{x,av@x} - q^{mean}_{x,av@(x+\Delta x)}\right) \Delta y_P \tag{5.5}$$
$$+ \left(q^{mean}_{y,av@y} - q^{mean}_{y,av@(y+\Delta y)}\right) \Delta x_P$$

where the subscript av represents the space-averaged (within the volume/surface-area) value, and the superscript *mean* represents the time-averaged (over the time interval Δt) value. As expected, the above algebraic formulation incorporates both the spatial and temporal variations of the volumetric term and the flux term.

5.1.3 Approximations

Approximations are required to convert the volumetric term and the flux term in the algebraic formulation (Eq. 5.5) to certain discrete values in space (within the volume/surface) and time (within the time interval Δt). Now, let us come up with a *simple, intuitive, and one of the best approximation,* for the discrete grid point-based and time instant-based representation of the algebraic formulation.

What should be the approximation to obtain the space-averaged value of the volumetric enthalpy stored $\overline{\mathcal{H}}_{av}$ and the volumetric heat generated $\overline{\mathcal{Q}}_{gen,av}$ inside the CV? Both $\overline{\mathcal{H}}$ and $\overline{\mathcal{Q}}_{gen}$ varies from point to point inside the CV; consisting of almost infinite number of points. Instead, if we want to approximate them by one-point value, which point in the CV will be a good representation of the average value? An intuitive answer would be centroid of the CV (represented by P), resulting in $\overline{\mathcal{H}}_{av} \approx \overline{\mathcal{H}}_P$ and $\overline{\mathcal{Q}}_{gen,av} \approx \overline{\mathcal{Q}}_{gen,P}$—become close to reality if the size/volume of the CV is *small enough* ($\Delta V \to 0$). Note that the symbol $\approx$ corresponds to approximate representation. Similar approximation for the average value at the various surfaces $av@face$, results in the conduction flux as $q^{mean}_{x,av@x} \approx q^{mean}_{x,w}$, $q^{mean}_{y,av@y} \approx q^{mean}_{y,s}$, $q^{mean}_{x,av@(x+\Delta x)} \approx q^{mean}_{x,e}$ and $q^{mean}_{y,av@(y+\Delta y)} \approx q^{mean}_{y,n}$; where the subscript w, s, e, and n

corresponds to the centroid of the west, south, east, and north surfaces of the CV, respectively. The centroids are shown in Fig. 5.3, with separate symbols as circle for the centroid P (of the CV) and square/triangle for the centroid f (e, w, n, and s) at the various faces.

The flux term also requires an approximation for the time-averaged (mean) value. Thus, which time instant over the interval Δt (in-between the old/previous time instant t and the new/present time instant $t + \Delta t$) will represent the mean value? Extending the centroid for the above space-averaged value to time, the answer is $t + \Delta t/2$. This leads to the approximated value of conduction flux as $q_{x/y,f}^{mean} \approx \left(q_{x/y,f}^{t} + q_{x/y,f}^{t+\Delta t} \right)/2$, called as *Crank-Nicolson method*. As mentioned in Chap. 1, a movie for temporal evolution of temperature is obtained such that the previous picture (at time instant t) is used to compute present picture (at time instant $t + \Delta t$) of temperature field. Thus, a much simpler representation of conduction flux variation (between previous t and present $t + \Delta t$ time) is to take it at t, i.e., $q_{x/y,f}^{mean} \approx q_{x/y,f}^{t}$. This is called as an *explicit method* and approaches to reality if the duration of *time step* is *small-enough* ($\Delta t \rightarrow 0$). The explicit method assumes that the previous time-instant value $q_f(t)$ prevails over the entire time step Δt except at $t + \Delta t$ where it suddenly changes to $q_f(t + \Delta t)$.

Since the final algebraic formulation should be in terms of temperature at the various grid points, the volumetric enthalpy is represented as $\overline{\mathcal{H}}_P = \rho c_P T_P$, and conduction flux is obtained from Fourier's law of heat conduction as $q_{x,f=e,w} = -k\,(\partial T/\partial x)_f$ and $q_{y,f=n,s} = -k\,(\partial T/\partial y)_f$. Thus, another approximation is needed for the algebraic formulation of the normal gradient of temperature at the various face centers. From the east face center (Fig. 5.3), a horizontal line (note that it is along x-direction and normal to the face center e) intersects the centroid P of the representative CV and the centroid E of the adjoining CV (across the east face). Thus, a simple and intuitive approximation of the gradient is $(\partial T/\partial x)_e \approx (T_E - T_P)/\delta x_e$ —approaches close to reality if the distance δx_e ($= x_E - x_P$) between the two adjoining centroids is small enough. Similar approximation can be applied to approximate heat fluxes at the other surfaces of the representative CV.

The above-discussed assumptions for an algebraic formulation are specifically two in number called here as first and second approximations; presented in separate subsections below.

5.1.3.1 First Approximation

It is an approximation to convert the functional variation of the volumetric term and flux term into a discrete representation in space and time—one point for the spatial variation, and one time instant for the temporal variation. The *one-point/space-averaged value* corresponds to centroid of the volume and the surfaces of the CV, for the volumetric term and the flux term, respectively. The *one time-instant/time-averaged value* for the temporal variation of a flux term corresponds to the either of the three time instants: previous t, present $t + \Delta t$, and mean $(t + \Delta t)/2$. The respective time instant leads explicit, implicit, and Crank-Nicolson method.

For a representative 2D CV with the cell center denoted as P (Fig. 5.4), the I-approximation for the volumetric term is given as

$$\begin{aligned}
\text{Volumetric enthalpy}: & \quad \overline{\mathcal{H}}_{av@vol} \approx \overline{\mathcal{H}}_P \; (= \rho c_p T_P) \\
\text{Volumetric heat generation Rate}: & \quad \overline{Q}_{gen,av@vol} \approx \overline{Q}_{gen,P}
\end{aligned} \tag{5.6}$$

whereas, for the flux term at the east face center, the I-approximation is given as

Conduction-flux at the east face-center

$$q_{x,av@x+\Delta x}^{mean} \begin{cases} \approx q_{x,e}^{t} & \text{for Explicit Method} \\ \approx q_{x,e}^{t+\Delta t} & \text{for Implicit Method} \\ \approx \left(q_{x,e}^{t} + q_{x,e}^{t+\Delta t} \right)/2 & \text{for Crank - Nicolson Method} \end{cases} \tag{5.7}$$

Similar expression can be obtained for the conduction flux at the other face centers; and a general equation for all the face centers is given as

$$\left. \begin{aligned}
q_{x,av@x+\Delta x}^{mean} & \approx q_{x,e}^{\chi} \\
q_{y,av@y+\Delta y}^{mean} & \approx q_{y,n}^{\chi} \\
q_{x,av@x}^{mean} & \approx q_{x,w}^{\chi} \\
q_{y,av@y}^{mean} & \approx q_{y,s}^{\chi}
\end{aligned} \right\} \begin{array}{ll} \chi = n & \text{for Explicit Method} \\ \chi = n+1 & \text{for Implicit Method} \\ \chi = n+1/2 & \text{for Crank - Nicolson Method} \end{array} \tag{5.8}$$

where n, $n + 1$, and $n + 1/2$ are the time levels, corresponding to the time instant t, $t + \Delta t$, and $t + \Delta t/2$, respectively. The time level consist of integer values with $n = 0$, 1 and 2 for $t = 0$, Δt and $2\Delta t$, respectively; and act as running indices for time. The first picture of a fluid dynamics movie corresponds to $n = 0$; whereas, after a time step/interval of Δt, the second picture corresponds to $n = 1$.

The conduction flux at all the faces of the CV is used later (Eqs. 5.10 and 5.11) to compute the mean rate of net conduction heat transfer into the CV Q_{cond}^{mean} ($\approx Q_{cond}^{\chi}$). For the calculation of the temperature of present time instant $t + \Delta t$, note that Q_{cond}^{mean} is computed by using the conduction flux/temperature of the previous time instant t in an explicit method. Whereas, the implicit method uses the conduction flux/temperature of the same time instant $t + \Delta t$ at which temperature is to be obtained. Computation of Q_{cond}^{mean} by a Crank-Nicolson method results in equal weightage (50%) to the conduction flux/temperature at time t and time $t + \Delta t$, corresponding to the explicit and implicit methods, respectively. The temporal accuracy of the explicit and implicit methods is first-order and that of the Crank-Nicolson method is second-order.

5.1.3.2 Second Approximation

It is an approximation to convert the differential formulation in the Fourier's law of heat conduction to an algebraic formulation. The $II-$approximation corresponds to discrete representation of *first-derivative* of temperature. A *second order central difference*-based II−approximation for the normal gradient of temperature is given as

$$q_{x,e}^{\chi} \approx -k\frac{T_E^{\chi} - T_P^{\chi}}{\delta x_e}, \quad q_{x,w}^{\chi} \approx -k\frac{T_P^{\chi} - T_W^{\chi}}{\delta x_w}$$
$$q_{y,n}^{\chi} \approx -k\frac{T_N^{\chi} - T_P^{\chi}}{\delta y_n}, \quad q_{y,s}^{\chi} \approx -k\frac{T_P^{\chi} - T_S^{\chi}}{\delta y_s} \tag{5.9}$$

5.1.4 Approximated Algebraic Formulation

The energy conservation law and the two approximations in the physical law-based FVM are shown in Fig. 5.4.

Using the I-approximation for the volumetric term and flux term (Eqs. 5.6 and 5.8), the algebraic formulation in Eq. 5.5 is approximated as

$$\rho c_p \frac{T_P^{n+1} - T_P^n}{\Delta t} \Delta x_P \Delta y_P = Q_{cond}^{\chi} + \overline{Q}_{gen,P} \Delta x_P \Delta y_P \tag{5.10}$$
$$\text{where } Q_{cond}^{\chi} = \left(q_{x,w}^{\chi} - q_{x,e}^{\chi}\right) \Delta y_P + \left(q_{y,s}^{\chi} - q_{y,n}^{\chi}\right) \Delta x_P$$

Using the II-approximation for the conduction flux (Eq. 5.9), the final algebraic formulation is given as

$$\rho c_p \frac{T_P^{n+1} - T_P^n}{\Delta t} \Delta x_P \Delta y_P = \left(k\frac{T_E^{\chi} - T_P^{\chi}}{\delta x_e} - k\frac{T_P^{\chi} - T_W^{\chi}}{\delta x_w}\right) \Delta y_P$$
$$+ \left(k\frac{T_N^{\chi} - T_P^{\chi}}{\delta y_n} - k\frac{T_P^{\chi} - T_S^{\chi}}{\delta y_s}\right) \Delta x_P + \overline{Q}_{gen,P} \Delta x_P \Delta y_P \tag{5.11}$$

Note that the derivation of the above LAE for temperature is a two step process with derivation of LAE consisting of conduction flux in the first step (Eq. 5.10) and LAE for temperature in the second step (Eq. 5.11). This is analogous to the derivation of PDE, with $-\nabla \cdot \overrightarrow{q}$ in the first step and $k\nabla^2 T$ in the second step; shown in Fig. 5.5.

Various simplified versions of the above general LAE are possible. For example, for 1D conduction (e.g., in the x-direction), substituting $\Delta y_P = 1$ and eliminating the conduction term in the y-direction, the above equation reduces to

$$\rho c_p \frac{T_P^{n+1} - T_P^n}{\Delta t} \Delta x_P = k\left(\frac{T_E^{\chi} - T_P^{\chi}}{\delta x_e} - \frac{T_P^{\chi} - T_W^{\chi}}{\delta x_w}\right) + \overline{Q}_{gen,P} \Delta x_P \tag{5.12}$$

Further simplification is possible for steady-state conditions—there is no change in the enthalpy stored inside the CV—with the LHS of the above equations equal to zero. Furthermore, there is no time instant and the temperature in this case corresponds to steady-state condition; thus, the superscript n corresponding to time level

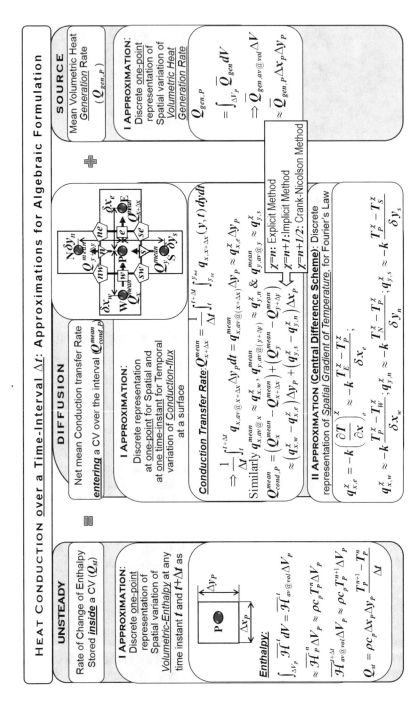

Fig. 5.4 Pictorial representation of the two approximations used in the physical law-based FVM for a 2D heat conduction

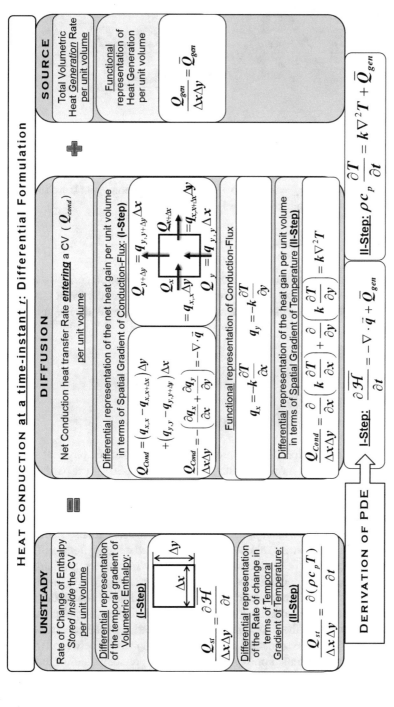

Fig. 5.5 Pictorial representation of physical law-based differential formulation for a 2D heat conduction

disappears from the RHS of the above equations. Thus, the general LAE for 2D and 1D steady-state conductions are given as

$$\left(\frac{T_E - T_P}{\delta x_e} - \frac{T_P - T_W}{\delta x_w}\right)\Delta y_P + \left(\frac{T_N - T_P}{\delta y_n} - \frac{T_P - T_S}{\delta y_s}\right)\Delta x_P$$
$$+ \frac{\overline{Q}_{gen,P}}{k}\Delta x_P \Delta y_P = 0 \tag{5.13}$$

$$\left(\frac{T_E - T_P}{\delta x_e} - \frac{T_P - T_W}{\delta x_w}\right) + \frac{\overline{Q}_{gen,P}}{k}\Delta x_P = 0 \tag{5.14}$$

If there is no volumetric heat generation and the grid points are equi-spaced ($\delta x_e = \delta x_w = \Delta x = \delta y_n = \delta y_s = \Delta y$), the above equations gets simplified for 2D and 1D steady-state conductions as as

$$T_P = \frac{T_E + T_W + T_N + T_S}{4} \text{ for 2D and } T_P = \frac{T_E + T_W}{2} \text{ for 1D}$$

The above algebraic equations for 1D steady-state conduction correctly represents the physical example, a person standing exactly in between the ice at $0\,°\text{C}$ and the fire at $100\,°\text{C}$ (Fig. 3.3), where the temperature experienced is the mean of the two temperatures.

5.1.5 Discussion

5.1.5.1 Mathematical Approximations

Although the presentation of the approximations above for the physical law-based FVM is different from those presented by Patankar (1980) for a partial differential equation-based FVM, mathematically the approximations for the two types of FVM are same. The mathematical approximations are of two types (Patankar 1980): *stepwise* and *piecewise linear*.

Considering the above *first approximation* for volumetric enthalpy in the unsteady term, the one-point/centroid value of the *approximated* enthalpy/temperature varies from one CV to the other adjoining CV in a *stepwise* manner. Similarly, for the spatial variation of conduction flux at the various surfaces of a CV, there is a stepwise variation of one-point/centroid value of the approximated conduction flux from one surface to the other adjoining surface. Whereas, for the temporal variation of conduction flux within the time interval Δt, the variation of the constant one time-instant value of the approximated heat flux is also stepwise in between two consecutive discrete time instants. For explicit method, the previous time-instant heat flux q_f^n prevails over the entire time step (in-between the time instant of t and $t + \Delta t$) except at $t + \Delta t$ where it suddenly changes to q_f^{n+1}; and the present time-instant heat flux q_f^{n+1} prevails over the entire time step, for the implicit method. Whereas, for the Crank-Nicolson method, a linear variation of heat flux is assumed over the time interval Δt to obtain an approximated heat flux $q_f^{t+\Delta t/2} \approx (q_f^{t+\Delta t} + q_f^t)/2$. Thus, the unsteady profile assumption is *stepwise* for explicit and implicit; and *piecewise linear* for Crank-Nicolson method.

Considering the above *second approximation* for Fourier's law of heat conduction, the temperature profile is assumed as linear between two consecutive grid points for the approximation of the normal temperature gradient. Thus, for $(\partial T/\partial x)_w$ and $(\partial T/\partial x)_e$, the approximation results in a change in the slope of the linear variation of the temperature between T_W and T_P as compared to that between T_P and T_E. Similarly, for a Crank-Nicolson method, there is a change in the slope of linear timewise variation of heat flux variation at any discrete time instant. Since there is a discontinuity in the slope of the variation, across any discrete time instant for the Crank-Nicolson method and any grid point for the central difference scheme-based normal gradient approximation, the variations are called as *piecewise linear*.

5.1.5.2   Explicit, Implicit, and Crank-Nicolson Method

For the explicit method, the mean rate of net conduction heat transfer into the CV Q_{cond}^{χ} (refer Eqs. 5.9 and 5.10) can be determined easily (independent of the unsteady term), as it consists of temperature of the previous time instant t. The simulation proceeds time step by step such that the information from the picture of temperature field of the previous time instant "t" is used to obtain the picture for the present time instant "$t + \Delta t$". Whereas, in an implicit and Crank-Nicolson method, the equation for Q_{cond}^{χ} consists of temperature of the present time step which is an unknown. This results in the coupling of the algebraic formulation for the conduction/diffusion and the unsteady term.

Later on, during the solution methodology below, the implication of above discussion will result in the solution of system of LAEs *independently* in an explicit method and in a *coupled* way in an implicit method; commonly solved in CFD by an indirect/iterative method. Considering a 25 LAEs obtained from the FVM for 2D conduction (corresponding to temperature at the 25 internal grid points in Fig. 5.3), this means that the number of unknown in each equation will be only one for the explicit method; and not more than five for the implicit and Crank-Nicolson method. The convenience (independent solution of the system of LAEs) in an explicit method is offset by a constraint in the time step Δt. The time step depends on the chosen grid size—Δt is restricted to be equal or less than a certain value Δt_{stab}; presented below (Eq. 5.40) as a stability criterion. For a larger Δt, the transient solution will quickly go unstable —the numbers for the temperature field as well as steady-state convergence parameter approaches towards infinity. If the stability criterion-based Δt_{stab} is very small in a problem, it will lead a large number of time marching and large computational time to obtain the final steady-state solution. The stability criterion-based time-step restriction is also needed for Crank-Nicolson but not for implicit method.

Thus, the explicit and Crank-Nicolson are conditionally stable; whereas, the implicit method is unconditionally stable.

At larger time step ($\Delta t > \Delta t_{stab}$), the unsteady profile assumption (discussed above) becomes highly inaccurate for explicit as well as Crank-Nicolson but not for implicit method. The inaccuracy leads to a numerical instability for explicit and physically unrealistic solution for Crank-Nicolson method (Patankar 1980). Thus, the unsteady profile assumption in an implicit method is closest to the reality for a larger time interval.

5.1.5.3   System of Linear Algebraic Equations

Note that the finite volume method, corresponding to the derivation of LAEs, is presented above for a representative CV. However, it is indeed applicable for all the CVs in the domain leading from the single to a system of LAEs. Thus, the same equation (Eq. 5.11) needs to be solved repeatedly to scroll through all the CVs/grid points inside the domain. However, as the neighbors of a CV change with the change in the CV under consideration, there is a change in the value of thermal conductivity as well as heat generation (if they are non-uniform in the domain), geometrical parameters (if the grid is non-uniform) and temperature of the neighboring cell centers.

Also note that the neighboring information is in-built—called as structured— in the Cartesian grid used here. This is because the subscript P, E, W, N, S for the temperature corresponds to the coordinate (x, y), $(x + \delta x_e, y)$, $(x - \delta x_w, y)$, $(x, y + \delta y_n)$, and $(x, y - \delta y_s)$; and represented later by running indices (i, j), $(i + 1, j)$, $(i - 1, j)$, $(i, j + 1)$, and $(i, j - 1)$, respectively. The running indices consists of integer values which acts like a tag or address of a grid point. It also corresponds to an element number of a matrix which is used as the data structure to store the temperature at the various grid points.

This recursive calculation involving geometrical parameters, thermo-physical properties, and temperature at different discrete time instants, along with an in-built neighboring information, is well suited for programming and then solving in a computer; discussed below. This is used to obtain the temperature at particular time instant $t + \Delta t$, using the values of previous time instant t. Thus, the present picture of temperature field is computed from the temperature in the previous picture, with a picture-by-picture (after a time interval of Δt) creation leading to a movie for the temperature field.

5.1.5.4   Stopping Criterion During the Unsteady Solution

Most of the unsteady-state solution finally approaches to a steady-state condition and the time step-by-step solution is stopped in a computer program by using a stopping criterion. Considering $n + 1$ as the present and n as the previous time level, and $T_{i,j}^{n+1}$ and $T_{i,j}^{n}$ as the respective time-instant temperature, the measure of unsteadiness (by the temporal derivative of the temperature) in the solution is approximated as

$$\left(\frac{\partial T}{\partial t}\right)_{i,j} \approx \frac{T_{i,j}^{n+1} - T_{i,j}^{n}}{\Delta t} \leq \epsilon_{st} \text{ (user-defined zero; say } 10^{-3} \text{ or } 10^{-5})$$

where ϵ_{st} is steady-state convergence tolerance, representing practically zero for a computer. It corresponds to the minimum accuracy of the steady-state convergence desired for the system of LAEs. Since the temperature field in a transient simulation approaches the steady-state asymptotically, ϵ_{st} is reduced upto a critical value of the convergence tolerance $\epsilon_{st,crit}$ below which there is almost no variation in the temperature field. However, for the above dimensional form of the temperature gradient, the $\epsilon_{st,crit}$ varies with a change in the thermophysical properties, physical conditions, and boundary conditions in a problem.

Thus, for almost same value of $\epsilon_{st,crit}$ (say 10^{-4}) applicable for all the problems, the non-dimensional form of the above temperature gradient is used as a measure of the *unsteadiness*; given as

$$\frac{\partial \theta}{\partial \tau} = \frac{l_c^2}{\alpha \Delta T_c}\left(\frac{\partial T}{\partial t}\right)_{i,j} \approx \frac{1}{Fo_{\Delta t}}\frac{T_{i,j}^{n+1} - T_{i,j}^{n}}{\Delta T_c} \leq \epsilon_{st}$$

where l_c is the characteristic length scale and $\Delta T_c (\equiv T_2 - T_1)$ is the characteristic temperature difference for a particular problem —leading to the non-dimensional temperature $\theta = (T - T_1)/(T_2 - T_1)$ and time $\tau = \alpha t/l_c^2$. Furthermore, $Fo_{\Delta t} \equiv \alpha \Delta t/l_c^2$ is Fourier number (based on the time step) or a non-dimensional time step. This unsteadiness $(\partial \theta/\partial \tau)$ is computed for all the grid points of the temperature and should be reduced simultaneously for all of them. For a conduction problem, a computer program can be developed with the unsteady results for temperature as dimensional T or non-dimensional θ. For the respective type of results, the *steady-state convergence criterion* is given as

$$\text{Results} \begin{cases} \text{Dimensional:} & \left|\dfrac{l_c^2}{\alpha \Delta T_c}\dfrac{T_{i,j}^{n+1} - T_{i,j}^{n}}{\Delta t}\right|_{max} \text{for all}_{(i,j)} \\[2em] \text{Non-Dimensionalz} & \left|\dfrac{\theta_{i,j}^{n+1} - \theta_{i,j}^{n}}{\Delta \tau}\right|_{max} \text{for all}_{(i,j)} \end{cases} \leq \epsilon_{st} \qquad (5.15)$$

where the unsteadiness (L.H.S of the above equation) is computed for all the grid points and the maximum of the grid-point values is considered above; to ensure that the criterion is ensured by all the grid points in the domain. Ideally, with increase in the time, a monotonic decay in the non-dimensional unsteadiness is expected; however, a transient simulation should to terminated if there is almost no change in its value or it increases monotonically.

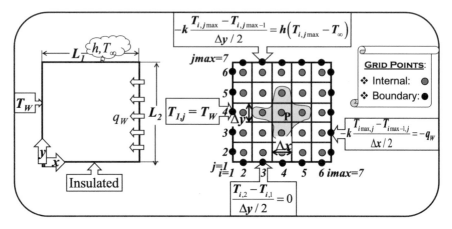

Fig. 5.6 Finite difference discretization of the various types of BCs, for 2D heat conduction in a plate

5.2   Finite Difference Method for Boundary Conditions

As discussed earlier, the FVM (presented in the previous section) is used to obtain LAE for the internal grid points, and FDM (presented in the previous chapter) is used to obtain LAE for the boundary grid points. For a 2D heat conduction in a plate, Fig. 5.6 shows a uniform grid and finite difference method-based discretization of the various types of BCs. Thus, the temperature at the boundary grid points are given as

$$
\left.
\begin{aligned}
T_{x=0} &= T_W \\
-k\left(\frac{\partial T}{\partial x}\right)_{x=L_1} &= -q_W \\
\left(\frac{\partial T}{\partial y}\right)_{y=0} &= 0 \\
-k\left(\frac{\partial T}{\partial y}\right)_{y=L2} &= h(T_{y=L_2} - T_\infty)
\end{aligned}
\right\}
\Rightarrow
\begin{aligned}
T_{1,j} &= T_W \\
T_{imax,j} &= T_{imax-1,j} + \frac{q_W \Delta x}{2k} \\
T_{i,1} &= T_{i,2} \\
T_{i,jmax} &= \frac{2kT_{i,jmax-1} + h\Delta y T_\infty}{(2k + h\Delta y)}
\end{aligned}
$$

(5.16)

where a first-order forward/backward difference method is used for the implementation of the non-Dirichlet BCs. Thus, for the problem and grid shown in Fig. 5.6, the FVM will lead to 25 LAEs for the 25 internal grid points and FDM will result in 20 LAE for the 20 boundary grid points; one LAE for each of the grid points.

5.3 Flux-based Solution Methodology on a Uniform Grid: Explicit Method

The physical law-based FVM/Algebraic formulation is presented above first for conduction flux (Eq. 5.9) and then for temperature (5.10); similar sequence can be followed in the solution methodology. This is presented here as two-step flux-based solution methodology, which results in conduction flux in the first step, and temperature in the second step. Whereas, for the coefficient of LAE-based solution methodology presented later, the final LAEs (Eq. 5.11 for all the CVs in the domain) are solved independently for the explicit method (and iteratively for the implicit method) in a single step. Note that the discrete field variable are obtained for both conduction flux and temperature in the flux-based methodology, and only temperature field is obtained in the coefficient of LAE-based methodology. Also note that the former methodology involves physical quantities, while the latter involves the mathematical coefficients in a computer program. The flux-based methodology is presented here for explicit method on a uniform grid, while the coefficient of LAE-based methodology will be presented below for a non-uniform grid. However, both the methodology will be presented for 1D as well as 2D conduction.

5.3.1 One-Dimensional Conduction

Solution methodology is presented for two cases of 1D heat conduction problems: (a) Dirichlet BC at both the boundaries, and (b) Dirichlet at the west boundary and non-Dirichlet (convective) BC at the east boundary of the domain. This is shown in Fig. 5.7, considering $imax = 7$ (5 internal and 2 boundary) grid points. Figure 5.7 shows a *uniform* grid generated first by dividing the 1D domain of length L into five CVs of same size and then defining the internal grid points ($i = 2 - 6$) at the centroid of the CVs. The grid points are shown in the figure by circles —black for boundary grid points ($i = 1$ & 7) and gray for the internal grid points ($i = 2 - 6$). Furthermore, the internal grid points are of two types: border ($i = 2$ & 6) and interior ($i = 3 - 5$); differentiated by the thicker boundary of the gray circle for border grid points in the figure. The figure also shows the face centers by square symbols. The different symbols and colors used here later help in the effective presentation of the implementation of the formulation into a computer program. For a representative CV, considering the running indices of the grid point P as $i = 4$, the figure shows the face center of the CV as w and e, width of the CV as Δx, and distance from the neighboring grid points as δx_e and δx_w.

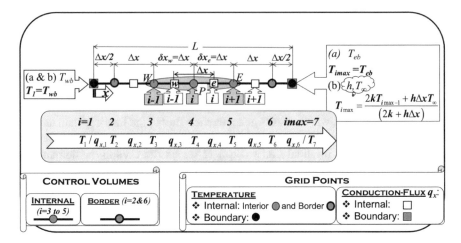

Fig. 5.7 Two different cases (**a** and **b**) of 1D heat conduction problems, along with the different types of CVs/grid points and discretized BCs on a uniform grid

5.3.1.1   Explicit method

Using the algebraic formulation for 1D unsteady-state heat conduction with *uniform heat generation*, the *explicit method*-based LAE for the representative CV (Fig. 5.7) is given (from Eqs. 5.9 and 5.10 with $\chi = n$) as

$$q_{x,e}^{n} = -k \left(\frac{T_{E}^{n} - T_{P}^{n}}{\delta x_{e}} \right) \text{ and } q_{x,w}^{n} = -k \left(\frac{T_{P}^{n} - T_{W}^{n}}{\delta x_{w}} \right) \tag{5.17}$$

$$\rho c_{p} \frac{T_{P}^{n+1} - T_{P}^{n}}{\Delta t} \Delta x = Q_{cond,P}^{n} + Q_{gen}$$

$$T_{P}^{n+1} = T_{P}^{n} + \frac{\Delta t}{\rho c_{p} \Delta x} \left(Q_{cond,P}^{n} + Q_{gen,P} \right) \tag{5.18}$$

$$\text{where } Q_{cond,P}^{n} = q_{w}^{n} - q_{e}^{n} \text{ and } Q_{gen} = \overline{Q}_{gen} \Delta x$$

5.3.1.2   Implementation Details

The implementation of the formulation, presented above, into a computer program/-code requires certain details—called implementation details. One of them is the *data structure* of the different variables encountered in the formulation. Matrix (row or column vector in case of 1D problems) is taken as the data structure to store the flow properties (temperature in case of the conduction heat transfer) at discrete grid points. While using the flux-based methodology, conduction flux q_x at the various face centers are also computed and stored in a matrix.

Note that only one running index i, corresponding to the cell centers, is used for the matrix of all the variables (Fig. 5.7); resulting in an efficient implementation. Thus, it avoids using two different running indices: i_1 for a variable defined at the cell centers (T here) and i_2 for a variable defined at the face centers (q_x here). This is achieved by following a *convention* where the running indices of a variable defined at a face corresponds to that of the CV on which it lies at the positive face—east for the 1D case here. Using the convention for a representative CV P with $i = 4$ in Fig. 5.7, $T_P = T_4$, $q_{x,e} = q_{x,4}$ and $q_{x,w} = q_{x,3}$ ($q_{x,w}$ becomes $q_{x,e}$ for $i = 3$). The figure also shows the specific grid points for T_i ($i = 1 - 7$) and $q_{x,i}$ ($i = 1 - 6$), where the subscript corresponds to the running indices or element number of a matrix.

Thermo-physical properties are mostly uniform in the domain. Geometrical parameters such as width of the CVs Δx, and distance between cell center δx, are non-uniform in general. However, for the uniform grid here, note from Fig. 5.7 that the width of all the CVs is Δx. Furthermore, it can be seen that the distance between any two consecutive circle (grid point) is $\delta x = \Delta x$ except at the end of the domain where it is $\Delta x/2$. Thus, δx is equal to Δx and $\Delta x/2$ during the computation of q_x at the internal and boundary grid points (for q_x), respectively. Thus, $q_{x,1} = -k(T_2 - T_1)/(\Delta x/2)$, $q_{x,6} = -k(T_7 - T_6)/(\Delta x/2)$ and $q_{x,i} = -k(T_{i+1} - T_i)/\Delta x$ for $i = 2 - 5$.

Another implementation detail is the initial and end value of the "*for loop*", used for recursive computation. This is used to implement an expression involving vectorized operation such as transient computation of the heat flux $q_{x,i}$ ($i = 1$ to $imax - 1$) and temperature T_i ($i = 2$ to $imax - 1$). Seven circles in Fig. 5.7 indicate that the variable T_i (at the cell centers) varies from $i = 1$ to $imax$, and six squares in the figure indicate that $q_{x,i}$ (defined at the face centers) varies from $i = 1$ to $imax - 1$. Note that heat flux is computed here at the internal as well as boundary grid points, and temperature is computed only at the internal grid points. The temperature of the boundary grid points are obtained from the discretized form of BCs, shown in Fig. 5.7.

A flux-based implementations of Eqs. 5.17 and 5.18 are done by replacing the subscript E, P, W, e and w by $i + 1$, i, $i - 1$, i and $i - 1$, respectively; the subscript and the running indices can be seen in Fig. 5.7. The implementation is shown as a two step process in Fig. 5.8. The figure shows that the temperatures of the present time level is calculated using the values of previous time level as follows:

First step: For the grid points at the face centers represented by square in step 1 of Fig. 5.8 , calculate the heat flux at previous time level n using the discrete representation of Fourier's law (Eq. 5.17) as

$$q^n_{x\,i,j} = \begin{cases} -k\dfrac{T^n_{i+1} - T^n_i}{\Delta x/2} & \text{for } i = 1 \text{ and } imax - 1 \\[2mm] -k\dfrac{T^n_{i+1} - T^n_i}{\Delta x} & \text{for } i = 2 \text{ to } imax - 2 \end{cases} \qquad (5.19)$$

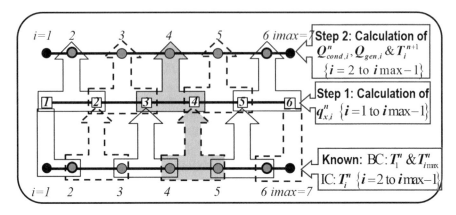

Fig. 5.8 Computational stencils to compute the conduction fluxes in the first step, and net mean conduction heat transfer rates entering the CVs in the second step, during the explicit method and flux-based implementation for the solution of 1D unsteady-state conduction

Second step: For the internal grid points seen in the step 2 of Fig. 5.8, calculate the net mean conduction heat transfer rate entering the internal CVs, and then the volumetric heat generation and temperature (Eq. 5.18) as

$$
\left.
\begin{aligned}
Q^n_{cond,i} &= (q^n_{x,i-1} - q^n_{x,i}) \\
Q_{gen} &= \overline{Q}_{gen}\,\Delta x \\
T^{n+1}_i &= T^n_i + \frac{\Delta t}{\rho c_p \Delta x}(Q^n_{cond,i} + Q_{gen})
\end{aligned}
\right\} \quad \text{for } i = 2 \text{ to } imax - 1 \quad (5.20)
$$

A geometrical representation of the grid points involved in above equations is shown as the *computational stencil* in Fig. 5.8. For example, applying these equations to grid point $i = 4$, the heat flux $q^n_{x,4}$ is a function of T^n_5 and T^n_4, and $Q^n_{cond,4}$ is a function of $q^n_{x,3}$ and $q^n_{x,4}$. The respective functional relationship for the steps 1 and 2 are shown by the shaded arrows in the figure. The computational stencil for all the other grid points are also shown (unshaded) in the figure.

5.3.1.3   Solution Algorithm

For the flux-based solution methodology, the solution algorithm for 1D unsteady-state conduction on a *uniform* grid—using *explicit method* —is given as

1. User input:

   (a) Thermo-physical property of the material: enter thermal conductivity, density and specific heat.
   (b) Geometrical parameters: enter the length of the domain L.

(c) Grid size: enter the maximum number of grid points $imax$.

(d) IC (initial condition): enter T_0

(e) BCs (Boundary conditions) parameters: enter the parameter such as T_{wb} and T_{eb} for case(a); and T_{wb}, h and T_∞ for case(b) in Fig. 5.7.

(f) Governing parameter: enter volumetric heat generation.

(g) Steady-state convergence tolerance: enter the value of ϵ_{st} (practically zero defined by the user).

2. Calculate geometric parameters ($\Delta x = L/(imax - 2)$) and stability criterion-based time step ($\Delta t = \Delta x^2/2\alpha$; based on Eq. 5.30) for explicit method. Use the time step slightly less than the computed value.

3. For the temperature $T_{i,j}$, apply IC at all the internal grid points and BCs at all those boundary grid points which are subjected to a Dirichlet BC.

4. Apply BC for the boundary grid points which are subjected to a non-Dirichlet BC. Then, the temperature at present time instant is considered as the previous time-instant temperature, $T_{old\,i,j} = T_{i,j}$ (for all the grid points), as the computations is to be continued further for the computation of the temperature of the present time instant $T_{i,j}$.

5. Computation of the conduction flux and the temperature at all the grid points:

   (a) Compute conduction flux at the face centers of the internal CVs (Eq. 5.19).

   (b) Compute the net mean conduction heat transfer rate entering the CVs, volumetric heat generation rate and then the temperature for the next time step (Eq. 5.20)—at the internal grid point.

6. Calculate a stopping parameter—non-dimensional steady-state convergence criterion (L.H.S of Eq. 5.15) —called here as *unsteadiness*.

7. If the unsteadiness in previous step is greater than steady-state tolerance ϵ_{st} (Eq. 5.15), go to step 4 for the next time-instant solution.

8. Otherwise, stop and post-process the steady-state results for analysis of the conduction problem. In an inherently unsteady problem, the condition in previous step will never be satisfied and a different stopping parameter is used.

The above algorithm presents only two matrices for temperature (T and T_{old}), and replaces the value of T_{old} by T (in step 4 above) if the solution has not reached to steady state. Thus, it avoids storage of all the discrete time-instant value of the temperature field—for an efficient management of the memory of the computer. This efficient memory storage technique is fine if the user is interested only in the final steady-state picture of the temperature field. However, if the user is interested to create a movie of the field then the temperature T obtained after every time step needs to be permanently stored.

A computer program for the flux-based methodology of CFD development is given below in Prog. 5.1, considering an example problem for the 1D unsteady heat conduction on a uniform grid.

Example 5.1: Flux-based methodology of CFD development and code verification for a 1D unsteady-state heat conduction problem.

Consider 1D conduction in a long steel ($\rho = 7750\,\text{kg/m}^3$, $c_p = 500\,\text{J/kg K}$ and $k = 16.2\,\text{W/m K}$) sheet of thickness $L = 1\,\text{cm}$. The sheet is initially at a uniform temperature of $30\,°\text{C}$ and is suddenly subjected to a constant temperature of $T_{wb} = 0\,°\text{C}$ on the west and $T_{eb} = 100\,°\text{C}$ on east boundary; case (a) in Fig. 5.7.

(a) Using the flux-based solution methodology, presented above, develop a computer program for the *explicit method* on a *uniform* 1D Cartesian grid.
(b) Present a CFD application of the code for a volumetric heat generation of 0 and $100\,\text{MW/m}^3$. Consider maximum number of grid points as $imax = 12$ and the steady-state convergence tolerance as $\epsilon_{st} = 10^{-4}$. Plot the steady-state temperature profiles with and without volumetric heat generation; and compare with the exact solution. The comparison of the present numerical with the exact solution is called as code verification. Also show the unsteady-state temperature profiles at different intermediate time instants.

Solution:

(a) Using the flux-based implementation for 1D unsteady heat conduction, an explicit method-based computer program for the present problem is presented in Prog. 5.1. The program uses a 99% of maximum time step for a stable solution in the explicit method-based 1D conduction problem, given as $\Delta t_{stab} = 0.5\Delta x^2/\alpha$ (Eq. 5.30; derivation presented below); $\Delta t_{stab} = 0.1196\,\text{sec}$ for the present problem.

Listing 5.1 Scilab code for 1D unsteady-state conduction, using solution algorithm (Sect. 5.3.1.3) for the explicit method and flux-based approach of CFD development on a uniform 1D-Cartesian grid .

```
1   //STEP-1:  User-Input
2   rho=7750.0;  cp=500.0;  k=16.2;  L=1.0;  imax=12;
3   T0=30;T_wb=0.0;T_eb=100.0;
4   Q_vol_gen=0;epsilon_st=0.0001;
5   //STEP-2:  Geometrical  Parameter  &  Stability
        criterion-based  time  step
6   alpha=k/(rho*cp);  DTc=T_eb-T_wb;
7   Dx  =  L/(imax-2);  Dt  =0.99*0.5*  (Dx*Dx/alpha);
8   Q_gen=Q_vol_gen*Dx;  //  Total  Heat  Generation
9   //STEP-3:  IC  and  Dirichlet  BCs
10  T(2:imax-1)=T0;  T(1)=T_wb;  T(imax)=T_eb;
11  //Time-Marching  for  Explicit  LAEs:  START
12  unsteadiness_nd=1;  n=0;
13  while  unsteadiness_nd>=epsilon_st
14  //Step  4:  Consider  the  temp.  as  the  previous
        value
15      n=n+1;  T_old=T;
16  //STEP  5:  Computation  of  conduction  flux  &  temp.
```

```
17          for  i =1: imax -1
18              if ( i ==1) | ( i == imax -1)  then
19            qx_old ( i ) =-k*( T_old ( i +1) -T_old ( i ) ) / ( Dx
                 /2.0) ;
20              else
21              qx_old ( i ) =-k*( T_old ( i +1) -T_old ( i ) ) / Dx ;
22              end
23          end
24          for  i =2: imax -1
25              Q_cond_old ( i ) = ( qx_old ( i -1) - qx_old ( i ) ) ;
26              T ( i ) = T_old ( i ) + ( Dt / ( rho * cp * Dx ) ) * (
                 Q_cond_old ( i ) + Q_gen ) ;
27          end
28    // STEP  6:  Steady - state  convergance  criterion  ( Eq
          . 5.15)
29          unsteadiness = max ( abs ( T - T_old ) ) / Dt ;
30          unsteadiness_nd = unsteadiness * L * L / ( alpha * DTc ) ;
31    printf ( " Time  step  no.  %5d ,  Unsteadiness_nd )  =  %8
          .4 e \ n " ,  n  ,  unsteadiness_nd ) ;
32    //  STEP  7:  End  of  While  loop
33    end
```

(b) The program is applied on the present problem for $\overline{Q}_{gen} = 0$ and $100\,MW/m^3$, and the results are shown in Fig. 5.9. The figure shows that the transient numerical result obtained on a uniform grid size of $imax = 12$, with an IC of $30\,°C$, and BC of $0\,°C$ on the west and $100\,°C$ on the east boundary. As time progresses, it can be seen that the initial temperature distribution monotonically approaches to a steady-state temperature profile which is linear and non-linear for without and with heat generation case, respectively. For the steady state, the figure shows an excellent agreement between the numerical and an analytical solution; given for the temperature profile and the location of the maxima in the profile as

$$T = T_{wb} + \frac{x}{L}(T_{eb} - T_{wb}) + \frac{\overline{Q}_{gen} x(L - x)}{2k}$$

$$x_{T=Tmax} = \frac{k(T_{eb} - T_{wb})}{L\overline{Q}_{gen}} + \frac{L}{2}$$

5.3.2 Two-Dimensional Conduction

Similar to the 1D conduction discussed above, here for 2D conduction, Unsteady-state solution methodology will be discussed for two cases of the conduction problem:

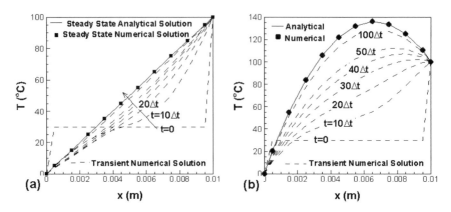

Fig. 5.9 CFD application and testing, of the *explicit method*-based 1D code on *uniform* Cartesian grid, for unsteady state heat conduction: transient temperature profile for (**a**) without and (**b**) with heat generation. The figure also shows comparison of the numerical with the analytical solution for the steady-state temperature profile

(a) Dirichlet BC on all the boundaries and (b) a different BC in each of the boundary. They are shown in Fig. 5.10, for $imax \times jmax = 7 \times 7$ (5×5 internal) grid points.

Figure 5.10 shows a *uniform* grid generated by dividing the 2D domain of dimension $L_1 \times L_2$ into 5×5 CVs of same size—obtained by drawing equi-spaced horizontal and vertical line. The figure shows filled circles for the grid points; and unfilled symbols as square for vertical (east/west) face centers and triangle for horizontal (north/south) face centers of the CVs. The circles are filled with a different color—gray for internal and black for boundary grid points. The different geometrical representations and colors used here, brings clarity later in the presentation of the implementation details for the formulation into a computer program.

A representative CV with grid point P is shown in the figure for the running indices $i = j = 4$. The figure also shows its face center, width of the CV and its distance from the neighboring grid points. Finally, a computational stencil enclosing the grid P and its neighbor —which are involved in Eq. 5.11—is shown in the figure. In general, a uniform grid corresponds to $\Delta x \neq \Delta y$ —resulting in rectangular CVs instead of the square ($\Delta x = \Delta y$) CVs. As discussed above for 1D conduction, here for 2D conduction on a uniform grid, the calculation at various *grid points for conduction flux*—represented by square and triangle symbol in Fig. 5.10—involves $\delta x_e = \delta x_w = \Delta x$ and $\delta y_n = \delta y_s = \Delta y$ for the internal grid points, and δx_e (or δx_w) $= \Delta x/2$ and δy_n (orδy_s) $= \Delta y/2$ for the boundary grid points of the *fluxes*. The two types of grid points are differentiated in the figure with a larger size of the squares/triangles for the boundary as compared to the internal grid points.

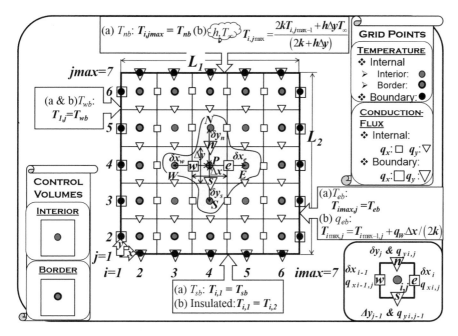

Fig. 5.10 Computational domain, discretized BCs, and different types of grid points for two different cases (**a** and **b**) of 2D heat conduction problems on a uniform grid.

5.3.2.1 Explicit method

Using the FVM for 2D unsteady-state heat conduction with *uniform* heat generation in a representative CV (Fig. 5.10), the explicit method-based LAE for the conduction flux (Eq. 5.9 with $\chi = n$) and temperature (Eq. 5.10 with $\chi = n$) are given as follows:

$$q_{x,e}^{n} = -k\left(\frac{T_E^n - T_P^n}{\delta x_e}\right), \qquad q_{x,w}^{n} = -k\left(\frac{T_P^n - T_W^n}{\delta x_w}\right), \qquad (5.21)$$

$$q_{y,n}^{n} = -k\left(\frac{T_N^n - T_P^n}{\delta y_n}\right) \text{ and } q_{y,s}^{n} = -k\left(\frac{T_P^n - T_S^n}{\delta y_s}\right)$$

$$\rho c_p \frac{T_P^{n+1} - T_P^n}{\Delta t}\Delta x \Delta y = Q_{cond,P}^n + Q_{gen}$$

$$T_P^{n+1} = T_P^n + \frac{\Delta t}{\rho c_p \Delta x \Delta y}\left(Q_{cond,P}^n + Q_{gen,P}\right) \qquad (5.22)$$

$$\text{where } Q_{cond,P}^n = (q_{x,w}^n - q_{x,e}^n)\Delta y + (q_{y,s}^n - q_{y,n}^n)\Delta x$$

$$\text{and } Q_{gen} = \overline{Q}_{gen}\Delta x \Delta y$$

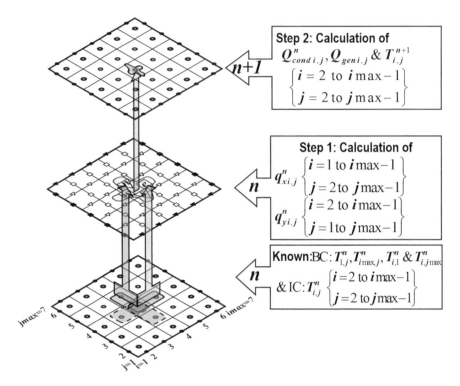

Fig. 5.11 Computational stencil to compute the conduction fluxes in the first step, and net mean conduction heat transfer rates entering a representative CV in the second step, during the explicit method and flux-based implementation for the solution of 2D unsteady-state conduction

5.3.2.2 Implementation Details

As discussed above for 1D, the flux-based implementation for 2D condution is also a two step process shown in Fig. 5.11. The temperatures of the present time level $T_{i,j}^{n+1}$ is calculated —using the previous time level values $T_{i,j}^n$ and boundary conditions—in two steps as follows:

First step: Calculate the conduction fluxes at the previous time level n (Eq. 5.21). This is done separately for the boundary and internal grid points of q_x and q_y (at the face centers of the CVs in Fig. 5.10) as

$$
q_{x\,i,j}^n = \left. \begin{array}{l} -k\dfrac{T_{i+1,j}^n - T_{i,j}^n}{\Delta x/2} \\[4mm] -k\dfrac{T_{i+1,j}^n - T_{i,j}^n}{\Delta x} \end{array} \right\} \begin{array}{l} \text{for } i = 1 \text{ and } imax - 1 \\[4mm] \text{for } i = 2 \text{ to } imax - 2 \end{array} \left. \right\} \text{ for } j = 2 \text{ to } jmax - 1
$$

$$(5.23)$$

$$q_{yi,j}^n = \begin{cases} -k\dfrac{T_{i,j+1}^n - T_{i,j}^n}{\Delta y/2} & \text{for } j = 1 \text{ and } jmax - 1 \\[2mm] -k\dfrac{T_{i,j+1}^n - T_{i,j}^n}{\Delta y} & \text{for } j = 2 \text{ to } jmax - 2 \end{cases} \Bigg\} \text{ for } i = 2 \text{ to } imax - 1$$

$$(5.24)$$

Second Step: Calculate the net mean conduction heat transfer rate entering the internal CVs, shown as the step 2 in Fig. 5.11; and then the volumetric heat generation and temperature (Eq. 5.22) as

$$\left. \begin{aligned} Q_{cond\,i,j}^n &= (q_{xi-1,j}^n - q_{xi,j}^n)\Delta y + (q_{yi,j-1}^n - q_{yi,j}^n)\Delta x \\ Q_{gen} &= \overline{Q}_{gen}\Delta x\Delta y \\ T_{i,j}^{n+1} &= T_{i,j}^n + \dfrac{\Delta t}{\rho c_p \Delta x \Delta y}(Q_{cond\,i,j}^n + Q_{gen}) \end{aligned} \right\} \quad \begin{aligned} &\text{for } i = 2 \text{ to } imax - 1 \\ &\text{for } j = 2 \text{ to } jmax - 1 \end{aligned}$$

$$(5.25)$$

The *computational stencil*, corresponding to a geometrical representation of the grid points involved in the two steps (Eqs. 5.23, 5.24 and 5.25), is shown in Fig. 5.11. For example, applying these equations to grid point (4, 4), the shaded arrow in the figure shows the computational stencil in two steps: first, for conduction flux $q_{x\,4,4}^n$ as a function of $T_{5,4}^n$ and $T_{4,4}^n$ as well as $q_{y\,4,4}^n$ as a function of $T_{4,5}^n$ and $T_{4,4}^n$; and second, $Q_{cond\,4,4}^n$ as a function of $q_{x\,4,4}^n$, $q_{x\,3,4}^n$, $q_{y\,4,4}^n$ and $q_{y\,4,3}^n$. The computational stencil is also applicable for all the other internal grid points.

Note that the computational stencil for the neighboring grid points is obtained by shifting the stencil—shown in the Fig. 5.11 —leftwards by Δx for $T_{3,4}^n$, rightwards by Δx for $T_{5,4}^n$, downward by Δy for $T_{4,3}^n$, and upward by Δy for $T_{4,5}^n$. Thus, the FVM results in the number of LAEs equal to the number of unknown temperatures at the internal grid points, *i.e.*, $(imax - 2) \times (jmax - 2)$. For example, considering the grid points shown in Fig. 5.10, the methods results in twenty-five LAEs with each LAE having only one unknown in the explicit method.

5.3.2.3 Solution Algorithm

The solution algorithm for the conduction flux-based solution methodology in 2D is same as presented in Sect. 5.3.1.3 for the 1D unsteady-state conduction, except a slight change due to increase in the dimensionality. Thus, the code developed and tested for 1D can be easily extended to 2D conduction. A computer program for the flux-based solution methodology of CFD development is given below in Prog. 5.4.1.3, considering an example problem for the 2D unsteady heat conduction on a uniform grid.

Example 5.2: Flux-based solution methodology of CFD development and application for a 2D unsteady-state heat conduction problem.
    Consider 2D conduction in a square shaped ($L_1 = 1$ m and $L_2 = 1$ m) *long* stainless-steel plate ($\rho = 7750$ kg/m^3, $c_p = 500$ J/kg K and $k = 16.2$ W/m K).

The plate is initially at uniform ambient temperature of 30 °C and is suddenly
subjected to a constant temperature of $T_{wb} = 100\,°C$ on the west boundary,
$T_{sb} = 200\,°C$ on the south boundary, $T_{eb} = 300\,°C$ on the east boundary, and
$T_{nb} = 400\,°C$ on the north boundary; case (a) in Fig. 5.10.

(a) Using the flux-based solution methodology of CFD development, presented
    above, develop a computer program for the *explicit method* on a *uniform*
    2D Cartesian grid.
(b) Present a CFD application of the code, for a volumetric heat generation of
    0 and $50\,kW/m^3$. Consider the maximum number of grid points as $imax \times
    jmax = 12 \times 12$, and the steady-state convergence tolerance as $\epsilon_{st} = 10^{-4}$.
    Plot the steady-state temperature contours, with and without volumetric heat
    generation.

Solution:

(a) Using the flux-based implementation for 1D unsteady heat conduction, an
    explicit method-based computer program for the present problem is pre-
    sented in Prog. 5.2. The program uses the maximum time step for a stable
    solution in the explicit method-based solution of the 2D conduction prob-
    lem, given as $\Delta t_{stab} = 0.25\Delta x^2/\alpha$ (Eq. 5.40; derivation presented below);
    $\Delta t_{stab} = 598$ sec for the present problem.

Listing 5.2 Scilab code for 2D unsteady-state conduction, using the explicit method and
flux-based approach of CFD development on a uniform 2D-Cartesian grid.

```
1   //STEP-1: User-Input
2   rho = 7750.0; cp = 500.0; k = 16.2;
3   L1=1;L2=1; imax=12;jmax=12;
4   T0=30;T_wb=100.0;T_sb=200;T_eb=300.0;T_nb=400;
5   Q_vol_gen=50000; epsilon_st=0.0001;
6   //STEP-2: Geometrical Parameter & Stability
        criterion-based time step
7   Dx = L1/(imax-2); Dy=L2/(jmax-2);
8   alpha=k/(rho*cp); DTc=T_nb-T_wb;
9   Dt =1*0.5*(alpha*((1/Dx^2)+(1/Dy^2)))^-1;//Eq
        .5.40
10  Q_gen=Q_vol_gen*Dx*Dy; // Total Heat Generation
11  //STEP-3: IC and Dirichlet BCs
12  T(2:imax-1,2:jmax-1)=T0;
13  T(1,1:jmax)=T_wb; T(imax,1:jmax)=T_eb;
14  T(1:imax,1)=T_sb;T(1:imax,jmax)=T_nb;
15  //Time-Marching for Explicit Unsteady-State LAEs
        :
16  unsteadiness_nd=1; n=0;
17  while unsteadiness_nd>=epsilon_st
18  //Step 4: Consider the temp. as the previous
        value
19      n=n+1; T_old=T;
```

```
20   //STEP 5: Computation of conduction flux & temp.
21       for  j=2:jmax-1
22       for  i=1:imax-1
23           if(i==1)|(i==imax-1)  then
24   qx_old(i,j)=-k*(T_old(i+1,j)-T_old(i,j))/(Dx
         /2.0);
25           else
26         qx_old(i,j)=-k*(T_old(i+1,j)-T_old(i,j))/
               Dx;
27           end
28       end
29       end
30       for  j=1:jmax-1
31       for  i=2:imax-1
32           if(j==1)|(j==jmax-1)  then
33   qy_old(i,j)=-k*(T_old(i,j+1)-T_old(i,j))/(Dy
         /2.0);
34           else
35         qy_old(i,j)=-k*(T_old(i,j+1)-T_old(i,j))/
               Dy;
36           end
37       end
38       end
39       for  j=2:jmax-1
40       for  i=2:imax-1
41   Q_cond_old(i,j)=((qx_old(i-1,j)-qx_old(i,j))*Dy)
         +((qy_old(i,j-1)-qy_old(i,j))*Dx);
42       T(i,j)=T_old(i,j)+(Dt/(rho*cp*Dx*Dy))*(
             Q_cond_old(i,j)+Q_gen);
43       end
44       end
45   //STEP 6: Steady-state convergance criterion
46       unsteadiness=max(abs(T-T_old))/Dt;
47       unsteadiness_nd=unsteadiness*L1*L1/(alpha*DTc
             );
48     printf("Time step no. %5d, Unsteadiness_nd) =
             %8.4e\n", n , unsteadiness_nd);
49   // STEP 7: While-end loop
50   end
```

(b) The program is applied on the present problem for $\overline{Q}_{gen} = 0$ and $50\,kW/m^3$, and the results are shown in Fig. 5.12. The figure shows that the steady-state numerical results obtained on a uniform grid size of $imax = 12$ and $jmax = 12$. For $\overline{Q}_{gen} = 0$, Fig. 5.12a shows a monotonic variation in the temperature; within the limits of the temperature imposed by the BCs. Whereas, for the heat generation case, Fig. 5.12b shows a local maximum of temperature (inside the domain) which is outside the range of temperature imposed by the BCs, $i.e.$, $T_{max} > 400\,°C$.

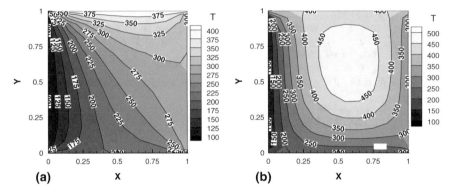

Fig. 5.12 CFD application of the *explicit method*-based 2D code on *uniform* Cartesian grid: steady-state temperature contour (line and flooded) for unsteady-state heat conduction (**a**) without and (**b**) with heat generation

5.4 Coefficients of LAE-based Solution Methodology on a Non-uniform Grid: Explicit and Implicit Method

5.4.1 One-Dimensional Conduction

For the two cases of 1D heat conduction problems, introduced in Fig. 5.7 above for a uniform grid, Fig. 5.13 is considered here for the presentation of solution methodology on a non-uniform grid.

5.4.1.1 Explicit and Implicit Methods

For 1D unsteady-state heat conduction, the FVM-based discretized equation (Eq. 5.12) with *uniform* heat generation is expressed in a general form as

$$\rho c_p \frac{T_P^{n+1} - T_P^n}{\Delta t} \Delta x_P = \lambda Q_{cond,P}^{n+1} + (1 - \lambda) Q_{cond,P}^n + \overline{Q}_{gen} \Delta x_P \qquad (5.26)$$

$$\text{where} \quad Q_{cond,P} = k \left(\frac{T_E - T_P}{\delta x_e} - \frac{T_P - T_W}{\delta x_w} \right)$$

where $\lambda = 0, 1$ and 0.5 corresponds to explicit, implicit and Crank-Nicolson method, respectively. Then, the above equation is expressed in the final LAE form as

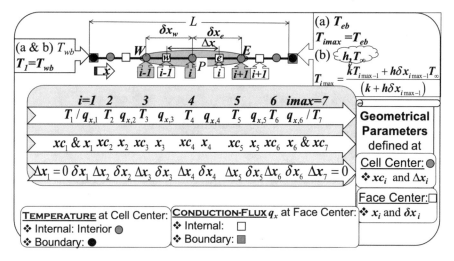

Fig. 5.13 Two different cases (**a** and **b**) of 1D heat conduction problems on a non-uniform Cartesian grid, along with the discretized BCs, different types of grid points, and discrete definition of the non-uniform geometrical parameters

$$a P \, T_P^{n+1} = a E \left[\lambda \, T_E^{n+1} + (1 - \lambda) \, T_E^{n} \right] + a W \left[\lambda \, T_W^{n+1} + (1 - \lambda) \, T_W^{n} \right] + b$$
$$\text{where } a E = k/\delta x_e, \;\; a W = k/\delta x_w, \;\; a P^0 = \rho c_p \Delta x_P / \Delta t$$
$$b = \left[a P^0 - (1 - \lambda) a E - (1 - \lambda) a W \right] T_P^n + \overline{Q}_{gen} \Delta x_P$$
$$a P = \lambda \, a E + \lambda \, a W + a P^0$$

$$(5.27)$$

Explicit method

Substituting $\lambda = 0$ in Eq. 5.27, the LAE for a representative CV is given as

$$a P \, T_P^{n+1} = a E \, T_E^{n} + a W \, T_W^{n} + b$$
$$\text{where } a E = k/\delta x_e, \;\; a W = k/\delta x_w, \;\; a P^0 = \rho c_p \Delta x_P / \Delta t \qquad (5.28)$$
$$b = \left[a P^0 - a E - a W \right] T_P^n + \overline{Q}_{gen} \Delta x_P \text{ and } a P = a P^0$$

where T_P^{n+1} can be obtained explicitly in terms of T_E^n, T_P^n and T_E^n; thus, this method is called explicit. For example, considering the case given in Fig. 5.13, the explicit method results in five LAEs —corresponding to the five internal grid points—with one unknown in each equation; thus, each LAE can be solved *explicitly/independently*.

The above LAE represents the unsteady diffusion phenomenon, as the value of temperature at a particular grid point T_P^{n+1} is *appropriately* influenced by the temperature of the grid points T_E^n, T_W^n and T_P^n. However, the coefficients $a E$, $a W$ and $a P^0 - a E - a W$ of the respective temperatures of previous time step in the LAE (Eq. 5.28) should have same sign, proposed by Patankar (1980) as a second rule. The rule for the LAE was proposed, with a physical reasoning on the nature of the diffusion phenomenon; the LAE ensures that an increase in the temperature T_E^n (or T_W^n

or T_P^n) results in an increase in the temperature T_P^{n+1} and vice-versa. The coefficient aE and aW consists of terms which are always positive. However, the coefficient $\left(aP^0 - aE - aW\right)$ of T_P^n (seen in the expression for b in Eq. 5.28) is conditionally positive; the condition is given as

$$
\begin{aligned}
aE + aW &\leq aP^0 \\
\frac{\alpha \Delta t}{\Delta x_P} \frac{\delta x_e + \delta x_w}{\delta x_e \delta x_w} &\leq 1 \quad \text{for non-uniform grid} \\
\alpha \Delta t / \Delta x^2 \text{ or Fo}_{\Delta t, \Delta x} &\leq 1/2 \quad \text{for uniform grid}(\delta x_e = \delta x_w = \Delta x)
\end{aligned}
\tag{5.29}
$$

where $\text{Fo}_{\Delta t, \Delta x}$, based on time step and grid size, is the *Fourier number*. The number represents a non-dimensional time step (should be less than half), as it is the ratio of the time step Δt to the diffusion time scale $\Delta x^2/\alpha$. The time scale corresponds to the time required for the transport of a diffusion mechanism-based disturbance over the distance Δx (Ferziger and Peric 2002). The above stability criterion for the diffusion is called as *grid Fourier number criterion*.

Thus, the limiting interval of time duration between two consecutive picture of temperature field—called as stable time step—is obtained from Eq. 5.29 as

$$
\Delta t_{stab} = \begin{cases} \dfrac{\Delta x_P \delta x_e \, \delta x_w}{\alpha \, (\delta x_e + \delta x_w)} & \text{for non-uniform grid} \\ \Delta x^2/2\alpha & \text{for uniform grid} \end{cases}
\tag{5.30}
$$

where $\Delta t > \Delta t_{stab}$ does not lead to the steady-state convergence, as the convergence parameter (LHS of Eq. 5.15) approaches towards infinity. A computer program for the computation of the above time step is presented above in Prog. 5.1 for uniform grid, with a Δt which is 99% of Δt_{stab}. For the non-uniform grid, the above equation is used to compute different time step for all the CVs in the domain, and the minimum of the time step is used in the unsteady computation. Note that the above equation was derived by Patankar (1980) from the *physical* reasoning-based second rule, rather than a detailed *mathematical* von Neumann stability analysis (Anderson (1995)).

Implicit method

Substituting $\lambda = 1$ in Eqs. 5.26 and 5.27, the LAE for a representative CV is given as

$$
\begin{aligned}
aP \, T_P^{n+1} &= aE \, T_E^{n+1} + aW \, T_W^{n+1} + b \\
&\text{where } aE = k/\delta x_e, \ aW = k/\delta x_w, \ aP^0 = \rho c_p \Delta x_P / \Delta t \\
&b = aP^0 T_P^n + \overline{Q}_{gen} \Delta x_P \text{ and } aP = aE + aW + aP^0
\end{aligned}
\tag{5.31}
$$

The method is called implicit because T_P^{n+1} is coupled with the unknowns T_E^{n+1} and T_W^{n+1}, leading to the *solution* of the system of LAEs; number of LAEs equal to the number of internal grid points (with unknown temperatures), *i.e.*, $imax - 2$. For

example, considering the case given in Fig. 5.13, the implicit-method results in five
LAEs with a total of five unknowns; and at least three unknowns in each equation.
Thus, the coupled system of LAEs needs to be solved *simultaneously/implicitly*.

5.4.1.2 Implementation Details on a Non-Uniform Grid

For a non-uniform grid, the grid points in Fig. 5.13 are non-uniformly spaced. This
results in a variation of width of the CVs Δx and distance between cell center δx;
thus, they are also stored in a matrix. They are computed from the coordinates of a
face center x as well as centroid xc of the CVs; also stored in a matrix. The x_i, xc_i,
Δx_i and δx_i are shown in Fig. 5.13, where the subscript corresponds to the running
indices or element number of a matrix. The implementation details for the uniform
grid above presents a convention, leading to the usage of the same running indices for
the conduction fluxes at the east (also north for 2D) face centers and the temperatures
at the cell centers. The convention is extended here for the geometrical parameters
of the non-uniform grid, shown in Fig. 5.13. The figure shows only one running
index i which are tagged as the running index for all the variables defined at a cell
center (T_P, xc_P and Δx_P) and east face center ($q_{x,e}$, x_e and δx_e), denoted by circle
and square symbols, respectively. This is seen in the figure as $T_P = T_4$, $xc_P = xc_4$,
$\Delta x_P = \Delta x_4$, $q_{x,e} = q_{x,4}$, $q_{x,w} = q_{x,3}$, $x_e = x_4$, $\delta x_e = \delta x_4$, $x_w = x_3$, and $\delta x_w = \delta x_3$,
for the representative CV with $i = 4$. If the thermo-physical property is non-uniform,
such as thermal conductivity is a function of temperature, it is also represented in a
matrix form. Finally, the coefficient of LAEs (aE, aW, aP, and b in Eq. 5.27) can
be expressed in a matrix form, as its value does not change (if the thermo-physical
property are not a function of temperature) during the solution of system of LAEs.

The coordinates of the face center of the CVs x_i (Fig. 5.13) are computed by
a non-uniform Cartesian grid generation method; presented below in an example
problem. The x_i's are used to compute the other geometrical parameters as follows
(refer Fig. 5.13):

$$\text{Face-center:} \quad x_i \qquad \text{for } i = 1 \text{ to } imax - 1 \qquad (5.32)$$

$$\text{Centroids}: \ xc_i = \frac{x_i + x_{i-1}}{2} \quad \text{for } i = 2 \text{ to } imax - 1;$$

$$xc_1 = x_1 \text{ and } \quad xc_{imax} = x_{imax-1}$$

$$\text{Width of CVs:} \ \ \Delta x_i = x_i - x_{i-1} \quad \text{for } i = 2 \text{ to } imax - 1$$

$$\text{Distance between}$$

$$\text{cell-centers:} \ \delta x_i = xc_{i+1} - xc_i \ \ \text{for } i = 1 \text{ to } imax - 1$$

Considering the representative CV P, corresponding to $i = 4$ in the Fig. 5.13, $\delta x_e =
\delta x_4 = xc_5 - xc_4$ and $(\delta x_w)_{i=4} = (\delta x_e)_{i=3} = \delta x_3 = xc_4 - xc_3$. A computer program
for the computation of above geometrical parameters is presented later in Prog. 5.3.

The single LAE for a representative CV in the Eqs. 5.28 and 5.31 is converted into
a system of LAEs by substituting all the subscripts to running indices: E, P, W, e,

and w are replaced by the running indices $i + 1$, i, $i - 1$, i, and $i - 1$, respectively. Furthermore, the coefficients aE, aW, aP, and b in the above system of LAEs are represented as a matrix, where $aW_i = aE_{i-1}$. This results in the final equation —implemented later in a computer program (Prog. 5.4 for the implicit method)—as

For Explicit method:

$$
\left.
\begin{aligned}
aP_i T_i^{n+1} &= aE_i\,T_{i+1}^n + aE_{i-1}\,T_{i-1}^n + b_i \\
\text{where } b_i &= \left[aP_i^0 - aE_i - aE_{i-1}\right] T_i^n + \overline{Q}_{gen}\Delta x_i \\
aP_i &= aP_i^0 = \rho c_p \Delta x_i / \Delta t
\end{aligned}
\right\} \text{ for } i = 2 \text{ to } imax - 1
$$

$$
\left.
aE_i = \frac{k}{\delta x_i}
\right\} \text{ for } i = 1 \text{ to } imax - 1
$$

$$(5.33)$$

For Implicit method:

$$
\left.
\begin{aligned}
aP_i T_i^{n+1} &= aE_i\,T_{i+1}^{n+1} + aE_{i-1}\,T_{i-1}^{n+1} + b_i \\
\text{where } aP_i^0 &= \rho c_p \Delta x_i / \Delta t, \\
b_i &= aP^0 T_i^n + \overline{Q}_{gen}\Delta x_i \\
aP_i &= aP^0 + aE_i + aE_{i-1}
\end{aligned}
\right\} \text{ for } i = 2 \text{ to } imax - 1
$$

$$
\left.
aE_i = \frac{k}{\delta x_i}
\right\} \text{ for } i = 1 \text{ to } imax - 1
$$

$$(5.34)$$

For both explicit and implicit methods, note that the LAE and all its coefficients (Eqs. 5.33 and 5.34) are computed at the cell centers; except aE_i which is computed at the face center. For the five internal grid points in Fig. 5.13, Eq. 5.34 is given in a Tri-Diagonal matrix form as

$$
\begin{bmatrix}
aP_2 & -aE_2 & 0 & 0 & 0 \\
-aE_2 & aP_3 & -aE_3 & 0 & 0 \\
0 & -aE_3 & aP_4 & -aE_4 & 0 \\
0 & 0 & -aE_4 & aP_5 & -aE_5 \\
0 & 0 & 0 & -aE_5 & aP_6
\end{bmatrix}
\begin{bmatrix}
T_2^{n+1} \\
T_3^{n+1} \\
T_4^{n+1} \\
T_5^{n+1} \\
T_6^{n+1}
\end{bmatrix}
=
\begin{bmatrix}
aE_1 T_1^{n+1} + b_2 \\
b_3 \\
b_4 \\
b_5 \\
aE_6 T_7^{n+1} + b_6
\end{bmatrix}
\quad (5.35)
$$

The coefficient of LAE-based implementation is shown as a one step process in Fig. 5.14, to calculate the temperatures of the present time level at the internal grid points T_i^{n+1} (independently for the explicit method, and iteratively for the implicit method) using the values of previous time level T_i^n and boundary conditions.

For the governing LAE in the *explicit method*, Eq. 5.33 shows that the temperature at a grid point for the present time level T_i^{n+1} is a function of the previous time level value of temperature of the same point T_i^n, and its east T_{i+1}^n and west T_{i-1}^n neighboring cell-center value. Thus, there is only one unknown in each equation and can be solved independently as the system of LAEs are uncoupled. The computational stencil for the LAEs is shown in Fig. 5.14a. For example, applying the equations to grid point 4, the temperature T_4^{n+1} is a function of T_3^n, T_4^n and T_5^n, shown by the shaded arrow

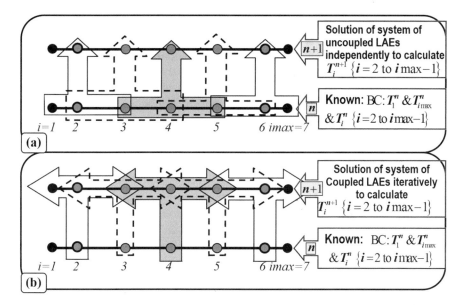

Fig. 5.14 Computational stencils to compute the temperature at the present time level T_i^{n+1}, during the coefficient of LAE-based implementation, for the (**a**) explicit and (**b**) implicit solution method in a 1D unsteady-state conduction

in the figure. The computational stencil for all the other grid points are also shown (unshaded) in the figure.

For the governing LAE in the implicit method, Eq. 5.34 shows that the temperature at an internal grid point for the present time level T_i^{n+1} is a function of the same time level value of its east T_{i+1}^{n+1} and west T_{i-1}^{n+1} neighboring grid, and the previous time level value of temperature at the same point T_i^n. Thus, there are a maximum of three unknowns in each LAE—resulting in a coupled system of LAEs—and can be solved by an iterative method; such as, Gauss-Seidel method presented in Sect. 4.2.1. The computational stencil for the LAEs is shown in Fig. 5.14b. For example, applying the equations to grid point 4, the temperature T_4^{n+1} is a function of T_3^{n+1}, T_4^n and T_5^{n+1}, shown by the shaded arrow in the figure. The computational stencil for all the other grid points are also shown (unshaded) in the figure.

5.4.1.3   Solution Algorithm

The solution algorithm for the coefficient of LAE-based solution methodology is similar to that presented in Sect. 5.3.1.3 for the flux-based approach, except that the *step-2* involves computation of the coefficient of the LAEs, and the *step-5* here involves the solution of the final LAE; Eq. 5.33 for the explicit method, and Eq. 5.34 for the implicit method. For the implicit method, note that the system of LAEs is solved directly by a Tri-Diagonal matrix algorithm (Patankar (1980)) or by an iterative method (such as Gauss-Seidel method). Furthermore, here for non-uniform grid, read the coordinates of the faces of the CVs $x(i)$ for $i = 1$ to $imax - 1$ from a file obtained from a grid generation procedure. Thereafter, calculate coordinate of the cell centers $xc(i)$, width of the CVs $\Delta x(i)$ and distance between cell centers $\delta x(i)$. Also calculate time step $\Delta t_{stab}(i)$ ($i = 2\,imax - 1$) using Eq. 5.30 and take Δt_{stab} as minimum value of $\Delta t_{stab}(i)$, for explicit method. For implicit method, enter any value of Δt.

For the coefficient of LAE-based approach of CFD development, a computer program is given below in Prog. 5.4 considering an example problem for 1D unsteady heat conduction on a non-uniform grid.

Example 5.3: Coefficient of LAE-based methodology of CFD development and code verification for a 1D unsteady-state heat conduction problem, on a non-uniform grid

Consider 1D conduction in a long steel ($\rho = 7750\,\text{kg/m}^3$, $c_p = 500\,\text{J/kg K}$ and $k = 16.2\,\text{W/m K}$) sheet of thickness $L = 1$ cm. The sheet is initially at a uniform temperature of 30∘C and is suddenly subjected to a constant temperature of $T_{wb} = 0\,^\circ\text{C}$ on the west and $h = 1000\,\text{W/m}^2\text{K}$ and $T_\infty = 100\,^\circ\text{C}$ on east boundary; case (b) in Fig. 5.13.

(a) Generate a non-uniform 1D Cartesian grid, using an algebraic method. The method involves a transformation of a uniform grid in a ξ-coordinate-based 1D computational domain of unit length, to a non-uniform grid in a x-coordinate-based physical domain of length L; using an algebraic equation, given (Hoffmann and Chiang 2000) as

$$x = L\frac{(\beta + 1)\,[(\beta + 1)\,/\,(\beta - 1)]^{(2\xi - 1)} - (\beta - 1)}{2\left\{1 + [(\beta + 1)\,/\,(\beta - 1)]^{(2\xi - 1)}\right\}} \qquad (5.36)$$

This equation results in a grid which is finest near the two ends of the domain, and gradually become coarser at the middle of the domain. It is called as equal *clustering* of grids at both the ends of the domain. Consider maximum number of grid points as $imax = 12$ and $\beta = 1.2$ (which controls the non-uniformity in the grid size).

(b) Using the coefficient of LAE-based solution methodology, presented above, develop a Gauss-Seidel method-based computer program for *implicit method* on a *non-uniform* 1D Cartesian grid.

(c) Present a CFD application of the code for a volumetric heat generation of 0 and 100 MW/m³. Consider the convergence tolerance as $\epsilon_{st} = 10^{-4}$ for steady state, and $\epsilon = 10^{-4}$ for the iterative solution. Plot the steady-state temperature profiles with and without volumetric heat generation, and compare with the exact solution. Also show the unsteady-state temperature profiles, at different intermediate time instants.

Solution:

(a) For the *non-uniform grid generation*, the *solution algorithm* is as follows:

1. User-Input: Enter length of the physical domain L, maximum number of grid points in the domain $imax$, and the parameter β to control the non-uniformity.
2. Consider uniform grid in a ξ—coordinate-based 1D computational domain of unit length, and compute the coordinates of the face centers ξ_i—uniformly spaced between 0 and 1.
3. Substitute the face-center coordinates ξ_i in the transformation equation (Eq. 5.36) to compute the x-coordinates x_i of the non-uniform spaced face centers (of the CVs in Fig. 5.13) in the x-coordinate-based 1D physical domain of length L
4. Compute coordinates of centroids of the CV in the physical plane xc_i, using Eq. 5.32. Thereafter, the equation is also used to compute the non-uniform width of the CVs Δx_i, and the distance between the cell centers δx_i (Fig. 5.13).

Using the above solution algorithm, a computer program for the non-uniform grid generation (with $imax = 7$) is presented in Prog. 5.4.

Listing 5.3 Scilab code for non-uniform 1D-Cartesian grid generation, using an algebraic method.

```
1   L=0.01;  imax=12;  Beta=1.2;  //STEP-1
2   Dxi=1/(imax-2);  xi=0:Dxi:1;  // STEP-2
3   Beta_p1=Beta+1;  Beta_m1=Beta-1;
4   Beta_p1_div_m1=(Beta_p1/Beta_m1)^(2*xi-1);
5   num=(Beta_p1*Beta_p1_div_m1)-Beta_m1;den=2*(1+
        Beta_p1_div_m1);
6   x=L*num./den;  // STEP-3
7   xc(2:imax-1)=(x(2:imax-1)+x(1:imax-2))/2;  //STEP
        -4
8   xc(1)=x(1);xc(imax)=x(imax-1);  //STEP-4
9   Dx(2:imax-1)=x(2:imax-1)-x(1:imax-2);// STEP-4
10  dx(1:imax-1)=xc(2:imax)-xc(1:imax-1);  // STEP-4
```

(b) Using the coefficient of LAE-based implementation and the solution algorithm for 1D unsteady heat conduction, an implicit method-based computer program for the present problem is presented in Prog. 5.4. The program uses a time step $\Delta t = 1$ sec for the implicit method; much larger that the time step $\Delta t_{stab} = 0.02438$ sec (obtained from Eq. 5.30 for the present non-uniform grid) needed for the explicit method. For the same grid size of $imax = 12$, note that the time step restriction $\Delta t_{stab} = 0.02438$ sec for the non-uniform grid here is much smaller than $\Delta t_{stab} = 0.1196$ sec for the uniform grid in the previous example problem.

Listing 5.4 Scilab code for 1D unsteady-state conduction, using the implicit method and coefficient of LAE-based approach of CFD development on a non-uniform 1D-Cartesian grid.

```
1   //STEP-1: User-Input
2   rho = 7750.0; cp = 500.0; k = 16.2;
3   T0=30; T_wb=0.0;T_inf=100.0;h=1000;
4   Q_vol_gen=0; epsilon_st=0.0001; Dt=1;
5   Q_gen=Q_vol_gen.*Dx; // Total Heat Generation
6   //STEP-2: Compute geometrical parameters for the
        non-uniform grid (already presented in Prog.
        5.3) and the coefficient of implicit LAEs
7   for i=1:imax-1
8       aE(i)=k/dx(i);
9   end
10  for i=2:imax-1
11      aP0(i)=rho*cp*Dx(i)/Dt;
12      aP(i)=aP0(i)+aE(i)+aE(i-1);
13  end
14  //STEP-3: IC and Dirichlet BCs
15  T(2:imax-1)=T0; T(1)=T_wb;
16  unsteadiness_nd=1; n=0;
17  alpha=k/(rho*cp); DTc=T_inf-T_wb;
18  //==== Time-Marching for Implicit  LAEs: START
19  while unsteadiness_nd>=epsilon_st
20      n=n+1;
21  //Step-4: Non-Dirichlet BCs and consider the
        temp. as the previous value
22      T(imax)=(k*T(imax-1))+(h*dx(imax-1)*T_inf);
23      T(imax)=T(imax)/(k+h*dx(imax-1));T_old=T;
24      for i=2:imax-1
25          b(i)=aP0(i)*T_old(i)+Q_gen(i);
26      end
27  //Step-5: Iterative solution (by GS method) at
        each time step
28      epsilon=0.0001; N=0; Error=1;
29      while Error>=epsilon
30       T(imax)=(k*T(imax-1))+(h*dx(imax-1)*T_inf);
31       T(imax)=T(imax)/(k+h*dx(imax-1));T_old_iter
             =T;
32          N=N+1;
33          for i=2:imax-1
34              T(i)=aE(i)*T(i+1)+aE(i-1)*T(i-1)+b(i
                    );
35              T(i)=T(i)/aP(i);
36          end
37          Error=max(abs(T-T_old_iter));
38      end
39  //STEP 6: Steady-state convergance criterion (Eq
        . 5.15)
40  unsteadiness=max(abs(T-T_old))/Dt;
41  unsteadiness_nd=unsteadiness*L*L/(alpha*DTc);
```

```
42   printf("Time step no. %5d, Unsteadiness_nd = %8
         .4e\n", n , unsteadiness_nd);
43   // STEP 7: While-end loop
44   end
```

(c) The program is applied for the present problem for $\overline{Q}_{gen} = 0$ and $100\ MW/m^3$ and the results are shown in Fig. 5.15. Since the symbols on the figure corresponds to the instantaneous temperature at the non-uniformly spaced $imax = 12$ grid points, the x-coordinate-based relative location of the symbols gives an idea on the nature of non-uniformity in the grid generation. Similar to the previous problem, as time progresses, it can be seen that the initial temperature distribution monotonically approaches to a steady-state temperature profile which is linear for without and non-linear for with heat generation case. Also notice from the triangle symbol at the right vertical axes of the figure that the temperature of the right boundary evolves with time in the present problem; in-contrast to the previous problem (Fig. 5.9). This is because the boundary temperature T_{imax} is a function of time wise evolving border temperature T_{imax-1} in the case of the non-Dirichlet BC at the east boundary here; the equation for the BC can be seen in Fig. 5.13 for the case (b).
For the steady state, the figure shows an excellent agreement between the numerical and an analytical solution; given as

$$T = T_{wb} + \frac{hx}{(k+hL)}(T_\infty - T_{wb}) + \frac{\overline{Q}_{gen}x}{2k}\left[\frac{(2k+hL)}{k+hL}L - x\right]$$

$$x_{T=Tmax} = \frac{hk(T_\infty - T_{wb})}{(k+hL)\overline{Q}_{gen}} + \frac{(2k+hL)}{k+hL}\frac{L}{2}$$

5.4.2   Two-Dimensional Conduction

For the two cases of 2D heat conduction problems, introduced in Fig. 5.10 above for a uniform grid, Fig. 5.16 is considered here for the presentation of solution methodology on a non-uniform grid.

5.4.2.1   Explicit and Implicit Methods

For 2D unsteady-state heat conduction, the FVM-based discretized equation (Eq. 5.10) with heat generation is expressed in a general form as

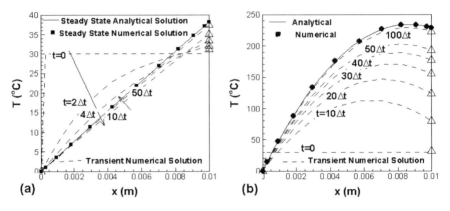

Fig. 5.15 CFD application and testing of the *implicit method*-based 1D code on *non-uniform* Cartesian grid, for unsteady-state heat conduction: transient temperature profile (**a**) without and (**b**) with heat generation. The figure also shows comparison of the numerical with the analytical solution for steady-state temperature profile

$$\rho c_p \frac{T_P^{n+1} - T_P^n}{\Delta t} \Delta x_P \Delta y_P = \lambda Q_{cond,P}^{n+1} + (1 - \lambda) Q_{cond,P}^n + \overline{Q}_{gen} \Delta x_P \Delta y_P$$

$$\text{where } Q_{cond,P} = k \left\{ \left(\frac{T_E - T_P}{\delta x_e} - \frac{T_P - T_W}{\delta x_w} \right) \Delta y_P \right.$$

$$\left. + \left(\frac{T_N - T_P}{\delta y_n} - \frac{T_P - T_S}{\delta y_s} \right) \Delta x_P \right\}$$

where $\lambda = 0, 1$ and 0.5 corresponds to explicit, implicit and Crank-Nicolson method, respectively. Then, the above equation is expressed in the final LAE form as

$$aP\, T_P^{n+1} = \quad aE \left[\lambda\, T_E^{n+1} + (1 - \lambda)\, T_E^n \right] + aW \left[\lambda\, T_W^{n+1} + (1 - \lambda)\, T_W^n \right]$$

$$+ aN \left[\lambda\, T_N^{n+1} + (1 - \lambda)\, T_N^n \right] + aS \left[\lambda\, T_S^{n+1} + (1 - \lambda)\, T_S^n \right] + b$$

where $aE = k\Delta y_P / \delta x_e,\ aW = k\Delta y_P / \delta x_w,\ aN = k\Delta x_P / \delta y_n,\ aS = k\Delta x_P / \delta y_s,$

$$aP^0 = \rho c_p \Delta x_P \Delta y_P / \Delta t, \quad aP = \lambda \sum_{NB} aNB + aP^0,$$

$$b = \left[aP^0 - (1 - \lambda) \sum_{NB} aNB \right] T_P^n + \overline{Q}_{gen} \Delta x_P \Delta y_P,$$

$$\sum_{NB} aNB = aE + aW + aN + aS$$

$$(5.37)$$

Explicit method:
Substituting $\lambda = 0$ in Eq. 5.37, the LAE for the representative CV is given as

$$aP\, T_P^{n+1} = aE\, T_E^n + aW\, T_W^n + aN\, T_N^n + aS\, T_S^n + b$$

where $aE = k\Delta y_P / \delta x_e,\ aW = k\Delta y_P / \delta x_w,\ aN = k\Delta x_P / \delta y_n,\ aS = k\Delta x_P / \delta y_s,$

$$b = \left(aP^0 - \sum_{NB} aNB \right) T_P^n + \overline{Q}_{gen} \Delta x_P \Delta y_P, \quad aP = aP^0 = \rho c_p \Delta x_P \Delta y_P / \Delta t$$

$$(5.38)$$

As discussed above for 1D conduction, the coefficients of all the temperatures of the previous time step in the above LAE should be positive to correctly capture of the conduction phenomenon. Thus, the condition for the coefficient $\left(aP^0 - \sum_{NB} aNB \right)$

of T_P^n (seen in the expression for b in Eq. 5.38) to be positive is given (stability criterion) as

$$aE + aW + aN + aS \leq aP^0$$

$$\alpha \Delta t \left\{ \frac{1}{\Delta x_P} \left(\frac{\delta x_e + \delta x_w}{\delta x_e \, \delta x_w} \right) + \frac{1}{\Delta y_P} \left(\frac{\delta y_n + \delta y_s}{\delta y_n \, \delta y_s} \right) \right\} \leq 1 \quad \text{Non-uniform grid}$$

$$\alpha \Delta t \left(\frac{1}{\Delta x^2} + \frac{1}{\Delta y^2} \right) \begin{matrix} (\delta x_e = \delta x_w = \Delta x_P = \Delta x) \\ (\delta y_n = \delta y_s = \Delta y_P = \Delta y) \end{matrix} \leq \frac{1}{2} \quad \text{Uniform grid}$$

$$\text{(5.39)}$$

A limiting/stable time step is given as

$$\Delta t_{stab} = \begin{cases} \dfrac{1}{\alpha} \left\{ \dfrac{1}{\Delta x_P} \left(\dfrac{\delta x_e + \delta x_w}{\delta x_e \, \delta x_w} \right) + \dfrac{1}{\Delta y_P} \left(\dfrac{\delta y_n + \delta y_s}{\delta y_n \, \delta y_s} \right) \right\}^{-1} & \text{Non-uniform grid} \\[4mm] \dfrac{1}{2\alpha} \left(\dfrac{1}{\Delta x^2} + \dfrac{1}{\Delta y^2} \right)^{-1} & \text{Uniform grid} \\[4mm] \Delta x^2 / 4\alpha & \text{if } \Delta x = \Delta y \end{cases}$$

$$\text{(5.40)}$$

For the uniform square grid, as compared to the $\Delta x^2/4\alpha$ above for the 2D, the time step is $\Delta x^2/2\alpha$ for the 1D conduction (Eq. 5.30). Thus, for the same uniform grid size, the stable time step becomes half for the 2D as compared to that for the 1D problem. The application of above time step is presented above in Prog. 5.2 for the 2D uniform grid, with a Δt which is 100% of Δt_{stab}. For non-uniform grid, the above equation results in a variation of the time step corresponding to the various CVs in the domain, and the minimum of the time step is used in the unsteady computation.

Implicit method:

Substituting $\lambda = 1$ in Eq. 5.37, the LAE for the representative CV is given as

$$aP \, T_P^{n+1} = aE \, T_E^{n+1} + aW \, T_W^{n+1} + aN \, T_N^{n+1} + aS \, T_S^{n+1} + b$$
$$\text{where } aE = k\Delta y_P/\delta x_e, \, aW = k\Delta y_P/\delta x_w, \, aN = k\Delta x_P/\delta y_n, \, aS = k\Delta x_P/\delta y_s,$$
$$b = aP^0 T_P^n + \overline{Q}_{gen} \Delta x_P \Delta y_P, \, aP = \sum_{NB} aNB + aP^0$$
$$aP^0 = \rho c_p \Delta x_P \Delta y_P / \Delta t$$

$$\text{(5.41)}$$

5.4.2.2 Implementation Details on a Non-Uniform Grid

A Cartesian structured grid is generated by the vertical and the horizontal lines which are equi-spaced for the uniform grid (Fig. 5.10), and non-uniformly spaced for the non-uniform grid (Fig. 5.16). Instead of a general definition of the matrix for the *geometrical parameters* as a 2D matrix for the 2D grid, Fig. 5.16 shows that they are defined here as 1D matrix to avoid unnecessary memory storage and computation. This is due to the characteristics of a Cartesian grid—the x-coordinate and y-coordinate remains constant on a vertical and horizontal line, respectively. For example, it can be seen in Fig. 5.16 that all the CVs between any two vertical lines

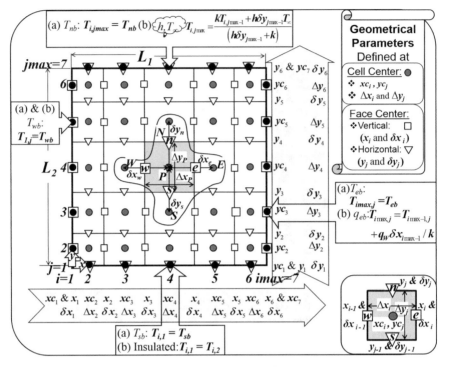

Fig. 5.16 Discrete definition of geometrical parameters for non-uniform Cartesian grid

have the same value of Δx; thus, Δx for the various 2D CVs is stored as Δx_i instead of $\Delta x_{i,j}$. Similarly, the CVs between any two horizontal lines have the same value of Δy; thus, Δy is stored as Δy_j instead of $\Delta y_{i,j}$.

As mentioned above, for implementation of a single running indices (i or j) for all the variables, note that the convention followed here is such that the running indices for the coordinates at positive (east for vertical and north for horizontal) face center corresponds to that of the centroid of the CV. For example, Fig. 5.10 shows that the running indices for the δx_e, δx_w, δy_n, and δy_s at the face centers of a CV P (i, j) is δx_i, δx_{i-1}, δy_j, and δy_{j-1} , respectively. The figure also shows the value of i/j for the various geometrical parameters defined as the 1D matrix.

The coordinates of the non uniformly spaced face centers of the CVs (Fig. 5.16) are computed by a non-uniform Cartesian grid generation method; presented below in an example problem (Example 5.4). This is used to compute the other geometrical parameters as follows:

$$
\begin{aligned}
\text{Face-Centers:} & \left.\begin{array}{l} x_i \\ y_j \end{array}\right\} && \begin{array}{l} \text{for } i = 1 \text{ to } imax - 1 \\ \text{for } j = 1 \text{ to } jmax - 1 \end{array} \\
\text{Centroids :} & \left.\begin{array}{l} xc_i = (x_i + x_{i-1})/2 \\ yc_j = (y_j + y_{j-1})/2 \end{array}\right\} && \begin{array}{l} \text{for } i = 2 \text{ to } imax - 1 \\ \text{for } j = 2 \text{ to } jmax - 1 \end{array} \\
& xc_1 = x_1, \; xc_{imax} = x_{imax-1} \; yc_1 = y_1, \; yc_{jmax} = y_{jmax-1} && \\
\text{Width of CVs:} & \left.\begin{array}{l} \Delta x_i = x_i - x_{i-1} \\ \Delta y_j = y_j - y_{j-1} \end{array}\right\} && \begin{array}{l} \text{for } i = 2 \text{ to } imax - 1 \\ \text{for } j = 2 \text{ to } jmax - 1 \end{array} \\
\begin{array}{l}\text{Distance between} \\ \text{cell-centers:}\end{array} & \left\{\begin{array}{l} \delta x_i = xc_{i+1} - xc_i \\ \delta y_j = yc_{j+1} - yc_j \end{array}\right\} && \begin{array}{l} \text{for } i = 1 \text{ to } imax - 1 \\ \text{for } j = 1 \text{ to } jmax - 1 \end{array}
\end{aligned}
\tag{5.42}
$$

A computer program for the computation of above geometrical parameters is presented below in Prog. 5.5.

The LAE for a representative CV in the Eq. 5.38 and 5.41 is converted into a system of LAEs by substituting all the subscripts to running indices: E, P, W, e, and w are replaced by the running indices $i + 1, i, i - 1, i$, and $i - 1$, respectively. Furthermore, the coefficients in the above system of LAEs is represented as a matrix, with $aW_{i,j} = aE_{i-1,j}$ and $aS_{i,j} = aN_{i,j-1}$. This results in the final equation—implemented in a computer program (Prog. 5.6 for the implicit method) below—as

For Explicit method:

$$
\left.\begin{aligned}
a P_{i,j} T_{i,j}^{n+1} &= a E_{i,j} T_{i+1,j}^n + a E_{i-1,j} T_{i-1,j}^n \\
&\quad + a N_{i,j} T_{i,j+1}^n + a N_{i,j-1} T_{i,j-1}^n + b_{i,j} \\
\text{where } a P_{i,j}^o &= \rho c_p \Delta x_i \Delta y_j / \Delta t, \; a P_{i,j} = a P_{i,j}^o \\
b_{i,j} &= \left[a P_{i,j}^0 - \sum_{NB} a N B_{i,j} \right] T_{i,j}^n + \overline{Q}_{gen} \Delta x_i \Delta y_j \\
\sum_{NB} a N B_{i,j} &= a E_{i,j} + a E_{i-1,j} + a N_{i,j} + a N_{i,j-1}
\end{aligned}\right\}
\begin{array}{l} \text{for } i = 2 \text{ to } imax - 1 \\ \text{for } j = 2 \text{ to } jmax - 1 \end{array}
\tag{5.43}
$$

For Implicit method:

$$
\left.\begin{aligned}
a P_{i,j} T_{i,j}^{n+1} &= a E_{i,j} T_{i+1,j}^{n+1} + a E_{i-1,j} T_{i-1,j}^{n+1} \\
&\quad + a N_{i,j} T_{i,j+1}^{n+1} + a N_{i,j-1} T_{i,j-1}^{n+1} + b_{i,j} \\
\text{where } a P_{i,j}^o &= \rho c_p \Delta x_i \Delta y_j / \Delta t, \\
a P_{i,j} &= \sum_{NB} a N B_{i,j} + a P_{i,j}^o \\
b_{i,j} &= a P_{i,j}^0 T_{i,j}^n + \overline{Q}_{gen} \Delta x_i \Delta y_j
\end{aligned}\right\}
\begin{array}{l} \text{for } i = 2 \text{ to } imax - 1 \\ \text{for } j = 2 \text{ to } jmax - 1 \end{array}
\tag{5.44}
$$

For both explicit and implicit methods, note that all the coefficients for the LAEs are computed at cell center, except $a E_{i,j}$ at the vertical (east) and $a N_{i,j}$ at the horizontal (north) face centers; given as

$$
\begin{aligned}
\text{Face-Centers:} & \quad \text{Vertical} \quad a E_{i,j} = \left.\frac{k \Delta y_j}{\delta x_i}\right\} && \begin{array}{l} \text{for } i = 1 \text{ to } imax - 1 \\ \text{for } j = 2 \text{ to } jmax - 1 \end{array} \\
& \quad \text{Horizontal} \quad a N_{i,j} = \left.\frac{k \Delta x_i}{\delta y_j}\right\} && \begin{array}{l} \text{for } i = 2 \text{ to } imax - 1 \\ \text{for } j = 1 \text{ to } jmax - 1 \end{array}
\end{aligned}
\tag{5.45}
$$

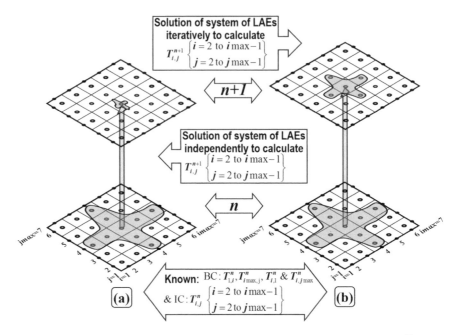

Fig. 5.17 Computational stencils to compute the temperature at the present time level $T_{i,j}^{n+1}$, during the coefficient of LAE-based implementation, for the (**a**) explicit and (**b**) implicit solution method in a 2D unsteady-state conduction

For the explicit method, there is only one unknown in each LAE (Eq. 5.43) and can be solved independently as the system of LAEs are uncoupled. The computational stencil for the LAEs is shown in Fig. 5.17a. For example, applying Eq. 5.43 to grid point (4, 4), the temperature $T_{4,4}^{n+1}$ is a function of $T_{3,4}^{n}$, $T_{5,4}^{n}$, $T_{4,4}^{n}$, $T_{4,3}^{n}$, and $T_{4,5}^{n}$, shown by the shaded arrow in the figure. Whereas, for the implicit method, Eq. 5.44 shows a maximum of five unknowns in each LAE and can be solved iteratively by a Gauss-Seidel method. The computational stencil for the LAEs is shown in Fig. 5.17b. For example, applying Eq. 5.44 to grid point (4, 4), the temperature $T_{4,4}^{n+1}$ is a function of $T_{3,4}^{n+1}$, $T_{5,4}^{n+1}$, $T_{4,4}^{n}$, $T_{4,3}^{n+1}$ and $T_{4,5}^{n+1}$, shown by the shaded arrow in the figure. The system of algebraic equations—corresponding to the internal grid point shown in Eq. 5.44 and Fig. 5.17b—results in a penta-diagonal matrix for the 2D conduction; as compared to the tri-diagonal matrix (Eq. 5.35).

For both the explicit and implicit methods, note that the computational stencil for the neighboring grid points is obtained by shifting the stencil—shown in the Fig. 5.17—left by Δx for $T_{3,4}^{n}$ and right by Δx for $T_{5,4}^{n}$; and below by Δy for $T_{4,3}^{n}$ and above by Δy for $T_{4,5}^{n}$. Furthermore, both the methods results in the number of LAEs equal to the number of unknown temperatures at the internal grid points, i.e., $(imax - 2) \times (jmax - 2)$. For example, considering the grid points given in Fig. 5.16, the methods results in twenty-five LAEs. In each LAE, there is only one unknown in explicit and at least five unknowns in implicit method. However, there are a total of twenty-five unknowns in the twenty-five LAEs for the implicit method. Thus, similar to the iterative solution of the FDM-based LAEs for the 2D steady-state conduction (presented in Sect. 4.2.2 and Example 4.7), here for the unsteady state, the system of LAEs needs to solved iteratively at each of the discrete present time instant $t + \Delta t$. However, the solution by an iterative method was only once earlier for the steady-state methodology, and after every time time interval Δt here for the implicit method-based unsteady-state methodology. Whereas, iterative method is not needed for explicit method-based unsteady-state methodology.

5.4.2.3 Solution Algorithm

The solution algorithm for the coefficient of LAE-based solution methodology for the 2D here is similar to that presented in Sect. 5.4.1.3 for the 1D unsteady-state conduction, except a slight change due to increase in the dimensionality. Thus, the code developed and tested for 1D can be easily extended for 2D conduction. A computer program for the coefficient of LAEs based approach of CFD development is given below in Prog. 5.6, considering an example problem for 2D unsteady heat conduction on a non-uniform grid.

Example 5.4: Coefficient of LAE-based methodology of CFD development and application for a 2D unsteady-state heat conduction problem, on a non-uniform grid

Consider 2D conduction in a square shaped ($L_1 = 1$ m and $L_2 = 1$ m) *long* stainless-steel plate ($\rho = 7750$ kg/m^3, $c_p = 500$ J/kg K and $k = 16.2$ W/m K). The plate is initially at a uniform ambient temperature of 30 °C and is suddenly subjected to a constant temperature of $T_{wb} = 100$ °C on the west boundary, insulated on the south boundary, constant incident heat flux of 10 kW/m^2 on the east boundary, and $h = 100$ W/m^{2}K and $T_\infty = 30$ °C on north boundary; case (b) in Fig. 5.16.

(a) Generate a non-uniform 2D Cartesian grid, using an algebraic method; given above in Eq. 5.36. However, the equation is also used in the y-direction to generate the 2D grid. Consider the non-uniformity control parameter as $\beta = 1.2$ and maximum number of grid points as 12×12.
(b) Using the coefficient of LAE-based solution methodology, presented above, develop a Gauss-Seidel method-based computer program for the *implicit method* on a *non-uniform* 2D Cartesian grid.
(c) Using the program, present a CFD application of the code for a volumetric heat generation of 0 and $5 kW/m^3$. Consider the convergence tolerance as $\epsilon_{st} = 10^{-4}$ for the steady state, and $\epsilon = 10^{-4}$ for the iterative solution. Plot the steady-state temperature contours with and without volumetric heat generation.

Solution:

(a) Using the solution algorithm, presented above in Example 5.3, a computer
    program for the non-uniform grid generation (with $imax = jmax = 12$) is
    presented below.

Listing 5.5 Scilab code for non-uniform 2D-Cartesian grid generation, using an algebraic
method.

```
1   L1=1; L2=1; imax=12; jmax=12; Beta=1.2; //STEP-1
2   xi=linspace(0,1,imax-1); // STEP-2
3   Beta_p1=Beta+1; Beta_m1=Beta-1;
4   Beta_p1_div_m1=(Beta_p1/Beta_m1)^(2*xi-1);
5   num=(Beta_p1*Beta_p1_div_m1)-Beta_m1; den=2*(1+
        Beta_p1_div_m1);
6   x=L1*num./den; y=x; // STEP-3
7   xc(2:imax-1)=(x(2:imax-1)+x(1:imax-2))/2; //STEP
        -4
8   xc(1)=x(1); xc(imax)=x(imax-1); yc=xc //STEP-4
9   [Xc Yc]=meshgrid(xc,yc); Zc=zeros(jmax,imax);
10  [X Y]=meshgrid(x,y); Z=zeros(jmax-1,imax-1);
11  // plotting
12  xset('window',1); surf(X,Y,Z); plot(Xc,Yc,'bo')
13  //STEP-4
14  Dx(2:imax-1)=x(2:imax-1)-x(1:imax-2); Dy=Dx;
15  dx(1:imax-1)=xc(2:imax)-xc(1:imax-1); dy=dx;
```

(b) Using the coefficient of LAE-based implementation for 2D unsteady heat
    conduction, an implicit method-based computer program for the present
    problem is presented below. The program uses a time step $\Delta t = 5000\,$s for
    the implicit method; much larger that the time step needed for the explicit
    method.

Listing 5.6 Scilab code for 2D unsteady-state conduction, using the implicit method and
coefficient of LAEs based approach of CFD development on a non-uniform 2D-Cartesian
grid.

```
1   //STEP-1: User-Input
2   rho = 7750.0; cp = 500.0; k = 16.2;
3   T0=30; T_wb=100.0; q_W=10000; T_inf=30.0; h=100;
4   Q_vol_gen=0; epsilon_st=0.0001; Dt=1000;
5   //STEP-2: Compute geometrical parameters for the
        non-uniform grid (already presented in Prog.
        5.5) and the coefficient of implicit LAEs
6   for j=2:jmax-1
7       for i=1:imax-1
8           aE(i,j)=k*Dy(j)/dx(i);
9       end
10  end
11  for j=1:jmax-1
12      for i=2:imax-1
```

```
13                aN(i,j)=k*Dx(i)/dy(j);
14       end
15   end
16   for  j=2:jmax-1
17       for  i=2:imax-1
18           aP0(i,j)=rho*cp*Dx(i)*Dy(j)/Dt;
19               aP(i,j)=aP0(i,j)+aE(i,j)+aE(i-1,j)+
                     aN(i,j)+aN(i,j-1);
20   end
21   end
22   //STEP-3:  IC  and  Dirichlet  BCs
23   T(2:imax-1,2:jmax-1)=T0;   T(1,1:jmax)=T_wb;
24   unsteadiness_nd=1; n=0;
25   alpha=k/(rho*cp); DTc=T_wb-T_inf;
26   //====  Time-Marching  for  Implicit  LAEs:  START
        ====
27   while  unsteadiness_nd >= epsilon_st
28        n=n+1;
29   //Step-4:  Non-Dirichlet  BCs  and  Consider  the
        temp. as the previous value
30       for  j=1:jmax
31       T(imax,j)=T(imax-1,j)+(q_W*dx(imax-1)/k);
32       end
33       for  i=1:imax
34           T(i,1)=T(i,2);
35   T(i,jmax)=((k*T(i,jmax-1))+(h*dy(jmax-1)*T_inf))
        /(k+h*dy(jmax-1));
36       end
37       T_old=T;
38       for  j=2:jmax-1
39           for  i=2:imax-1
40   b(i,j)=aP0(i,j)*T_old(i,j)+Q_vol_gen*Dx(i)*Dy(j)
        ;
41       end
42       end
43   //Step-5:  Iterative  solution  (by  GS  method)  at
        each time step
44        epsilon=0.0001; N=0; Error=1;
45   while  Error >= epsilon
46        N=N+1;
47       //Non-Dirichlet  BCs
48       for  j=1:jmax
49       T(imax,j)=T(imax-1,j)+(q_W*dx(imax-1)/k);
50       end
51       for  i=1:imax
52           T(i,1)=T(i,2);
53   T(i,jmax)=((k*T(i,jmax-1))+(h*dy(jmax-1)*T_inf))
        /(k+h*dy(jmax-1));
54       end
55       T_old_iter=T;
56       for  j=2:jmax-1
```

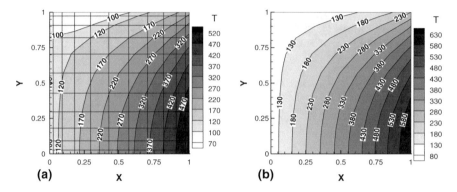

Fig. 5.18 CFD application of the *implicit method*-based 2D code on a *non-uniform* Cartesian grid: steady-state temperature contour (line and flooded) for unsteady-state heat conduction (**a**) without and (**b**) with heat generation

```
57        for  i=2:imax-1
58  T(i,j)=aE(i,j)*T(i+1,j)+aE(i-1,j)*T(i-1,j)+aN(i,
       j)*T(i,j+1)+aN(i,j-1)*T(i,j-1)+b(i,j);
59      T(i,j)=T(i,j)/aP(i,j);
60      end
61      end
62      Error=max(abs(T-T_old_iter));
63  end
64  //STEP 6:  Steady-state convergance criterion (Eq
       , 5.15)
65      unsteadiness=max(abs(T-T_old))/Dt;
66      unsteadiness_nd=unsteadiness*L1*L1/(alpha*DTc
         );
67  printf("Time step no. %5d, Unsteadiness_nd = %8
       .4e\n", n , unsteadiness_nd);
68  // STEP 7: While-end loop
69  end
```

(c) The program is applied for the present problem for $\overline{Q}_{gen} = 0$ and $5\,kW/m^3$ and the results are shown in Fig. 5.18. The figure shows that the steady-state numerical result obtained on a non-uniform grid size of $imax = 12$ and $jmax = 12$.

Problems

5.1 Consider the example problem on 2D conduction, *Example 4.7*, presented earlier with the FDM method-based *steady*-state solution methodology. Here, solve this problem using the FVM, *explicit method*, and flux-based *unsteady*-state solution methodology. For the computational domain and BCs presented in Example 4.7, modify the computer program presented in *Example 5.2* (Prog. 5.2) and obtain the steady-state temperature distribution on a grid size of 12×12. Present a *code verification*, by comparing the agreement between the numerical and the analytical (Eq. 4.42 and Prog. 4.3) results. In this regard, plot and discuss the analytically and numerically obtained variation of temperature along the vertical centerline (vertical line passing through the centroid) of the square domain. Consider the convergence tolerance as $\epsilon_{st} = 10^{-4}$ for the steady state.

5.2 Repeat the previous problem, for the *implicit method* and the coefficient of LAE-based solution methodology, using the code presented in *Example 5.4* for the non-uniform grid (Prog. 5.5 and 5.6). Consider the convergence tolerance $\epsilon = 10^{-4}$ for the iterative solution of the LAEs.

5.3 Consider the example problem on 1D conduction, *Example 5.1*, and modify the code (Prog. 5.1) to solve the problem on a non-uniform grid (Prog. 5.3).

5.4 Consider the example problem on 2D conduction, *Example 5.2*, and modify the code (Prog. 5.2) to solve the problem on a non-uniform grid (Prog. 5.5). However, instead of all the BCs as Dirichlet, solve the problem for the various types of thermal BCs; presented as case (b) in Fig. 5.16. Use the BCs related details given in *Example 5.4*. Present a CFD application of the code, with the results similar to the *Example 5.2*.

5.5 Consider the example problem on 1D conduction, *Example 5.3*, and develop a code as well as present the results for the *explicit method* and the coefficient of LAE-based solution methodology.

5.6 Consider the example problem on 2D conduction, *Example 5.4*, and develop a code for the *explicit method* and the coefficient of LAE-based solution methodology. Apply the code for the example problem and present the temperature contours. Thereafter, incorporate the implementation of the second-order numerical differentiation (Sect. 4.3) in the code, to compute the variation of the local heat flux along the east and west boundaries (q_x vs. y-coordinate), and along the north and south boundaries (x-coordinate vs. q_y), of the plate. Plot as well as discuss the *steady-state* heat flux profile at the various boundaries. Also incorporate the trapezoidal rule (Sect. 4.4) in the code, for the computation of the total rate of heat transfer from the different boundaries. Considering the heat transfer rates under the *steady-state* condition, present an application of the energy conservation law for the complete domain as the CV.

References

Anderson, J. D. (1995). *Computational fluid dynamics: The basics with applications*. New York: McGraw Hill.

Ferziger, J. H., & Peric, M. (2002). *Computational Methods for Fluid Dynamics* (3rd ed.). Berlin: Springer.

Hoffmann, K. A., & Chiang, S. T. (2000). *Computational fluid dynamics: Volume 1, 2 and 3*, Engineering Education System, Kansas.

Incropera, F. P., & Dewitt, D. P. (1996). *Fundamentals of heat and mass transfer* (4th ed.). New York: Wiley.

Patankar, S. V. (1980). *Numerical heat transfer and fluid flow*. New York: Hemisphere Publishing Corporation.

Chapter 6
Computational Heat Advection

Advection is a transport mechanism due to a bulk motion of the molecule. For energy transport due to the other mechanism as diffusion, the unsteady computation of non-uniform spatial distribution of conduction flux as well as temperature is presented in the last chapter. Instead of the conduction flux $\overrightarrow{q}$ $(= -k\nabla T)$, the distribution will correspond to enthalpy flux $\overrightarrow{h}$ $(= \overrightarrow{m} c_p T)$ for advection in this chapter, and to both the fluxes for convection in the next chapter. For a particular domain and boundary conditions, the instantaneous transported variables (flux as a vector and temperature as a scalar field) are driven by thermal diffusivity α for conduction and mass flux $\overrightarrow{m}$ $(= \rho \overrightarrow{u})$ as well as specific heat c_p for advection. Other than the thermo-physical property (k, ρ, and c_p), the driver of the energy transport process is velocity field $\overrightarrow{u}$ which is also a flow property and an unknown quantity obtained by solving the coupled equations for mass and momentum conservation. The coupling of the energy with the mass and momentum transport is avoided in this chapter by considering a prescribed continuity satisfying steady-state velocity field.

Onset of fluid flow initiates the advection, whereas conduction is present with as well as without flow. Thus, both conduction and advection mode of heat transfer occurs in a non-isothermal fluid flow called as the convective heat transfer. As discussed in Chap. 3, with the help of an example of 1D non-isothermal flow between ice and fire (Fig. 3.3), pure advection is an approximation where the conduction is assumed negligible—corresponding to a situation where the velocity of the flow is extremely high. Advection is more of a hypothetical phenomenon, and convection is the more commonly encountered physical phenomenon. Thus, although this chapter on pure heat advection corresponds to a hypothetical situation, it is extremely helpful in a component-by-component development as well as testing of the present product—the CFD software. After presenting the conduction/diffusion as the first component in the last chapter, the second component advection is presented in this chapter. The combination of these two components will lead to convection, which will be presented in the next chapter.

© The Author(s), under exclusive license to Springer Nature Switzerland AG 2022   157
A. Sharma, *Introduction to Computational Fluid Dynamics*,
https://doi.org/10.1007/978-3-030-72884-7_6

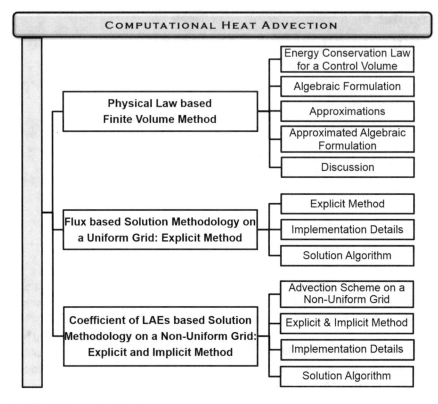

Fig. 6.1 Mind map for Chap. 6

The mind map for this chapter is similar to the previous chapter and is shown in Fig. 6.1. However, only 2D advection heat transfer is considered here as compared to both 1D and 2D conductions in the last chapter.

6.1 Physical Law-based Finite Volume Method

Similar to the previous chapter, a physical law for a finite-time interval Δt-based FVM is presented below for the algebraic formulation.

6.1.1 Energy Conservation Law for a Control Volume

Figure 6.2 shows the application of the two forms of energy conservation law to an arbitrary CV, for heat advection. Over a time interval Δt (encountered in CFD), the

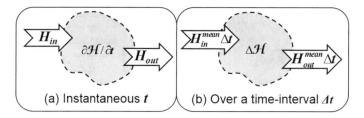

Fig. 6.2 Two different forms of conservation of energy for a CV, subjected to heat advection

law presented in the previous chapter for conduction is given here for a CV subjected to the heat advection (Fig. 6.2b) as

> Over a time interval Δt, the amount of advected thermal energy that enters a CV $H_{in}^{mean}\Delta t$ minus the amount of advected thermal energy that leaves the CV $H_{out}^{mean}\Delta t$ must be equal to the increase of enthalpy $\Delta\mathcal{H}$ stored within the CV.
>
> $$\left(H_{in}^{mean} - H_{out}^{mean}\right)\Delta t = \Delta\mathcal{H}\left[=\mathcal{H}^{t⇕∪§∪+\Delta∪} - \mathcal{H}^{t}\right]$$
>
> where $H^{mean} = \frac{1}{\Delta t}\int_{t}^{t+\Delta t} H dt$ and $\Delta\mathcal{H} = \int_{t}^{t+\Delta t}\frac{\partial\mathcal{H}}{\partial t}dt$ (6.1)

where H is the enthalpy flow rate and the superscript *mean* represents the time-averaged value over the time interval Δt. The above conservation statement, for the heat advection, corresponds to a law of conservation of enthalpy for a fluid which is not experiencing a change in phase.

6.1.2 Algebraic Formulation

Figures 6.2b and 6.3 show that the application of the time-interval-based conservation of energy for a CV involves two variables: enthalpy $\mathcal{H}$ and enthalpy flow rate H. Note that the enthalpy ($\mathcal{H} = Mc_pT_P$) involves the mass of the fluid inside a CV $M (= \rho\Delta V_P)$, and enthalpy flow rate $H (= Mc_pT)$ involves the mass flow rate M $(= \vec{m}\cdot\vec{\Delta S})$ at the various surfaces $\vec{\Delta S}$ of a CV. Also note that enthalpy $\mathcal{H}$ includes volume ΔV and enthalpy flow rate H includes surface area ΔS of a CV; the former is a volumetric term while the latter is a flux term. The corresponding volumetric and flux variables are defined as

Volumetric-Enthalpy: $\overline{\mathcal{H}} \equiv \dfrac{\mathcal{H}}{\Delta V} = \rho c_p T = f(\vec{x}, t)$ $\vec{x} \in \Delta V$

Enthalpy-Flux: $\vec{h} \equiv \dfrac{H}{\Delta S} = \vec{m}c_p T = f(\vec{x}, t)$ $\vec{x} \in \Delta S$

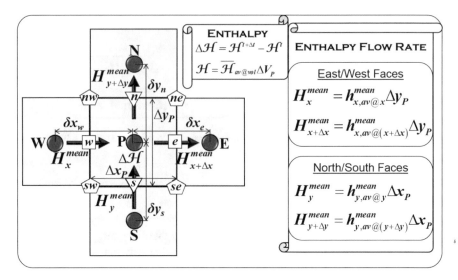

Fig. 6.3 Application of the time-interval-based law of conservation of energy (for heat advection) to a representative internal CV in a 2D Cartesian computational domain, with a non-uniform grid

where ΔV represents the volume and ΔS is the surface area (for *one* of the surface) of the CV. The variables are represented above as a *function* of spatial (within the volume/surface area) and temporal (within the time duration for unsteady heat transfer) coordinate.

Similar to the previous chapter, accounting the spatial variation of the $\mathcal{H}$ as the volumetric phenomenon and h as the surface phenomenon, an algebraic formulation is finally obtained from Eq. 6.1 (for a 2D Cartesian CV, shown in Fig. 6.3) as

$$\frac{\left(\overline{\mathcal{H}}^{t+\Delta t}_{\text{av@vol}} - \overline{\mathcal{H}}^{t}_{\text{av@vol}}\right) \Delta x_P \Delta y_P}{\Delta t} + Q^{\text{mean}}_{\text{adv}} = 0$$

$$\text{where } Q^{\text{mean}}_{\text{adv}} = \left(h^{\text{mean}}_{x,\text{av@}(x+\Delta x)} - h^{\text{mean}}_{x,\text{av@}x}\right) \Delta y_P \qquad (6.2)$$

$$+ \left(h^{\text{mean}}_{y,\text{av@}(y+\Delta y)} - h^{\text{mean}}_{y,\text{av@}y}\right) \Delta x_P$$

$$\overline{\mathcal{H}}_{\text{av@vol}} \Delta V_P = \int_{\Delta V_P} \overline{\mathcal{H}}^t \, dV, \ h^{\text{mean}}_{x/y,\text{av@face}} = \frac{1}{\Delta S} \int_{\Delta S} h^{\text{mean}}_{x/y,\text{face}} \, dS$$

where the subscript av represents the space-averaged (within the volume/surface area) value, and the superscript mean represents the time-averaged (over the time interval Δt) value. Also $h_{x/y,\text{face}}$ represents $h_{x,\text{face}}$ for the east/west and $h_{y,\text{face}}$ for the north/south surface. Furthermore, the subscript face corresponds to the various faces of the CV and ΔS is their surface area (Δx_P for the north/$y + \Delta y$ and south/y surfaces, and Δy_P for the east/$x + \Delta x$ and west/x surfaces; Fig. 6.3). Also ΔV_P is the volume of the CV; $\Delta x_P \Delta y_P$ for the 2D CV.

6.1.3 *Approximations*

The *approximation* for the volumetric enthalpy $\overline{\mathcal{H}}_{av@vol} = \overline{\mathcal{H}}_P \ (= \rho c_p T_P)$ is already presented in the previous chapter, and that for the enthalpy flux will be presented here.

6.1.3.1 First Approximation

The first approximation of the enthalpy flux here is quite similar to that for the conduction flux in the previous chapter (Eq. 5.8), with the average value of the enthalpy flux at the various faces (x, y, $x + \Delta x$, and $y + \Delta y$) represented by its value at the centroid (e, w, n, and s) given as

$$
\left.
\begin{aligned}
h^{mean}_{x,av@(x+\Delta x)} &\approx h^{\chi}_{x,e} \\
h^{mean}_{y,av@(y+\Delta y)} &\approx h^{\chi}_{y,n} \\
h^{mean}_{x,av@x} &\approx h^{\chi}_{x,w} \\
h^{mean}_{y,av@y} &\approx h^{\chi}_{y,s}
\end{aligned}
\right\}
\begin{aligned}
\chi &= n & &\text{for Explicit Method} \\
\chi &= n+1 & &\text{for Implicit Method} \\
\chi &= n+1/2 & &\text{for Crank-Nicolson Method}
\end{aligned}
\tag{6.3}
$$

where $\overrightarrow{h}^{n+1/2} \approx \left(\overrightarrow{h}^{n} + \overrightarrow{h}^{n+1} \right)/2$ for the Crank-Nicolson method. Thus, the first approximation is a general representation of a flux (conduction in the previous and advection in this chapter) acting on a surface in terms of its value at the one-point (centroid) and the one time-instant value.

6.1.3.2 Second Approximation: Advection Scheme

In contrast to the first approximation, the second approximation for the enthalpy flux is quite different from that for the conduction flux. The approximation for the flux term gets modified from the gradient of the temperature for the conduction flux to the value of the temperature for enthalpy flux. Substituting $\overrightarrow{h} = \overrightarrow{m} c_p T$ in Eq. 6.3, the enthalpy flux in the advection term is given as

$$
h^{\chi}_{x,e} = m_{x,e} c_p T^{\chi}_e, \ h^{\chi}_{x,w} = m_{x,w} c_p T^{\chi}_w, \ h^{\chi}_{y,n} = m_{y,n} c_p T^{\chi}_n \text{ and } h^{\chi}_{y,s} = m_{y,s} c_p T^{\chi}_s
\tag{6.4}
$$

where the temperature at a face center is obtained using a second approximation— called here as *advection scheme*, presented below. Note that the mass flux is considered as a known quantity in this as well as the next chapter.

Second approximation for the advection term is an *interpolation or extrapolation procedure* to obtain the temperature at a face center in terms of neighboring cell-center values. This is called as *convection scheme* in most of the published literature. However, since it is an approximation for the advection term (and the combination of advection and conduction is called as convection here), the second approximation

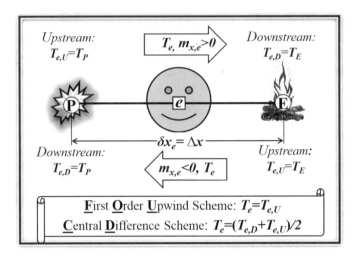

Fig. 6.4 Representation of FOU and CD advection scheme at the east face of representative fluid CV, where the cell center of the CV "P" is represented by ice and its east neighbor "E" is represented by fire. For the two different directions of mass flux at the east face center (represented by a smiling face), upstream cell center U and downstream cell center D are also shown

will be called as an *advection scheme* in this text book. Instead of only one diffusion scheme (central difference method) as the II-approximation for the conduction flux in the previous chapter, various advection schemes will be presented below. More options are given for the advection as compared to the diffusion scheme, as the advection term is numerically less well behaved in terms of accuracy and stability of the solution.

An advection scheme incorporates the physical behavior of advection, discussed in Chap. 3 and shown in Fig. 3.3. This is shown in Fig. 6.4, with a representation of ice at $0\,°C$ for the cell center P and fire at $100\,°C$ for the east neighbor E, and the temperature T_e experienced by the person at east face center e. This is shown for the First Order Upwind (FOU) and Central Difference (CD) scheme. For the the FOU scheme, the figures show a direction of mass-flux-based constant extrapolation procedure, with $T_e \approx T_P$ (ice temperature) if $m_{x,e} > 0$ and $T_e \approx T_E$ (fire temperature) if $m_{x,e} < 0$. Whereas, for the CD scheme, the figure shows that T_e is independent of the direction of mass flux, and is expressed by a linearly interpolated temperature of the first cell center on either side of the east face. As the face in the figure is exactly in between the neighboring cell centers (for the uniform grid considered here), it corresponds to the average temperature $T_e \approx (T_E + T_P)/2$.

Note that P is considered as the upstream cell center U, and E as the downstream cell center D, if the direction of mass flux at the east face center is positive ($m_{x,e} > 0$), vice versa for the negative mass flux ($m_{x,e} < 0$). Thus, using subscript U for upstream and D for downstream cell center, the general expression—incorporating both the directions of mass flux—for temperature at a face center f (e, w, n, s) of a CV is given as

$$\textbf{FOUScheme}: \quad T_f \approx T_{f,U} \tag{6.5}$$

$$\textbf{CDScheme}: \quad T_f \approx \frac{1}{2}T_{f,U} + \frac{1}{2}T_{f,D} \tag{6.6}$$

where for both the directions of mass flux at the east face center e, $T_{e,U}$ and $T_{e,D}$ are shown in Fig. 6.4.

There are further neighboring cell centers other than that represented by the ice T_P and the fire T_E. They are shown in Fig. 6.5 as T_W and T_{EE} (EE for east-of-east), the respective temperature is at UU (upstream-of-upstream) and DD (downstream-of-downstream) of the face center e if $m_{x,e} > 0$, i.e., $T_{e,UU} = T_W$ and $T_{e,DD} = T_{EE}$, and vice versa $T_{e,UU} = T_{EE}$ and $T_{e,DD} = T_W$ for $m_{x,e} < 0$. All the neighboring cell centers are maintained at a different temperature and will affect the temperature T_e. Thus, other than the *constant extrapolation* for the FOU and *linear interpolation* for the CD scheme (Fig. 6.4), one can have other local profile assumption as *linear extrapolation* and *quadratic interpolation* leading to Second-Order Upwind (SOU) scheme and Quadratic Upwind Interpolation for Convective Kinematics (QUICK) scheme, respectively. The QUICK scheme was proposed by Leonard (1979).

Similar to the derivation for finite difference equation in Sect. 4.1.2, the expressions for the temperature at a face center can be derived by two different approaches: first, Taylor series expansion and second, locally approximating the variation of temperature between the cell centers as linear for the SOU and quadratic for the QUICK scheme. The latter approach is presented in Fig. 6.5 to derive the expressions for the SOU and QUICK schemes. The figures show the derivations with the local variation as linear (between the cell centers $T_{e,U}$ and $T_{e,UU}$) for the SOU scheme, and quadratic (between the cell centers $T_{e,D}$, $T_{e,U}$, and $T_{e,UU}$) for the QUICK scheme. The derivations are shown for the temperature at the east face center; similar expression is obtained for the other face centers. The resulting general equation for SOU/QUICK advection scheme, at a face center f (e, w, n, s) of a CV, is given as

$$\textbf{SOUScheme}: \quad T_f \approx \frac{3}{2}T_{f,U} - \frac{1}{2}T_{f,UU} \tag{6.7}$$

$$\textbf{QUICKScheme}: \quad T_f \approx \frac{3}{8}T_{f,D} + \frac{6}{8}T_{f,U} - \frac{1}{8}T_{f,UU}$$

where for both the directions of mass flux at the east face center $f = e$, $T_{e,D}$, $T_{e,U}$, and $T_{e,UU}$ are shown in Fig. 6.5.

For the various advection schemes (Figs. 6.4 and 6.5), after substituting the appropriate cell center D, U and UU (in $T_{e,D}$, $T_{e,U}$, and $T_{e,UU}$) for $m_{x,e} > 0$ and $m_{x,e} < 0$, the resulting T_e is shown in Fig. 6.6, as T_e^+ for $m_{x,e} > 0$ and T_e^- for $m_{x,e} < 0$.

Although the advection schemes are shown in Fig. 6.6 for the east face of a CV, similar expressions for T_f^+ and T_f^- are obtained for the other face centers of a CV. They are presented in Table 6.1 as a generic expression for all the advection schemes, with w_1, w_2, and w_3 as the weights for the downstream, upstream, and

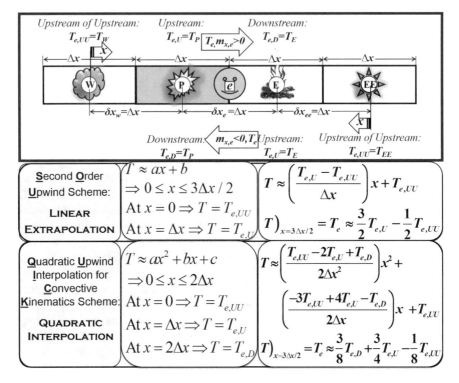

Fig. 6.5 Representation and derivation of SOU and QUICK advection schemes at the east face of representative fluid CV. Upstream and downstream cell centers are shown, for the two different directions of mass flux at the east face center (represented by a smiling face). The origin of x−coordinate is at W for $m_{x,e} > 0$ and at EE for $m_{x,e} < 0$, to derive a general equation

upstream-of-upstream cell centers of a face center f, respectively. The table also shows the value of the weights for the various advection schemes, which is same for all the face centers. It can be seen that the equations for the temperature at the face center are weighted average of the neighboring cell-center values, with the sum of the weights equal to 100%. The weights of the temperature at the downstream and upstream cell center are, respectively, seen as 0 and 100% for the FOU and 50 and 50% for the CD scheme. Whereas the weights of the downstream, upstream, and upstream-of-upstream cell-center temperature are, respectively, seen as 0, 150, and −50% for the SOU scheme, and 37.5, 75, and −12.5% for the QUICK scheme. The weights physically correspond to the percentage of the neighboring cell-center temperature/information reaching to the face center of interest. Although the heat transfer is two dimensional, note that the *approximation/scheme* is locally (at each face center) *one dimensional*.

Finally, for any face center f, a general expression for the product of mass flux and temperature (needed in Eq. 6.4 for the second approximation) is given as

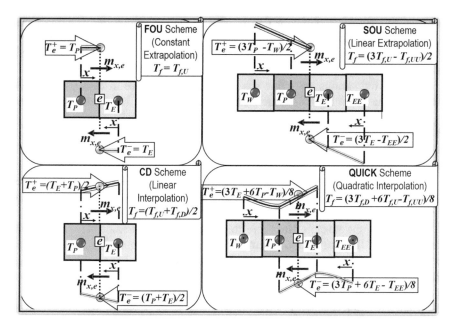

Fig. 6.6 Schematic representation of the various advection schemes to calculate the temperature at the east face center, considering both positive and negative directions of the mass flux at the east face

Table 6.1 Equations for temperature at the various face centers T_f, for both positive and negative directions of flow at the face center. The value of the weights of the temperature at the downstream, upstream, and upstream-of-upstream (of the various face centers) cell center is also presented for the FOU, CD, SOU, and QUICK advection schemes on a uniform grid

f	$T_f^{\pm} = w_1 T_{f,D}^{\pm} + w_2 T_{f,U}^{\pm} + w_3 T_{f,UU}^{\pm}$		Scheme	Weights		
$\downarrow$	T_f^+ (for $m_{x/y,f} > 0$)	T_f^- (for $m_{x/y,f} < 0$)		w_1	w_2	w_3
e	$w_1 T_E + w_2 T_P + w_3 T_W$	$w_1 T_P + w_2 T_E + w_3 T_{EE}$	FOU	0	1	0
w	$w_1 T_P + w_2 T_W + w_3 T_{WW}$	$w_1 T_W + w_2 T_P + w_3 T_E$	CD	$\dfrac{1}{2}$	$\dfrac{1}{2}$	
n	$w_1 T_N + w_2 T_P + w_3 T_S$	$w_1 T_P + w_2 T_N + w_3 T_{NN}$	SOU	0	$\dfrac{3}{2}$	$-\dfrac{1}{2}$
s	$w_1 T_P + w_2 T_S + w_3 T_{SS}$	$w_1 T_S + w_2 T_P + w_3 T_N$	QUICK	$\dfrac{3}{8}$	$\dfrac{6}{8}$	$-\dfrac{1}{8}$

$$\left. m_{x/y,f} T_f = m_{x/y,f}^+ T_f^+ + m_{x/y,f}^- T_f^- \right\} \; f = e, w, n, s$$

$$\text{where } T_f = w_1 T_D + w_2 T_U + w_3 T_{UU}$$

$$\left. \begin{aligned} m_{x/y,f}^+ &\equiv \max(m_{x/y,f}, 0) = \left| m_{x/y,f} \right| \text{ if } m_{x/y,f} > 0 \\ m_{x/y,f}^- &\equiv \min(m_{x/y,f}, 0) = -\left| m_{x/y,f} \right| \text{ if } m_{x/y,f} < 0 \end{aligned} \right\} \text{ otherwise } 0 \qquad (6.8)$$

where $m_{x/y,f}$ represents $m_{x,f}$ for the east/west face center and $m_{y,f}$ for the north/-south face center. Furthermore, the temperature T_f^+ is for positive direction and T_f^- is for negative direction of the fluid flow at a face center f, shown in Table 6.1 for the various faces. For the incorporation of the role of direction of mass flux on the different advection schemes, note that the above equation is presented as a single-line mathematical expression. This is done to avoid an "if" statement in a computer program, as it takes more computational time as compared to the single-line expression. Also note that only one of the two terms on the RHS of Eq. 6.8 will be non-zero.

Note that the various advection schemes, presented above in Figs. 6.4, 6.5, 6.6, are for all the adjoining cell centers uniformly spaced by the distance Δx in the $x-$direction (and Δy in the $y-$direction). This uniform spacing can be seen in Fig. 6.6, for any two adjoining cell centers inside the domains. However, at the boundary of the domain, the spacing between the boundary cell centers and the border cell centers is reduced by half, with $\Delta x/2$ at the east and west boundaries, and $\Delta y/2$ at the north and south boundaries. This resulted in a slightly different expression for the computation of conduction flux at the boundary face centers, as compared to that for the internal face centers, in the previous chapter (Eqs. 5.23 and 5.24).

Whereas, for the advection flux here, this results in a modification of the expression for SOU and QUICK advection schemes at the *border face center*—if the *direction of mass flux is inward* (inflow from the boundary of the domain). The border face centers are shown in Fig. 6.7 as inward arrows enclosing *black* filled triangle and square symbols. They are those face center whose at least one of the immediate neighboring face center is at the boundary of the domain. The faces passing through the border face centers are shown by the dashed lines in the figure. One such border face center and its associated border cell center and the neighboring cell centers are shown separately by a zoomed region in Fig. 6.7, with the distance δx_w (between UU and U cell centers) as half of the distance δx_e (between U and D), $\delta x_w = \delta x_e/2 = \Delta x/2$; as compared to same distances $\delta x_w = \delta x_e$ for the CV at P in Fig. 6.5. Since there is a change in the computational stencil used in the derivation of the schemes, the equations get modified and are derived (the derivation is similar to that shown in Fig. 6.5) as

For a border face centers with an inward flow:
$$\textbf{SOU Scheme}: T_f \approx 2T_{f,U} - T_{f,UU}$$
$$\textbf{QUICK Scheme}: T_f \approx \frac{1}{3}T_{f,D} + T_{f,U} - \frac{1}{3}T_{f,UU} \tag{6.9}$$

The Cartesian grid generation is such that one of the face center of a *border CV* coincides with the grid point at the boundary of the domain. The face centers are called here as *boundary face center*, shown by *gray* filled triangle/square in Fig. 6.7. The application of an advection scheme at the boundary face center is presented later in the solution methodology.

Thus, for an efficient implementation of the advection schemes, three different types of face centers are proposed: interior, border, and boundary, shown by unfilled, black filled, and gray filled symbols, respectively, in Fig. 6.7. The weights for an advection scheme are as shown in Table 6.1, for the interior and border face centers, except for the border face center with the SOU and QUICK schemes, if the mass flux is inward $- m_x > 0$ at the west, $m_x < 0$ at the east, $m_y > 0$ at the south, and $m_y < 0$ at the north border face center. The modified weights for the exceptional case are shown in Fig. 6.7.

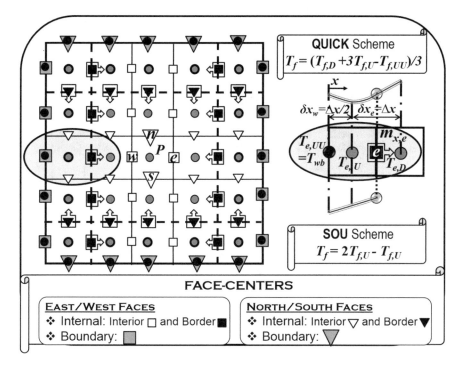

Fig. 6.7 Pictorial representation of a special case of implementation of higher order advection schemes, for the face centers parallel to the boundary of the domain (represented by filled square/-triangle) where the direction of mass flux is inward from the boundary of the domain (represented by arrows enclosing the face center)

Using the FOU scheme for the II-approximation, Eq. 6.4 for the enthalpy fluxes is approximated as

$$
\begin{aligned}
h_{x,e}^{\chi} &\approx c_p\left(m_{x,e}^{+} T_P^{\chi} + m_{x,e}^{-} T_E^{\chi}\right), \quad h_{x,w}^{\chi} \approx c_p\left(m_{x,w}^{+} T_W^{\chi} + m_{x,w}^{-} T_P^{\chi}\right), \\
h_{y,n}^{\chi} &\approx c_p\left(m_{y,n}^{+} T_P^{\chi} + m_{y,n}^{-} T_N^{\chi}\right) \,\&\, h_{y,s}^{\chi} \approx c_p\left(m_{y,s}^{+} T_S^{\chi} + m_{y,s}^{-} T_P^{\chi}\right)
\end{aligned}
\tag{6.10}
$$

6.1.4 Approximated Algebraic Formulation

The time-interval-based energy conservation law for heat advection and the two approximations in the physical law-based FVM are shown in Fig. 6.8. The corresponding approximated algebraic formulation is presented below for the *FOU upwind scheme*.

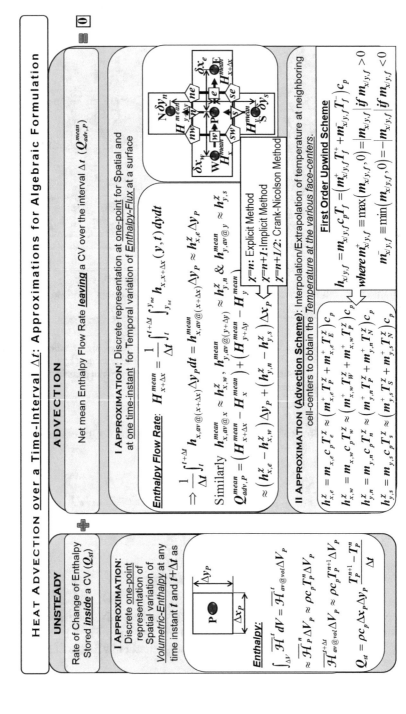

Fig. 6.8 Pictorial representation of the two approximations used in the physical law-based FVM for a 2D heat advection

Using the I−approximation for the volumetric term and flux term (Eqs. 5.6 and 6.3), the algebraic formulation in Eq. 6.2 is given as

$$\rho c_p \frac{T_P^{n+1} - T_P^n}{\Delta t} \Delta x_P \Delta y_P + Q_{adv}^\chi = 0$$

$$\text{where } Q_{adv}^\chi = \left(h_{x,e}^\chi - h_{x,w}^\chi \right) \Delta y_P + \left(h_{y,n}^\chi - h_{y,s}^\chi \right) \Delta x_P \tag{6.11}$$

Using the II−approximation for the enthalpy flux (FOU scheme; Eq. 6.10) and substituting $\overline{\mathcal{H}} = \rho c_p T$, the final algebraic formulation is given as

$$\rho \frac{T_P^{n+1} - T_P^n}{\Delta t} \Delta V_P + \left\{ \left[m_{x,e}^+ T_P^\chi + m_e^- T_E^\chi \right] - \left[m_{x,w}^+ T_W^\chi + m_{x,w}^- T_P^\chi \right] \right\} \Delta y_P$$

$$+ \left\{ \left[m_{y,n}^+ T_P^\chi + m_{y,n}^- T_N^\chi \right] - \left[m_{y,s}^+ T_S^\chi + m_{y,s}^- T_P^\chi \right] \right\} \Delta x_P = 0 \tag{6.12}$$

Similar to the algebraic formulation for the heat conduction, note that the above formulation is also a two-step process, with the derivation of the LAE consisting of enthalpy flux (Eq. 6.11) in the first step and the LAE for temperature (Eq. 6.12) in the second step. This is analogous to the derivation of PDE $\nabla \cdot \overrightarrow{h}$ in the first step and $\nabla \cdot T$ in the second step, as shown in Fig. 6.9.

6.1.5   Discussion

The first approximation is a discrete—one point and one time instant—representation of the spatial as well as temporal variation of the advection flux/diffusion flux terms. The second approximation corresponds to a discrete representation of normal *gradient* of temperature for *diffusion* in the previous chapter, and *value* of temperature for the *advection* here, at the various face centers of the CVs. The first as well as second approximation for the diffusion and advection flux—in the heat transport— corresponds to conduction in the previous chapter, and enthalpy in the present chapter. Note that they also correspond to viscous stress for diffusion, and momentum flux for advection, in the momentum transport, which will be discussed in the following chapters. Furthermore, note that the mass flux is a known quantity in this as well as next chapter. Whereas it is determined by solving continuity and momentum equation for a real-world convective heat transfer problem.

As introduced in the previous chapter, the steady-state stopping criterion during the unsteady solution is represented as the *non-dimensional* temperature gradient less than steady-state convergence tolerance ϵ_{st} (Eq. 5.15). For an unsteady-state advection problem, considering the characteristic velocity scale as u_c and the length scale as L_c, the non-dimensional temperature gradient is given as

$$\frac{\partial \theta}{\partial \tau} = \frac{l_c}{u_c \Delta T_c} \left(\frac{\partial T}{\partial t} \right)_{i,j} \approx \frac{1}{C_{L_c}} \frac{T_{i,j}^{n+1} - T_{i,j}^n}{\Delta T_c} \leq \epsilon_{st}$$

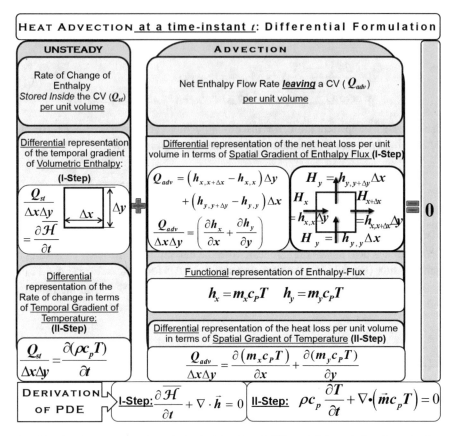

Fig. 6.9 Pictorial representation of physical law-based differential formulation for a 2D heat advection

where the non-dimensional temperature $\theta = (T - T_1)/(T_2 - T_1)$ and time $\tau = u_c t/l_c$. Furthermore, $C_{L_c} \equiv u_c \Delta t/l_c$ is the *Courant number* (based on the length scale) or a non-dimensional time step. This measure of the unsteadiness is computed for all the grid points of the temperature and should be reduced simultaneously for all of them. For the advection problem, a computer program can be developed with the unsteady results for temperature as dimensional T or non-dimensional θ. For the respective type of results, the steady-state convergence criterion is given as

$$\text{Results} \begin{cases} \text{Dimensional:} & \left| \dfrac{l_c}{u_c \Delta T_c} \dfrac{T_{i,j}^{n+1} - T_{i,j}^n}{\Delta t} \right|_{max} \text{for all } (i,j) \\[3ex] \text{Non-Dimensional:} & \left| \dfrac{\theta_{i,j}^{n+1} - \theta_{i,j}^n}{\Delta \tau} \right|_{max} \text{for all } (i,j) \end{cases} \leq \epsilon_{st} \quad (6.13)$$

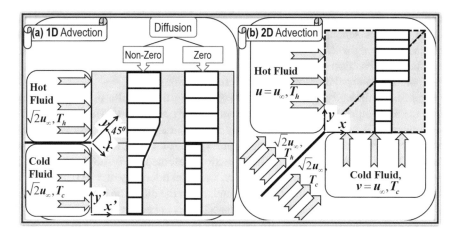

Fig. 6.10 Contact of upper hot and lower cold layer of equal velocity, representing one- and two-dimensional advection phenomenon, with zero diffusion

6.2 Flux-based Solution Methodology on a Uniform Grid: Explicit Method

2D unsteady-state solution methodology will be discussed for a contact of two parallel streams of equal velocity $\sqrt{2}u_\infty$ but unequal temperature (Patankar 1980), shown in Fig. 6.10. If the *diffusion is non-zero*, a mixing layer with a gradual change in the temperature (from the hot to the cold layer) is shown in Fig. 6.10a. Furthermore, there will be an increase in the cross-stream width of the mixing layer in the downstream direction. The figure also shows *zero-diffusion* and pure advection case with no mixing layer, maintaining the step change in the temperature in the downstream direction. For the uniform flow in $x'y'$ coordinate system in Fig. 6.10a (x' is parallel and y' is perpendicular to the flow direction with $u' = \sqrt{2}u_\infty$ and $v' = 0$), the pure advection problem is 1D with the solution as $T(x') = T_h$ or T_c. The problem becomes 2D for another coordinate direction xy, which is inclined at an angle of 45^0 from the direction of the horizontal free stream flow in Fig. 6.10a.

The 2D heat advection problem is shown in Fig. 6.10b, where the direction of the hot and cold stream is along the diagonal of a square computational domain. It represents advection of a step profile in a uniform flow oblique to the grid lines. It can be seen that the temperature BC corresponds to that of the hot layer T_h on the west boundary, and the cold layer T_c on the south boundary. Furthermore, there is a uniform velocity $u = v = u_\infty$ and uniform mass flux $m_x = m_y = \rho u_\infty$ inside the domain. Although the 2D pure heat advection problem is hypothetical, the persistence of the temperature discontinuity (along the diagonal) is a severe test case for any advection scheme.

For the 1D advection problem in the $x'y'$-coordinate, all the advection schemes give the correct results (Ferziger and Peric 2002). However, for the 2D advection problem in the xy-coordinate, FOU produces a smeared step profile and SOU as well as QUICK scheme results in an unbounded solutions, presented below. The results, as *exact for 1D while approximate for 2D advection*, are due to the fact that the 2D advection phenomenon is numerically modeled here by a locally 1D extrapolation/interpolation method (advection scheme).

The mathematical model for the advection (Fig. 6.9) results in first derivative in space as compared to the second derivative for conduction/diffusion (Fig. 5.5) phenomenon. Thus, for the 2D conduction in the Cartesian coordinate, the BCs are prescribed at two constant values of x and y, i.e., at $x = 0$, $x = L_1$, $y = 0$, and $y = L_2$ (Fig. 5.6). Whereas for the 2D advection presented in Fig. 6.10b, the BCs are prescribed at one constant value of x and y, *i.e.*, inlet boundary at $x = 0$ and $y = 0$. Note that the BCs are given all over the boundary for the diffusion as compared to the upstream boundary for the advection transport. This is because the information (of change of temperature at a point) travel equally in all the directions for the diffusion, and only in one (upstream to downstream) direction for the advection transport phenomenon.

Thus, for the present 2D advection problem, the BCs are prescribed only at the upstream/inflow boundary ($x = 0$ for the west and $y = 0$ for the south inlet boundary in Fig. 6.11), and the temperature at the downstream/outflow boundary is obtained by the application of an advection scheme. This is ensured by the extrapolation (FOU and SOU) but not by the interpolation (CD and QUICK) method for the advection scheme. This is because the FOU/SOU scheme is a function of only upstream values, and the CD/QUICK scheme also depends on downstream values. Considering the application of an advection scheme at the outlet boundary face center, shown as squares at $i$max $= 7$ and triangles at $j$max $= 7$ in Fig. 6.11, it can be seen that the cell centers which are upstream (of the outlet boundary) are inside and the downstream cell centers correspond to outside the domain. Since the downstream cell centers are not available neither there is an outlet BC, a special treatment is needed for the CD/QUICK scheme at the outlet boundary face centers; here, *FOU scheme is used*. This is shown in Fig. 6.11 with $T_{7,j} = T_{6,j}$ and $T_{i,7} = T_{i,6}$, resulting from the application of the FOU scheme for the face centers at the east and north outlet boundary of the domain, respectively. The FOU scheme-based computation of the temperature for the outlet boundary grid points ($T_{7,j} = T_{6,j}$ and $T_{i,7} = T_{i,6}$) also corresponds to the FDM-based backward difference discretization of a Neumann (insulated) BC at the outlet boundaries; $\partial T/\partial x = 0$ at the east and $\partial T/\partial y = 0$ at the north boundary of the domain in Fig. 6.11. The BC corresponds to a fully developed flow BC at the outlet, for the advection combined with diffusion, presented as convection in the next chapter. Thus, the implementation of the outflow BC (encountered in the convection later but not for advection here) is same as the application of FOU scheme at the outlet boundary (for advection here).

Interestingly, the application of Dirichlet BC at the inlet boundaries, seen as $T_{1,j} = T_h$ at the west and $T_{i,1} = T_c$ at the south boundary in Fig. 6.11, also turns out to be the same as the application of the FOU scheme at the inlet boundaries. Other than the Dirichlet BC at the inlet boundary, and the Neumann BC at the outlet boundary, the BC for the flow (in fluid dynamics) corresponds to no slip and free

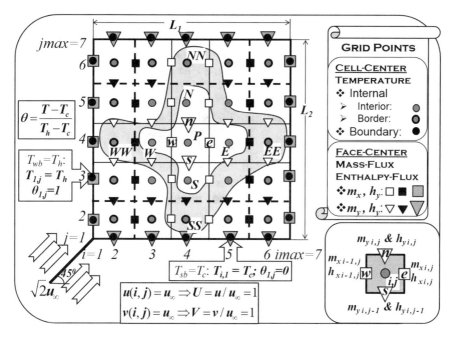

Fig. 6.11 Computational domain, discretized inlet BCs, and different types of grid points for a 2D heat advection problem

slip where application of any advection scheme at the boundary does not matter, as the mass flux (and thus the advection flux) at the boundary is zero.

Thus, for the advection here, and the convection and fluid dynamics in the later chapters, the *FOU scheme is used here at all the boundary face centers* during the computation of advection flux at the boundary. Whereas, at the internal face centers, appropriate advection scheme (FOU/CD/SOU/QUICK as per the users' choice) is applied using the equations shown in Table 6.1 and Fig. 6.7.

6.2.1 Explicit Method

Using the physical law-based FVM for 2D unsteady-state heat advection in a representative CV (Fig. 6.11), the explicit-method-based LAE for the enthalpy flux (Eq. 6.8) is given as

$$h_{x/y,f}^n = c_p \left[m_{x/y,f}^+ T_f^+ + m_{x/y,f}^- T_f^- \right] \text{ where } T_f = w_1 T_D + w_2 T_U + w_3 T_{UU}$$

$$h_{x,e}^n = c_p \left[m_{x,e}^+ (w_1 T_E^n + w_2 T_P^n + w_3 T_W^n) + m_{x,e}^- (w_1 T_P^n + w_2 T_E^n + w_3 T_{EE}^n) \right]$$

$$h_{x,w}^n = c_p \left[m_{x,w}^+ (w_1 T_P + w_2 T_W + w_3 T_{WW}) + m_{x,w}^- (w_1 T_W + w_2 T_P + w_3 T_E) \right]$$

$$h_{y,n}^n = c_p \left[m_{y,n}^+ (w_1 T_N + w_2 T_P + w_3 T_S) + m_{y,n}^- (w_1 T_P + w_2 T_N + w_3 T_{NN}) \right]$$

$$h_{y,s}^n = c_p \left[m_{x,s}^+ (w_1 T_P + w_2 T_S + w_3 T_{SS}) + m_{x,s}^- (w_1 T_S + w_2 T_P + w_3 T_N) \right]$$

$$(6.14)$$

where $T_{f=e,w,n,s}^{\pm}$ are substituted from Table 6.1. The table also shows the value of the weights for the various advection schemes on a uniform grid. Substituting $\chi = n$ in Eq. 6.11, the final discretized equation for the temperature on a uniform grid is given as

$$T_P^{n+1} = T_P^n - \frac{\Delta t}{\rho c_p \Delta x \Delta y} Q_{adv,P}^n$$

$$\text{where } Q_{adv,P}^n = (h_{x,e}^n - h_{x,w}^n) \Delta y + (h_{y,n}^n - h_{y,s}^n) \Delta x \quad (6.15)$$

6.2.2 Implementation Details

As discussed in the previous chapter for 2D unsteady-state conduction, the flux-based implementation for 2D advection is also a two-step process, shown in Fig. 6.12. Similar to conduction flux in the previous chapter, the grid point for enthalpy flux is also defined at the face centers and uses the same data structure. Using the previous time level values $T_{i,j}^n$ and the boundary conditions, the temperatures of the present time level $T_{i,j}^{n+1}$ are computed in two steps as follows:

First step: Calculate the enthalpy flux at the previous time level n (Eq. 6.14), using the FOU scheme at the boundary face centers and a user-defined advection scheme at the internal face centers, given as

$$h_{x\,i,j}^n = \begin{cases} c_p \left[m_{x,i,j}^+ T_{i,j}^n + m_{x,i,j}^- T_{i+1,j}^n \right] \} & \text{for } i = 1 \text{ and } (i\max - 1) \\ c_p \left[m_{x,i,j}^+ (w_1 T_{i+1,j}^n + w_2 T_{i,j}^n + w_3 T_{i-1,j}^n) \\ + m_{x,i,j}^- (w_1 T_{i,j}^n + w_2 T_{i+1,j}^n + w_3 T_{i+2,j}^n) \right] \end{cases} \text{ for } i = 2 \text{ to } (i\max - 2)$$

$$\implies \text{ for } j = 2 \text{ to } (j\max - 1)$$

$$(6.16)$$

$$h_{y\,i,j}^n = \begin{cases} c_p \left[m_{y,i,j}^+ T_{i,j}^n + m_{y,i,j}^- T_{i,j+1}^n \right] \} & \text{for } j = 1 \text{ and } (j\max - 1) \\ c_p \left[m_{y,i,j}^+ (w_1 T_{i,j+1}^n + w_2 T_{i,j}^n + w_3 T_{i,j-1}^n) \\ + m_{y,i,j}^- (w_1 T_{i,j}^n + w_2 T_{i,j+1}^n + w_3 T_{i,j+2}^n) \right] \end{cases} \text{ for } j = 2 \text{ to } (j\max - 2)$$

$$\implies \text{ for } i = 2 \text{ to } (i\max - 1)$$

$$(6.17)$$

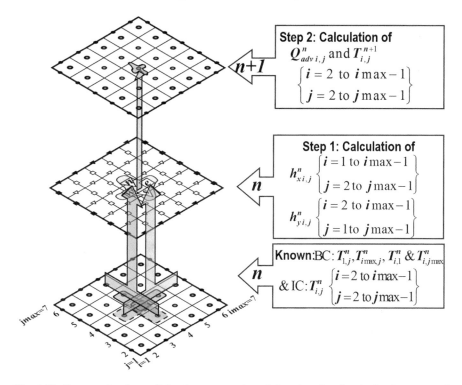

Step 2: Calculation of
$Q^n_{adv\,i,j}$ and $T^{n+1}_{i,j}$
$$\begin{cases} i = 2 \text{ to } i\max - 1 \\ j = 2 \text{ to } j\max - 1 \end{cases}$$

Step 1: Calculation of
$h^n_{x\,i,j}$ $\begin{cases} i = 1 \text{ to } i\max - 1 \\ j = 2 \text{ to } j\max - 1 \end{cases}$
$h^n_{y\,i,j}$ $\begin{cases} i = 2 \text{ to } i\max - 1 \\ j = 1 \text{ to } j\max - 1 \end{cases}$

Known: BC: $T^n_{1,j}, T^n_{i\max j}, T^n_{i,1}$ & $T^n_{i,j\max}$
& IC: $T^n_{i,j}$ $\begin{cases} i = 2 \text{ to } i\max - 1 \\ j = 2 \text{ to } j\max - 1 \end{cases}$

Fig. 6.12 Computational stencil for the computation of the advection flux in the first step and total heat efflux/lost by advection in the second step of a flux-based implementation details, for the explicit-method-based solution of the 2D unsteady-state advection

where the implementation of the BC at the inlet boundary is presented above as the application of the FOU scheme at the inlet boundary face centers. Whereas the implementation of the FOU scheme at the outlet boundary corresponds to the application of Neumann/outflow BC at the outlet boundary. For the implementation of the various advection schemes at internal (interior as well as border) face center, the weights w_1, w_2, and w_3 given in Table 6.1 are used. However, for the border face center, if the mass flux is inward to the domain (Fig. 6.7), the weights get modified for the SOU ($w_2 = 2$ and $w_3 = -1$) and QUICK ($w_1 = -1/3$, $w_2 = 1$, and $w_3 = 1/3$) scheme.

Second Step: Calculate the total heat efflux/lost by advection at the internal cell centers, and then the temperature (Eq. 6.15) as

$$\left.\begin{aligned} Q^n_{adv,i,j} &= (h^n_{x\,i,j} - h^n_{x\,i-1,j})\Delta y + (h^n_{y\,i,j} - h^n_{y\,i,j-1})\Delta x \\ T^{n+1}_{i,j} &= T^n_{i,j} - \frac{\Delta t}{\rho c_p \Delta x \Delta y} Q^n_{adv\,i,j} \end{aligned}\right\} \begin{aligned} &\text{for } i = 2 \text{ to } i\max - 1 \\ &\text{for } j = 2 \text{ to } j\max - 1 \end{aligned}$$

$$(6.18)$$

Corresponding to a geometrical representation of the grid points involved in Eqs. 6.16, 6.17, and 6.18, the computational stencil is shown in Fig. 6.12. For example, applying these equations to the grid point $(4, 4)$, the shaded arrow in the figure shows the computational stencil for the above two steps: first, $h^n_{x\,4,4}$ as a function of $T^n_{6,4}$, $T^n_{5,4}$, $T^n_{4,4}$, and $T^n_{3,4}$ (two upstream and two downstream), and $h^n_{y\,4,4}$ as a function of $T^n_{4,6}$, $T^n_{4,5}$, $T^n_{4,4}$, and $T^n_{4,3}$; and second, $Q^n_{\text{adv}\,4,4}$ as a function of $h^n_{x\,4,4}$, $h^n_{x\,3,4}$, $h^n_{y\,4,4}$, and $h^n_{y\,4,3}$. The computational stencil is also applicable for all the other face centers and cell centers, and is shown for the maximum number (four) of neighboring cell-center temperature needed for an advection scheme.

6.2.3 Solution Algorithm

For the explicit-method and flux-based solution methodology on a uniform grid, the algorithm for 2D unsteady-state heat advection is as follows:

1. User input:

   (a) Thermo-physical property of the fluid: enter density.
   (b) Geometrical parameters: enter the length of the domain L_1 and L_2.
   (c) Grid: enter the maximum number of grid points $i$max and $j$max.
   (d) IC (initial condition): enter T_0.
   (e) BCs (Boundary conditions): enter the parameter corresponding to the BCs (such as T_{wb}, T_{sb} T_{eb}, and T_{nb}, for the problem shown in Fig. 6.11).
   (f) Governing parameter: enter the velocity field in the domain.
   (g) Steady-state convergence tolerance: enter the value of ϵ_{st} (practically zero defined by the user).
   (h) Select the advection scheme for the solution, and enter the weights for the advection scheme.

2. Calculate geometric parameters ($\Delta x = L_1/(i\text{max} - 2)$ and $\Delta y = L_2/(j\text{max} - 2)$) and stability criterion-based time step ($\Delta t = 1/[(|u|\,\Delta x) + (|v|\,\Delta y)]$, based on Eq. 6.29 below) for explicit method. Use the time step slightly less than the computed value.
3. Apply IC at all the internal grid points and BCs at all the boundary grid points.
4. Calculate mass flux.
5. Apply BC for the boundary grid points which are subjected to a non-Dirichlet BC (FOU leading to this BC at the outlet boundary). Then, the temperature at present time instant is considered as the previous time-instant temperature, $T_{\text{old}\,i,j} = T_{i,j}$ (for all the grid points), as the computations is to be continued further for the computation of the temperature of the present time instant $T_{i,j}$.
6. Computation of advection flux and temperature:

(a) Compute advection flux at all the face centers in the domain (Eqs. 6.16 and 6.17).
(b) Compute total heat efflux/lost by advection and then the temperature for the next time step (Eq. 6.18)—at internal grid points.

7. Calculate a stopping parameter—steady-state convergence criterion (L.H.S of Eq. 6.13)—called here as unsteadiness.
8. If the unsteadiness is greater than steady-state tolerance ϵ_{st} (Eq. 6.13), go to step 5 for the next time-instant solution.
9. Otherwise, stop and post-process the steady-state results for analysis of the advection problem. In an inherently unsteady problem, the condition in previous step will never be satisfied and a different stopping parameter is used.

A computer program for the enthalpy-flux-based approach of CFD development is given below in Prog. 6.1, considering an example problem for the 2D unsteady heat advection on a uniform grid.

Example 6.1: Flux-based solution methodology of CFD development and testing for a 2D unsteady-state heat advection problem.

Consider a flow of two parallel streams of equal velocity $\sqrt{2}u_\infty$, inclined at an angle of $45°$ from $x-$axis, but unequal temperature (hot above at T_h and cold below at T_c), shown in Fig. 6.10b. This is solved as a 2D advection problem in a Cartesian (x, y) computational domain of size $L_1 = 10$ m and $L_2 = 10$ m, shown in Fig. 6.11. Consider the BCs at the inlet boundary as $T_{wb} = T_h = 100\,°C$ and $T_{sb} = T_c = 0\,°C$, prescribed velocity as $u = v = u_\infty = 5$ m/s and thermophysical properties of the fluid as $1000\,kg/m^3$ for density and $4180\,J/kg.K$ for specific heat.

(a) Using the flux-based approach of CFD development presented above, develop a *non-dimensional* computer program for the explicit method on a uniform grid.
(b) Present a CFD application of the code for the FOU, SOU, and QUICK scheme, on the grid size $imax \times jmax$ as (i) 12×12 and (ii) 52×52, and the steady-state convergence tolerance as $\epsilon_{st} = 10^{-4}$. For the various schemes, plot the steady-state *non-dimensional* temperature profiles along the vertical centerline, and compare the numerical result with the exact solution.

Solution:

(a) For the development of the non-dimensional code, consider the characteristic scales as $u_c = u_\infty$ for velocity, $l_c = L_1$ for length, and $\Delta T_c = (T_{wb} - T_{sb})$ for temperature. Thus, the independent and dependent non-dimensional variables are given as

$$\text{Coordinates:} \qquad X = \frac{x}{L_1}, \quad Y = \frac{y}{L_1}, \quad \tau = \frac{u_\infty t}{L_1},$$

$$\text{Flow-Properties:} \; U = \frac{u}{u_\infty}, \quad V = \frac{v}{u_\infty}, \quad \theta = \frac{T - T_{sb}}{T_{wb} - T_{sb}}$$

The non-dimensional computational setup corresponds to the size of the domain $L_1^* \times L_2^*$ as 1×1, the prescribed velocity as $U = V = 1$, the temperature BCs as $\theta = 1$ on the west, and $\theta = 0$ in the south boundary (Fig. 6.11), and initial condition considered as $\theta = 0.5$. The setup results in a non-dimensional grid size and time step, with the non-dimensional form of steady-state convergence criterion for $\partial\theta/\partial\tau$ (Eq. 6.13).

Using the flux-based implementation for the 2D unsteady heat advection, an explicit-method-based computer program for the present problem is presented below. It uses the maximum non-dimensional time step for a stable solution in the explicit method, obtained by using Eq. 6.29. Although the code uses the dimensional value of the density and specific heat, note that the non-dimensional results are independent of their values.

Listing 6.1 Scilab code for 2D unsteady state advection, using the explicit method and flux based approach of CFD development on a uniform 2D Cartesian grid.

```
1   //STEP-1: User-Input
2   rho = 1000.0;  cp = 4180.0;
3   L1=1.0;  L2=1.0;    // Non-Dim. Domain Size
4   imax=12;  jmax=12;    //Grid Size
5   T0=0.5;  T_wb=1.0;  T_sb=0.0;    //Non-Dim. Temp.
6   u=1;  v=1;  epsilon_st=0.0001;  // Non-Dim.
        Velocity
7   disp("SELECT THE ADVECTION SCHEME (1/2/3)");
8   printf("1. FOU \n");  printf("2. SOU \n");  printf
        ("3. QUICK\n");
9   scheme = input("ENTER THE ADVECTION SCHEME : ");
10  //*FUNCTION: Weights for Advection Schemes*
11  function [w]= weights(scheme,k)
12      if (scheme==1) then
13          w(1)=0;  w(2)=1;  w(3)=0;
14      elseif (scheme==2) then
15          if (k==2) then
16              w(1)=0;  w(2)=2;  w(3)=-1;
17          else
18              w(1)=0;  w(2)=3/2;  w(3)=-1/2;
19          end
20      elseif (scheme==3) then
21          if (k==2) then
22              w(1)=1/3;  w(2)=1;  w(3)=-1/3;
23          else
24              w(1)=3/8;  w(2)=6/8;  w(3)=-1/8;
25          end
26      end
```

```
27  endfunction
28  //STEP-2: Geometrical Parameter
29  //and Stability criterion based time-step
30  Dx = L1/(imax-2); Dy = L2/(jmax-2);
31  Dt=((abs(u)/Dx)+(abs(v)/Dx))^-1; // For FOU
        Scheme
32  if (scheme==2) then
33      Dt=(2/3)*Dt;   // For SOU Scheme
34  elseif (scheme==3) then
35      Dt=(4/9)*Dt;   // For QUICK Scheme
36  else
37  end
38  //STEP-3: IC and BCs
39  T(2:imax-1,2:jmax-1)=T0;
40  T(1,1:jmax)=T_wb; T(1:imax,1)=T_sb;
41  //STEP-4: Calculate mass flux
42  mx=rho*u;my=rho*v;
43  mx_p=max(mx,0); mx_m=min(mx,0);
44  my_p=max(my,0); my_m=min(my,0);
45  // Function to compute face-center temperature
46  function [Tf_p, Tf_m]=Temp_f(w,T1,T2,T3,T4)
47      Tf_p=w(1)*T3+w(2)*T2+w(3)*T1;
48      Tf_m=w(1)*T2+w(2)*T3+w(3)*T4;
49  endfunction
50  //Time-Marching for Explicit Unsteady State LAEs
        :
51  unsteadiness_nd=1; n=0
52  while unsteadiness_nd>=epsilon_st
53      n=n+1
54  // Step 5: Non-Dirichlet BC: FOU scheme leading
        to dT/Dx=0 for east and DT/dY for north
        outlet boundary
55  T(imax,1:jmax)=T(imax-1,1:jmax); T(1:imax,jmax)=
        T(1:imax,jmax-1);
56  //Step 5: Consider the temp. as the previous
        value
57      T_old=T;
58  // STEP6: Computation of enthalpy-flux and temp
        .
59      for j=2:jmax-1
60          for i=1:imax-1
61              if (i==1)|(i==imax-1) then //FOU@bdy.
62                  Te_p=T_old(i,j); Te_m=T_old(i+1,j);
63              else //Any scheme for internal
64                  [w]=weights(scheme,i);
65  [Te_p Te_m]=Temp_f(w,T_old(i-1,j), T_old(i,j),
        T_old(i+1,j), T_old(i+2,j));
66              end
67              hx_old(i,j)=cp*(mx_p*Te_p+mx_m*Te_m);
68          end
69      end
```

```
70        for  j=1:jmax-1
71          for  i=2:imax-1
72              if  (j==1)|(j==jmax-1)  then  //FOU@bdy.
73                  Tn_p=T_old(i,j);  Tn_m=T_old(i,j+1);
74              else  //Any  scheme  for  internal
75                  [w]=weights(scheme,j);
76   [Tn_p Tn_m]  =  Temp_f(w,T_old(i,j-1),  T_old(i,j),
        T_old(i,j+1),  T_old(i,j+2));
77              end
78              hy_old(i,j)=cp*(my_p*Tn_p+my_m*Tn_m);
79          end
80        end
81        for  j=2:jmax-1
82          for  i=2:imax-1
83   Q_adv_old(i,j)=((hx_old(i,j)-hx_old(i-1,j))*Dy)
        +((hy_old(i,j)-hy_old(i,j-1))*Dx);
84   T(i,j)=T_old(i,j)-(Dt/(rho*cp*Dx*Dy))*Q_adv_old(
        i,j);
85          end
86        end
87   //STEP  7:  Steady  state  convergance  criterion
88   unsteadiness_nd=max(abs(T-T_old))/Dt;//STEP  7:
        Steady  state  converegnce  criterion
89   printf("Time  step  no.  %5d,  unsteadiness_nd)  =  %8
        .4e\n",  n  ,  unsteadiness_nd);
90   //  STEP  8:  While-end  loop
91   end
```

(b) The program is applied for the present problem and the results are shown
    in Fig. 6.13, for the FOU, SOU, and QUICK schemes. The figure shows a
    comparison between the numerical results and the step profile from the exact
    solution. As compared to the FOU scheme, the figure shows that the results
    obtained by the SOU and the QUICK scheme are more accurate; however,
    they are unbounded—not bounded within the range of variation of boundary
    temperature from 0 to 1. Whereas FOU unconditionally satisfies the bound-
    edness criteria. With grid refinement for the SOU and QUICK schemes, the
    figure shows that the unboundedness in the solution gets reduced.
    While FOU results in almost monotonic steady-state convergence, the SOU
    and QUICK schemes result in an oscillatory convergence; the convergence
    corresponds to the temporal variation in the value of the unsteadiness
    during time marching. For the steady-state convergence in a coarser grid
    size of 12×12, the number of time step/interval required is 18, 28, and
    201 for the FOU, SOU, and QUICK, for the time step as $(\Delta t_{stab})_{FOU}$,
    $2(\Delta t_{stab})_{FOU}/3$, and $4(\Delta t_{stab})_{FOU}/9$, respectively. Whereas, for the finer
    grid size of 52×52, the number and the value of time step needed for con-
    vergence are, respectively, found as 98 and $(\Delta t_{stab})_{FOU}$ for the FOU, 160 and
    $0.57(\Delta t_{stab})_{FOU}$ for the SOU, and 2725 and $0.05(\Delta t_{stab})_{FOU}$ for the QUICK
    scheme. With the grid refinement, note the stable time step (required for the

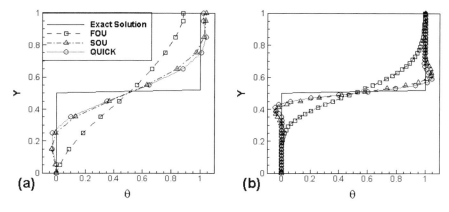

Fig. 6.13 CFD application of the *explicit-method*-based 2D unsteady advection code on a *uniform* Cartesian grid: comparison of the present results (on various advection schemes) with the exact solution for steady-state *non-dimensional* temperature profiles along the vertical centerline, on a grid size of (**a**) 12 × 12 and (**b**) 52 × 52

timewise convergence of the unsteady results) decreases slightly for the SOU and substantially for the QUICK scheme, as compared to the same value for the FOU scheme.

FOU introduces excessive numerical diffusion resulting in poor accuracy. It nearly solves an advection-diffusion equation with an effective diffusivity in a direction normal to the flow as $\Gamma = u_\infty \sin\theta\cos\theta(\Delta x\cos\theta + \Delta y\sin\theta)$ (Ferziger and Peric 2002), where u_∞ is the magnitude of the flow velocity and θ is the direction of the flow from the $x-$coordinate. FOU is robust as compared to SOU/QUICK, as it has better convergence characteristics, it allows larger time step to obtain converged results, for the explicit method. Whereas, with grid refinement, the figures show that SOU/QUICK as compared to FOU scheme approaches to the exact solution more rapidly along with a reduction in the unboundedness of the numerical solution.

6.3 Coefficients of LAEs-Based Solution Methodology on a Non-Uniform Grid: Explicit and Implicit Method

For a non-uniform grid, introduced in the previous chapter in Fig. 5.16, the methodology is presented below.

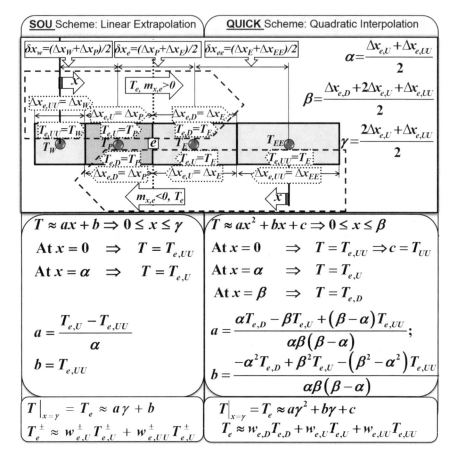

Fig. 6.14 For a non-uniform grid, representation and derivation of SOU and QUICK advection schemes at the east face of representative fluid CV. Upstream and downstream cell centers are shown, for the two different directions of mass flux at the east face center. The origin of x−coordinate is at W for $m_{x,e} > 0$ and at EE for $m_{x,e} < 0$ to derive a general equation

6.3.1 Advection Scheme on a Non-Uniform Grid

Using a procedure similar to that for the uniform grid in Figs. 6.5 and 6.14 shows the derivation of the SOU and QUICK schemes for a non-uniform grid. The figure shows that the derivation involves α, β, and γ, which are expressed in terms of non-uniform width of the CVs which are downstream ($\Delta x_{e,D}$), upstream ($\Delta x_{e,U}$), and upstream-of-upstream ($\Delta x_{e,UU}$) of the east face center e. They are used to determine the constants a, b, and c, which are substituted to the expression for T_e in the figure, and obtain the general equation for advection scheme at the east face center of a CV as

Table 6.2 General equations for the weights w of temperature at the cell centers which are downstream D, upstream U, and upstream-of-upstream UU of a face center f. For the various advection schemes on a non-uniform grid, the equation is also presented for the east and discussed for the north face center

$w_f \downarrow$	CD	SOU	QUICK
$w_{f,1}^{\pm}$	$\dfrac{\Delta s_{f,U}}{\Delta s_{f,D} + \Delta s_{f,U}}$	0	$\dfrac{\Delta s_{f,U}\left(2\Delta s_{f,U} + \Delta s_{f,UU}\right)}{\left(\Delta s_{f,D} + \Delta s_{f,U}\right)\left(\Delta s_{f,D} + 2\Delta s_{f,U} + \Delta s_{f,UU}\right)}$
$w_{f,2}^{\pm}$	$\dfrac{\Delta s_{f,D}}{\Delta s_{f,D} + \Delta s_{f,U}}$	$\dfrac{2\Delta s_{f,U} + \Delta s_{f,UU}}{\Delta s_{f,U} + \Delta s_{f,UU}}$	$\dfrac{\Delta s_{f,D}\left(2\Delta s_{f,U} + \Delta s_{f,UU}\right)}{\left(\Delta s_{f,D} + \Delta s_{f,U}\right)\left(\Delta s_{f,U} + \Delta s_{f,UU}\right)}$
$w_{f,3}^{\pm}$	0	$-\dfrac{\Delta s_{f,U}}{\Delta s_{f,U} + \Delta s_{f,UU}}$	$-\dfrac{\Delta s_{f,D}\Delta s_{f,U}}{\left(\Delta s_{f,U} + \Delta s_{f,UU}\right)\left(\Delta s_{f,D} + 2\Delta s_{f,U} + \Delta s_{f,UU}\right)}$
		$w_{f,1}^{\pm} = 0$, $w_{f,2}^{\pm} = 1$ and $w_{f,3}^{\pm} = 0$ **for the FOU Scheme**	
		$\Delta s = \Delta x$ for $f = e, w$ and $\Delta s = \Delta y$ for $f = n, s$; shown below for $f = e$	
$w_{e,1}^{+}$	$\dfrac{\Delta x_P}{\Delta x_E + \Delta x_P}$	0	$\dfrac{\Delta x_P\left(2\Delta x_P + \Delta x_W\right)}{\left(\Delta x_E + \Delta x_P\right)\left(\Delta x_E + 2\Delta x_P + \Delta x_W\right)}$
$w_{e,1}^{-}$	$\dfrac{\Delta x_E}{\Delta x_P + \Delta x_E}$	0	$\dfrac{\Delta x_E\left(2\Delta x_E + \Delta x_{EE}\right)}{\left(\Delta x_P + \Delta x_E\right)\left(\Delta x_P + 2\Delta x_E + \Delta x_{EE}\right)}$
$w_{e,2}^{+}$	$\dfrac{\Delta x_E}{\Delta x_E + \Delta x_P}$	$\dfrac{2\Delta x_P + \Delta x_W}{\Delta x_P + \Delta x_W}$	$\dfrac{\Delta x_E\left(2\Delta x_P + \Delta x_W\right)}{\left(\Delta x_E + \Delta x_P\right)\left(\Delta x_P + \Delta x_W\right)}$
$w_{e,2}^{-}$	$\dfrac{\Delta x_P}{\Delta x_P + \Delta x_E}$	$\dfrac{2\Delta x_E + \Delta x_{EE}}{\Delta x_E + \Delta x_{EE}}$	$\dfrac{\Delta x_P\left(2\Delta x_E + \Delta x_{EE}\right)}{\left(\Delta x_P + \Delta x_E\right)\left(\Delta x_E + \Delta x_{EE}\right)}$
$w_{e,3}^{+}$	0	$-\dfrac{\Delta x_P}{\Delta x_P + \Delta x_W}$	$-\dfrac{\Delta x_E \Delta x_P}{\left(\Delta x_P + \Delta x_W\right)\left(\Delta x_E + 2\Delta x_P + \Delta x_W\right)}$
$w_{e,3}^{-}$	0	$-\dfrac{\Delta x_E}{\Delta x_E + \Delta x_{EE}}$	$-\dfrac{\Delta x_P \Delta x_E}{\left(\Delta x_E + \Delta x_{EE}\right)\left(\Delta x_P + 2\Delta x_E + \Delta x_{EE}\right)}$

The above equation for $w_e^{\pm}$ is used to obtain similar equation for $w_n^{\pm}$,

by replacing x, EE, E, and W with y, NN, N and S, respectively.

$$T_e^{\pm} = w_{e,1}^{\pm} T_{e,D}^{\pm} + w_{e,2}^{\pm} T_{e,U}^{\pm} + w_{e,3}^{\pm} T_{e,UU}^{\pm} \tag{6.19}$$

where $w_{e,1}^{\pm}$, $w_{e,2}^{\pm}$, and $w_{e,3}^{\pm}$ are the weights for the downstream, upstream, and upstream-of-upstream cell center, respectively. The weights are shown in Table 6.2 in terms of non-uniform width of the CVs, instead of being constant for uniform grid (Table 6.1). Furthermore, the weights for the non-uniform grid in the above equation as compared to the uniform grid (Table 6.1) consist of additional subscript e (for the east face center) and superscript $\pm$ (+ corresponds to the positive and $-$corresponds to the negative $x-$direction of the flow at the east face center, $m_e > 0$ and $m_e < 0$, respectively).

The additional subscript and superscript for the weights are due to the fact that the non-uniform width of the CVs makes the weights dependent on the face center ($w_{e,1} \neq w_{w,1} \neq w_{n,1} \neq w_{s,1}$) as well as the flow direction ($w_{f1}^{+} \neq w_{f1}^{-}$), shown in Table 6.2. The width of a CV is shown in the table as a generic variable Δs, with Δx for east and west face centers, and Δy for north and south face centers of a CV. Furthermore, substituting $\Delta s_D = \Delta s_U = \Delta s_{UU} = \Delta s$ for the weights in Table 6.2,

Table 6.3 Equations for temperature at the various face centers T_f, for both positive and negative directions of flow at the face center, applicable for the FOU, CD, SOU, and QUICK advection schemes on a non-uniform grid

	$T_f^\pm = w_{f,1}^\pm T_{f,D}^\pm + w_{f,2}^\pm T_{f,U}^\pm + w_{f,3}^\pm T_{f,UU}^\pm$	
$f \downarrow$	T_f^+ (for $m_{x/y,f} > 0$)	T_f^- (for $m_{x/y,f} < 0$)
e	$T_e^+ = w_{e,1}^+ T_E + w_{e,2}^+ T_P + w_{e,3}^+ T_W$	$T_e^- = w_{e,1}^- T_P + w_{e,2}^- T_E + w_{e,3}^- T_{EE}$
w	$T_w^+ = w_{w,1}^+ T_P + w_{w,2}^+ T_W + w_{w,3}^+ T_{WW}$	$T_w^- = w_{w,1}^- T_W + w_{w,2}^- T_P + w_{w,3}^- T_E$
n	$T_n^+ = w_{n,1}^+ T_N + w_{n,2}^+ T_P + w_{n,3}^+ T_S$	$T_n^- = w_{n,1}^- T_P + w_{n,2}^- T_N + w_{n,3}^- T_{NN}$
s	$T_s^+ = w_{s,1}^+ T_P + w_{s,2}^+ T_S + w_{s,3}^+ T_{SS}$	$T_s^- = w_{s,1}^- T_S + w_{s,2}^- T_P + w_{s,3}^- T_N$

note that the weights for the non-uniform grid degenerate to that for the uniform grid (Table 6.1). Also note that there is no change in the FOU scheme for the non-uniform as compared to uniform grid; thus, not shown in Table 6.2.

Equation 6.19 is generalized for any face center f as

$$T_f^\pm = w_{f,1}^\pm T_{f,D}^\pm + w_{f,2}^\pm T_{f,U}^\pm + w_{f,3}^\pm T_{f,UU}^\pm$$

where the subscripts D, U, and UU at the east face center are E, P, and W for $m_e > 0$, and P, E, and EE for $m_e < 0$ (Fig. 6.14), resulting in the equation for T_e^+ and T_e^-, respectively. Similarly, the subscript for the temperature at the cell centers neighboring the other face centers ($f = w$, n and s) can be obtained. This results in a specific equation for temperature at a face center $T_{f=e,w,n,s}^\pm$, shown in Table 6.3.

6.3.2 Explicit and Implicit Method

For two-dimensional unsteady-state heat advection, the FVM-based discretized equation (Eq. 6.15) is expressed in a general form as

$$\rho c_p \frac{T_P^{n+1} - T_P^n}{\Delta t} \Delta x_P \Delta y_P + \lambda Q_{adv,P}^{n+1} + (1 - \lambda) Q_{adv,P}^n = 0 \tag{6.20}$$
$$\text{where } Q_{adv,P} = c_p \left\{ \left(m_{x,e} T_e - m_{x,w} T_w \right) \Delta y_P + \left(m_{y,n} T_n - m_{y,s} T_s \right) \Delta x_P \right\}$$

Furthermore, $\lambda = 0$, 1, and 0.5 correspond to explicit, implicit, and Crank-Nicolson method, respectively. For the *implicit* and the *Crank-Nicolson* method, the above equation may lead to matrices which are not *diagonally dominant*, and an iterative method may fail to converge for all the advection scheme presented above, except the FOU scheme. Thus, the implicit part of the above equation Q_{adv}^{n+1} is modified for the CD, SOU, and QUICK schemes — called here as **H**igher **O**rder **S**cheme (HOS) — as

$$Q_{adv,P}^{n+1} = Q_{adv,P}^{FOU,n+1} + \left(Q_{adv,P}^{HOS,N} - Q_{adv,P}^{FOU,N} \right) \tag{6.21}$$

where the *term inside the bracket* is computed using the values from the previous iteration N, while the implicit system of LAEs (consisting of neighboring cell-center temperatures at the present time level $n + 1$) is obtained by the FOU scheme. Once an iterative method converges at every time step, the FOU contribution in the above equation cancels out and it results in the solution by the HOS.

This method, called as *deferred correction approach,* was first reported by Khosla and Rubin (1974). This approach reduces the size of the computation stencil, storage requirements, and the efforts required to solve the LAEs. Furthermore, this approach usually converges at approximately the rate obtained by the FOU scheme (Ferziger and Peric 2002). The improvement in the convergence characteristics results in a reduction of the computational time for the iterative solution of the system of LAEs. Moreover, *once the iteration converges, the accuracy of the solution corresponds to that of the other not the FOU scheme*. Although the deferred correction approach is introduced here for the advection term, note that it is also used during the application of a higher order scheme for any other term (such as diffusion) in the governing equations.

Substituting, Q_{adv}^{n+1} from Eq. 6.21 to 6.20, we get

$$\rho c_p \frac{T_P^{n+1} - T_P^n}{\Delta t} \Delta x_P \Delta y_P + \lambda \left(Q_{adv,P}^{FOU,n+1} + Q_{adv,P}^{d,N} \right) + (1 - \lambda) Q_{adv}^n = 0$$
$$\text{where } Q_{adv,P} = c_p \left\{ \left(m_{x,e} T_e - m_{x,w} T_w \right) \Delta y_P + \left(m_{y,n} T_n - m_{y,s} T_s \right) \Delta x_P \right\} \tag{6.22}$$
$$Q_{adv,P}^{d,N} = Q_{adv,P}^{HOS,N} - Q_{adv,P}^{FOU,N}$$

where the application of the deferred correction approach results in the implicit treatment of only the nearest cell centers (E, W, N, and S) and the terms corresponding to the other cell centers (WW, EE, NN, and SS) are taken from the previous iteration N. Substituting the product of mass flow rate and temperature at a face center $m_f T_f$ from Eqs. 6.8 to 6.22 for $Q_{adv,P}$, we get

$$Q_{adv,P} \approx \left\{ \left[m_{x,e}^+ T_e^+ + m_{x,e}^- T_e^- \right] - \left[m_{x,w}^+ T_w^+ + m_{x,w}^- T_w^- \right] \right\} c_p \Delta y_P$$
$$+ \left\{ \left[m_{y,n}^+ T_n^+ + m_{y,n}^- T_n^- \right] - \left[m_{y,s}^+ T_s^+ + m_{y,s}^- T_s^- \right] \right\} c_p \Delta x_P$$

where the face-center temperature $T_f^\pm$ is substituted in terms of appropriate cell-center value for the FOU and HOS to obtain $Q_{adv,P}^{FOU}$ and $Q_{adv,P}^{HOS}$, respectively. Substituting the resulting equation to Eq. 6.22 and using $m_f = m_f^+ + m_f^-$ (refer Eq. 6.8), the final form of the LAE is given as

$$a P \, T_P^{n+1} = a E \left[\lambda \, T_E^{n+1} + (1 - \lambda) \, T_E^n \right] + a W \left[\lambda \, T_W^{n+1} + (1 - \lambda) \, T_W^n \right]$$
$$+ a N \left[\lambda \, T_N^{n+1} + (1 - \lambda) \, T_N^n \right] + a S \left[\lambda \, T_S^{n+1} + (1 - \lambda) \, T_S^n \right] + b$$

where $a E = -m_{x,e}^- \Delta y_P$, $a W = m_{x,w}^+ \Delta y_P$, $a N = -m_{y,n}^- \Delta x$, $a S = m_{y,s}^+ \Delta x_P$

$$a P^0 = \rho \Delta x_P \Delta y_P / \Delta t, \quad a P = \lambda \left(\sum_{NB} a N B + S_{m,P} \right) + a P^0$$

$$b = \left[a P^0 - (1 - \lambda) \left(\sum_{NB} a N B + S_{m,P} \right) \right] T_P^n - \lambda Q_{\text{adv}, P}^{d,N} - (1 - \lambda) Q_{\text{adv}, P}^{d,n}$$

$$\sum_{NB} a N B = a E + a W + a N + a S$$

$$S_{m,P} = \left(m_{x,e} - m_{x,w} \right) \Delta y_P + \left(m_{y,n} - m_{y,s} \right) \Delta x_P \approx 0$$

$$Q_{\text{adv}, P}^d = \left[\left(m_{x,e}^+ T_e^{d,+} + m_{x,e}^- T_e^{d,-} \right) - \left(m_{x,w}^+ T_w^{d,+} + m_{x,w}^- T_w^{d,-} \right) \right] \Delta y_P$$
$$+ \left[\left(m_{y,n}^+ T_n^{d,+} + m_{y,n}^- T_n^{d,-} \right) - \left(m_{y,s}^+ T_s^{d,+} + m_{y,s}^- T_s^{d,-} \right) \right] \Delta x_P$$

$$T_{f=e,w,n,s}^{d,\pm} = T_f^{\text{HOS},\pm} - T_f^{\text{FOU},\pm} = T_f^{\text{HOS},\pm} - T_U$$

$$(6.23)$$

where the temperature at a face center $T_f^{\text{HOS},\pm}$ is obtained by the higher order (CD, SOU, and QUICK) scheme, using Tables 6.3 and 6.2. Thereafter, the deferred temperature $T_{f=e,w,n,s}^{d,\pm} = T_f^{\text{HOS},\pm} - T_U$ is obtained. However, since FOU is used at all the boundary face centers irrespective of the advection scheme chosen by a user (discussed in Sect. 6.2), note that the deferred temperatures are zero for all the boundary face centers. This is given for the west, east, south, and north border CVs in Fig. 6.11 as

$$\text{Border CVs} : \begin{cases} \text{West}(i = 2) : T_w^{d,\pm} = 0; \ \text{East}(i = i\text{max} - 1) : T_e^{d,\pm} = 0 \\ \text{South}(j = 2) : T_s^{d,\pm} = 0; \ \text{North}(j = j\text{max} - 1) : T_n^{d,\pm} = 0 \end{cases}$$

$$(6.24)$$

Thus, $T_f^{d,\pm}$ are non-zero only for the internal face centers, and for the advection schemes other than the FOU scheme. Thus, the weights (Table 6.2) are used to obtain the temperature at internal face centers only.

Note that Eq. 6.23 is a general equation for all the advection schemes, with $Q_{\text{adv}, P}^d = 0$ for the FOU scheme. The generic equation is presented for the various solution methods—explicit, implicit, and Crank-Nicolson, although the *deferred correction approach is not relevant for the explicit method*. The transient deferred advection out-flux $Q_{\text{adv}, P}^d$ is taken at previous time level n for the explicit method ($Q_{\text{adv}, P}^{d,n}$), and at previous iteration N for the implicit method ($Q_{\text{adv}, P}^{d,N}$). At the boundary face centers, the deferred temperature $T_f^{d,\pm}$ is zero (Eq. 6.24). Since the velocity field considered for the advection term satisfies the mass conservation law for almost all the problems in CFD, the mass source $S_{m,P} \approx 0$ in Eq. 6.23.

6.3.2.1 Explicit Method

Substituting $\lambda = 0$ in Eq. 6.23, the LAE for a representative CV is given as

$$a P \, T_P^{n+1} = a E \, T_E^n + a W \, T_W^n + a N \, T_N^n + a S \, T_S^n + b$$

where $a E = -m_{x,e}^- \Delta y_P$, $a W = m_{x,w}^+ \Delta y_P$, $a N = -m_{y,n}^- \Delta x_P$, $a S = m_{y,s}^+ \Delta x_P$

$$b = \left(a P^0 - \sum_{NB} a N B + S_{m,P} \right) T_P^n - Q_{\text{adv}, P}^{d,n} \quad \text{and} \quad a P = a P^0 = \rho \Delta x_P \Delta y_P / \Delta t$$

$$(6.25)$$

where $Q_{\text{adv}, P}^d$ is given in Eq. 6.23.

For the derivation of stability criterion for the explicit method, an expanded form of the above equation is expressed in terms of all the neighboring cell-center coefficients (Table 6.3) as

$$aP\,T_P^{n+1} = aEE\,T_{EE}^n + aE\,T_E^n + aW\,T_W^n + aWW\,T_{WW}^n$$
$$+ aNN\,T_{NN}^n + aN\,T_N^n + aS\,T_S^n + aSS\,T_{SS}^n + b$$

where $aEE = -w_{e,3}m_{x,e}^-\Delta y_P$, $aE = \left(-w_{e,1}m_{x,e}^+ - w_{e,2}m_{x,e}^- + w_{w,3}m_{x,w}^-\right)\Delta y_P$

$aWW = w_{w,3}m_{x,w}^+\Delta y_P$, $aW = \left(w_{w,1}m_{x,w}^- + w_{w,2}m_{x,w}^+ - w_{e,3}m_{x,e}^+\right)\Delta y_P$

$aNN = -w_{n,3}m_{y,n}^-\Delta x_P$, $aN = \left(-w_{n,1}m_{y,n}^+ - w_{n,2}m_{y,n}^- + w_{s,3}m_{y,s}^-\right)\Delta x_P$

$aSS = w_{s,3}m_{y,s}^+\Delta x_P$, $aS = \left(w_{s,1}m_{y,s}^- + w_{s,2}m_{y,s}^+ - w_{n,3}m_{y,n}^+\right)\Delta x_P$

$b = \left(aP^0 - \sum_{NB} aNB - S_{m,P}\right)T_P^n$ and $aP = aP^0 = \rho\Delta x_P\Delta y_P/\Delta t$

$$(6.26)$$

As discussed in the previous chapter, the coefficients of all the coefficients in above LAE should be positive—to obtain a stable steady-state solution with the explicit method. Considering the sign of the weights ($w_{f,1}^\pm$ as well as $w_{f,2}^\pm$ are positive and $w_{f,3}^\pm$ is negative, shown in Table 6.2) and the mass fluxes $m_f^\pm$ (Eq. 6.8), the various products of the weights and mass fluxes in the above coefficients for the LAE ensure all the coefficients as positive only for the FOU scheme. Thus, a derivation for the stability criterion is presented below for the FOU scheme only.

Since the weight $w_{f,3}$ is negative (Tables 6.1 and 6.2), and m_f^+ is a positive and $m_{x/y,f}^-$ is a negative value (Eq. 6.8), the negative values of $-w_{e,3}m_{x,e}^-$, $w_{w,3}m_{x,w}^+$, $-w_{n,3}m_{y,n}^-$, and $w_{s,3}m_{y,s}^+$ result in the negative values of the coefficients aEE, aWW, aNN, and aSS, respectively (Eq. 6.26). The weight $w_{f,3}$ is non-zero for the SOU and QUICK schemes; thus these schemes do not ensure all the coefficients as positive. Whereas, for the CD difference scheme with a positive $w_{f,1}$, the first term in Eq. 6.26 for aE, aW, aN, and aS consists of $-w_{e,1}m_{x,e}^+$, $w_{w,1}m_{x,w}^-$, $-w_{n,1}m_{y,n}^+$, and $w_{s,1}m_{y,s}^-$, respectively; thus, the coefficients will be negative (unconditionally unstable). The FOU scheme, with $w_{f,1} = w_{f,3} = 0$, $w_{f,2} = 1$, and $aEE = aWW = aNN = aSS = 0$, ensures all the coefficients of Eq. 6.26 as positive.

For the FOU scheme, the condition for the coefficient of T_P^n (seen in the expression for b in Eq. 6.25) to be positive—to correctly capture the advection phenomenon—is given as

$$aE + aW + aN + aS + S_{m,P} \le aP^0 \qquad (6.27)$$

$$\left(-m_{x,e}^- + m_{x,w}^+\right)\frac{\Delta t}{\rho\Delta x_P} + \left(-m_{y,n}^- + m_{y,s}^+\right)\frac{\Delta t}{\rho\Delta y_P} \le 1$$

where the mass source $S_{m,P} \approx 0$ (Eq. 6.23). Furthermore, the mass flux at the face centers $m_{x,f=e,w}^+ = -m_{x,f=e,w}^- = \rho\left|u_f\right|$ and $m_{y,f=n,s}^+ = -m_{y,f=n,s}^- = \rho\left|v_f\right|$ (Eq. 6.8). Considering the sum of the mass flux (within the bracket) in the above equation, only one of the mass flux in a particular direction will be non-zero, since $m_{x,e}^+$ ($m_{x,e} < 0$) and $m_{x,w}^-$ ($m_{x,w} > 0$) do not occur simultaneously for almost all the CVs. Thus, substituting $-m_{x,e}^- + m_{x,w}^+ = \rho\left|u\right|$ and $-m_{y,n}^- + m_{y,s}^+ = \rho\left|v\right|$, the above equation is given as

$$\frac{|u|\,\Delta t}{\Delta x_P} + \frac{|v|\,\Delta t}{\Delta y_P} \le 1 \quad \text{for 2D advection}$$

$$\frac{|u|\,\Delta t}{\Delta x_P} \le 1 \quad \text{for 1D advection} \tag{6.28}$$

where the term on the L.H.S of the above equation corresponds to a *Courant number* $C\,(\equiv |u|\,\Delta t/\Delta x)$. The number represents a non-dimensional time step (should be less than one), as it is the ratio of the time step Δt to the advection time scale $\Delta x/|u|$. The time scale corresponds to the time required for an *advection mechanism-based transport* of a disturbance over the distance Δx (Ferziger and Peric 2002). The Courant number should be less than 1 for the FOU scheme in a 1D advection problem. The restriction for the time step means that a fluid particle cannot move more than the width of the CV (Eq. 6.28 for 1D advection) in a single time step. The above stability criterion for the advection is called as **C**ourant-**F**riedrichs-**L**ewy(CFL) *condition*.

A limiting/stable time step is given as

$$\Delta t_{stab} = \left(\frac{|u|}{\Delta x_P} + \frac{|v|}{\Delta y_P}\right)^{-1} \quad \text{for the FOU Scheme} \tag{6.29}$$

The application of the above time step is presented above in Prog. 6.1 for a uniform grid, with $\Delta t = \Delta t_{stab}$. The above equation, for the non-uniform grid, should be used to compute a different time step for each CV, and then the minimum of the time steps should be used in the computation. For the SOU and QUICK advection schemes, the above equation for the FOU scheme can be used as a guideline for the stable time step Δt_{stab}; *necessary but not the sufficient condition*. An actual computation involves a trial-and-error method, starting with $\Delta t_{stab}|_{\text{FOU}}$ as the first guess. If the computations do not converge to the steady state, a slightly smaller value of the time step is used until convergence is achieved. Whereas, for the CD scheme, the explicit method fails in the steady-state convergence for the pure advection problem. Thus, the *explicit method is unconditionally unstable for the CD scheme* (Anderson 1995) and is conditionally stable for the FOU/SOU/QUICK scheme. Even for the *unconditionally stable implicit method*, most of the iterative solvers fail to converge with the CD scheme (Ferziger and Peric 2002).

6.3.2.2 Implicit Method

Substituting $\lambda = 1$ in Eq. 6.23, the LAE for a representative CV is given as

$$aP\,T_P^{n+1} = aE\,T_E^{n+1} + aW\,T_W^{n+1} + aN\,T_N^{n+1} + aS\,T_S^{n+1} + b$$
$$\text{where } aE = -m_{x,e}^-\Delta y_P,\ aW = m_{x,w}^+\Delta y_P,\ aN = -m_{y,n}^-\Delta x_P,\ aS = m_{y,s}^+\Delta x_P$$
$$b = aP^0 T_P^n - Q_{adv,P}^{d,N},\ aP = \sum_{NB} aNB + aP^0 \text{ and } aP^0 = \rho\Delta x_P \Delta y_P/\Delta t \tag{6.30}$$

where the deferred-correction-approach-based $Q_{adv,P}^d$ is given in Eq. 6.23.

6.3.3 Implementation Details

As discussed in the previous chapter for the non-uniform Cartesian grid, all the CVs between any two vertical and horizontal lines have the same value of Δx and Δy, respectively; thus, a computationally efficient 1D instead of 2D matrix is used, i.e., Δx_i ($i = 2$ to $i\max - 1$) and Δy_j ($j = 2$ to $j\max - 1$). Furthermore, the weights for the east and west face centers which are dependent on Δx_i can also be represented as a 1D matrix $w_{e,i}^{\pm}$ for the east face center; similarly, $w_{n,j}^{\pm}$ for the north face center. Here, instead of computing the weights at the east and west face centers separately, a single running indices is used for the east face center (using the convention discussed in the last chapter for the heat flux)—with $w_{w,i}^{\pm} = w_{e,i-1}^{\pm}$ for a cell center i. However, since there are three weights in the general equation for an advection scheme, the weights are represented as $w_{e,i,k}^{\pm}$ with $k = 1$, 2, and 3. This is shown in Table 6.4, obtained from the equations for the weights shown in Table 6.2.

The single LAE for a representative CV in Eq. 6.25 for the explicit method, and Eq. 6.30 for the implicit method, is converted into a system of LAEs by substituting all the subscripts to running indices: E, P, W, e, and w are replaced by the running indices $i + 1$, i, $i - 1$, i, and $i - 1$, respectively. Furthermore, the coefficients (aE, aW, aN, aS, aP, and aP^0) in the above system of LAEs are represented as a matrix. This is done using a relationship between aW ($m_w^+ \Delta y_P = (m_w - m_w^-)\Delta y_P$) and aE ($-m_e^- \Delta y_P$) as $aW_{i,j} = aE_{i-1,j} + m_{x\,i-1,j}\Delta y_j$; similarly, $aS_{i,j} = aN_{i-1,j} + m_{y\,i-1,j}\Delta x_i$. This results in the final equation – implemented in a computer program (Prog. 6.2) for implicit method given below:

$$
\left.
\begin{aligned}
&a P_{i,j} T_{i,j}^{n+1} \\
&= a E_{i,j}\, T_{i+1,j}^{\chi} + (a E_{i-1,j} + m_{x\,i-1,j}\Delta y_j)\, T_{i-1,j}^{\chi} + \\
&\quad a N_{i,j}\, T_{i,j+1}^{\chi} + (a N_{i,j-1} + m_{y\,i,j-1}\Delta x_i)\, T_{i,j-1}^{\chi} + b_{i,j} \\
&\text{where } a P_{i,j}^o = \rho \Delta x_i \Delta y_j / \Delta t, \\
&\textbf{Explicit :} \chi = n,\ a P_{i,j} = a P_{i,j}^o \\
&\qquad b = \left[a P^0 - \textstyle\sum_{NB} a N B_{i,j}\right] T_{i,j}^n - Q_{\text{adv}\,i,j}^{d,n} \\
&\textbf{Implicit:} \chi = n+1,\ a P_{i,j} = \textstyle\sum_{NB} a N B_{i,j} + a P_{i,j}^o, \\
&\qquad b = a P^0 T_P^n - Q_{\text{adv}\,i,j}^{d,N}
\end{aligned}
\right\}
\quad
\begin{aligned}
&\text{for } i = 2\text{ to } i\max - 1 \\
&\text{for } j = 2\text{ to } j\max - 1
\end{aligned}
$$

$$(6.31)$$

Furthermore,

$$
\sum_{NB} a N B_{i,j} = a E_{i,j} + a E_{i-1,j} + m_{x\,i-1,j}\Delta y_j + a N_{i,j} + a N_{i,j-1} + m_{y\,i,j-1}\Delta x_i
$$

$$
\begin{aligned}
Q_{\text{adv}\,i,j}^{d} &= \left[\left(m_{x\,i,j}^+ T_{e\,i,j}^{d,+} + m_{x\,i,j}^- T_{e\,i,j}^{d,-}\right) - \left(m_{x\,i-1,j}^+ T_{e\,i-1,j}^{d,+} + m_{x\,i-1,j}^- T_{e\,i-1,j}^{d,-}\right)\right]\Delta y_j \\
&+ \left[\left(m_{y\,i,j}^+ T_{n\,i,j}^{d,+} + m_{y\,i,j}^- T_{n\,i,j}^{d,-}\right) - \left(m_{y\,i,j-1}^+ T_{n\,i,j-1}^{d,+} + m_{y\,i,j-1}^- T_{n\,i,j-1}^{d,-}\right)\right]\Delta x_i
\end{aligned}
$$

$$(6.32)$$

where the deferred temperature is computed at the face center. For both explicit and implicit methods, (i, j) for $T_{e,i,j}^{d,\pm}$, $m_{x\,i,j}^{\pm}$, and $aE_{i,j}$ correspond to the east face center; and (i, j) for $T_{n,i,j}^{d,\pm}$, $m_{y\,i,j}^{\pm}$, and $aN_{i,j}$ correspond to the north face center. The (i, j) for all the other variables in Eq. 6.31 correspond to that for the cell centers. For the values at the face centers, the coefficient for the LAEs (Eq. 6.23) and the deferred temperature $T_f^{d,\pm} = T_f^{\pm} - T_U$ (Eq. 6.24 for $T_f^{d,\pm}$ and Table 6.3 for $T_f^{\pm}$) are implemented as

$$
\text{Face-Centers:} \quad
\begin{aligned}
&\text{Vertical} \quad aE_{i,j} = -m_{x,i,j}^{-}\Delta y_j \quad
\begin{array}{l}
\text{for } i = 1 \text{ to } i\max - 1 \\
\text{for } j = 2 \text{ to } j\max - 1
\end{array} \\
&\text{Horizontal} \quad aN_{i,j} = -m_{y,i,j}^{-}\Delta x_i \quad
\begin{array}{l}
\text{for } i = 2 \text{ to } i\max - 1 \\
\text{for } j = 1 \text{ to } j\max - 1
\end{array}
\end{aligned}
\tag{6.33}
$$

$$
\left.
\begin{aligned}
&T_{e,i,j}^{d,+} = 0,\ T_{e,i,j}^{d,-} = 0 \text{ (FOU Scheme)} \quad\quad\quad\quad \text{for } i = 1\ \&\ i\max - 1 \\
\end{aligned}
\right\}
$$

$$
\left.
\begin{aligned}
T_{e,i,j}^{d,+} &= w_{e,i,1}^{+}T_{i+1,j} + (w_{e,i,2}^{+} - 1)T_{i,j} + w_{e,i,3}^{+}T_{i-1,j} \\
T_{e,i,j}^{d,-} &= w_{e,i,1}^{-}T_{i,j} + (w_{e,i,2}^{-} - 1)T_{i+1,j} + w_{e,i,3}^{-}T_{i+2,j}
\end{aligned}
\right\} \text{ for } i = 2 \text{ to } i\max - 2
$$

$$
\Longrightarrow \text{ for } j = 2 \text{ to } j\max - 1
\tag{6.34}
$$

$$
\left.
\begin{aligned}
&T_{n,i,j}^{d,+} = 0,\ T_{n,i,j}^{d,-} = 0 \text{ (FOU Scheme)} \quad\quad\quad\quad \text{for } j = 1\ \&\ j\max - 1 \\
\end{aligned}
\right\}
$$

$$
\left.
\begin{aligned}
T_{n,i,j}^{d,+} &= w_{n,j,1}^{+}T_{i,j+1} + (w_{n,j,2}^{+} - 1)T_{i,j} + w_{n,j,3}^{+}T_{i,j} \\
T_{n,i,j}^{d,-} &= w_{n,j,1}^{-}T_{i,j} + (w_{n,j,2}^{-} - 1)T_{i,j+1} + w_{n,j,3}^{-}T_{i,j+2}
\end{aligned}
\right\} \text{ for } j = 2 \text{ to } j\max - 2
$$

$$
\Longrightarrow \text{ for } i = 2 \text{ to } i\max - 1
\tag{6.35}
$$

where the implementation for $w_{e,i,k}^{\pm}$ and $w_{n,j,k}^{\pm}$ is presented in Table 6.4.

Note from the above equations that the weights are needed at *internal* (not boundary) face centers. Also note that the FOU scheme is used at all the boundary face centers irrespective of an advection scheme chosen by a user.

For the unsteady-state solution method, discussed above, note that the *maximum number* of neighboring cell-center temperature involved (corresponds to the QUICK scheme) in a LAE is eight: four nearest neighbors (E, W, N, and S) and four other neighbors (WW, EE, NN, and SS). Only the temperatures at the four nearest neighbors are treated implicitly, and the temperatures for all the eight neighbors are taken from the previous iteration N, for the implicit method. Whereas, for the explicit method, all the eight neighboring cell-center temperatures are treated explicitly. This can be seen in Eqs. 6.31 and 6.32. The resulting computational stencil is shown in Fig. 6.15(a) for the explicit method, and Fig. 6.15(b) for the implicit method. For example, applying the equations to grid point (4, 4), the shaded arrow in the figure for *explicit method* shows that the temperature $T_{4,4}^{n+1}$ is a function of all the eight neighboring temperatures at previous time level n ($T_{2,4}^{n}$, $T_{3,4}^{n}$, $T_{5,4}^{n}$, $T_{6,4}^{n}$, $T_{4,4}^{n}$, $T_{4,2}^{n}$, $T_{4,3}^{n}$, $T_{4,5}^{n}$, and $T_{4,6}^{n}$).

Whereas, for implicit method, the shaded arrow in Fig. 6.15b shows that the temperature $T_{4,4}^{n+1}$ is a function of a maximum of 12 neighboring values—four nearest

Table 6.4 For the implementation of the various advection schemes on a non-uniform Cartesian grid, equations for the weights ($w_{e,i,k}$ and $w_{n,j,k}$) of the temperature at the cell centers which are downstream ($k = 1$), upstream ($k = 2$), and upstream-of-upstream ($k = 3$) of the east and north face center

$w_e \downarrow$		Advection scheme	
	CD	SOU	QUICK
$w_{e,i,1}^+$	$\dfrac{\Delta x_i}{\Delta x_{i+1} + \Delta x_i}$	0	$\dfrac{\Delta x_i\,(2\Delta x_i + \Delta x_{i-1})}{(\Delta x_{i+1} + \Delta x_i)\,(\Delta x_{i+1} + 2\Delta x_i + \Delta x_{i-1})}$
$w_{e,i,1}^-$	$\dfrac{\Delta x_{i+1}}{\Delta x_i + \Delta x_{i+1}}$	0	$\dfrac{\Delta x_{i+1}\,(2\Delta x_{i+1} + \Delta x_{i+2})}{(\Delta x_i + \Delta x_{i+1})\,(\Delta x_i + 2\Delta x_{i+1} + \Delta x_{i+2})}$
$w_{e,i,2}^+$	$\dfrac{\Delta x_{i+1}}{\Delta x_{i+1} + \Delta x_i}$	$\dfrac{2\Delta x_i + \Delta x_{i-1}}{\Delta x_i + \Delta x_{i-1}}$	$\dfrac{\Delta x_{i+1}\,(2\Delta x_i + \Delta x_{i-1})}{(\Delta x_{i+1} + \Delta x_i)\,(\Delta x_i + \Delta x_{i-1})}$
$w_{e,i,2}^-$	$\dfrac{\Delta x_i}{\Delta x_i + \Delta x_{i+1}}$	$\dfrac{2\Delta x_{i+1} + \Delta x_{i+2}}{\Delta x_{i+1} + \Delta x_{i+2}}$	$\dfrac{\Delta x_i\,(2\Delta x_{i+1} + \Delta x_{i+2})}{(\Delta x_i + \Delta x_{i+1})\,(\Delta x_{i+1} + \Delta x_{i+2})}$
$w_{e,i,3}^+$	0	$-\dfrac{\Delta x_i}{\Delta x_i + \Delta x_{i-1}}$	$-\dfrac{\Delta x_{i+1}\Delta x_i}{(\Delta x_i + \Delta x_{i-1})\,(\Delta x_{i+1} + 2\Delta x_i + \Delta x_{i-1})}$
$w_{e,i,3}^-$	0	$-\dfrac{\Delta x_{i+1}}{\Delta x_{i+1} + \Delta x_{i+2}}$	$-\dfrac{\Delta x_i\Delta x_{i+1}}{(\Delta x_{i+1} + \Delta x_{i+2})\,(\Delta x_i + 2\Delta x_{i+1} + \Delta x_{i+2})}$

The above equation for $w_{e,i}^{\pm}$ is used to obtain similar equation for $w_{n,j}^{\pm}$, by replacing x, $i + 2$, $i + 1$, and $i - 1$ with y, $j + 2$, $j + 1$, and $j - 1$, respectively.

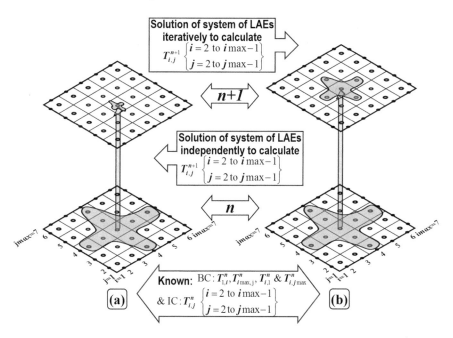

Fig. 6.15 Computational stencils to compute the temperature at the present time level $T_{i,j}^{n+1}$, during the coefficient of LAEs-based implementation, for the (**a**) explicit and (**b**) implicit solution method in a 2D unsteady-state advection

neighbors at $n + 1$ ($T_{3,4}^{n+1}$, $T_{5,4}^{n+1}$, $T_{4,3}^{n+1}$, and $T_{4,5}^{n+1}$) and all the eight neighbors at n. This is solved iteratively by an iterative (such as Gauss-Seidel) method, using the temperature of the eight neighbors at previous time level n for the first iteration and at previous iteration level N, for the solution of the subsequent iteration $N + 1$, within a time step. Using the deferred correction approach for the implicit method, the system of algebraic equations (corresponding to the internal grid point shown in Fig. 6.15b and Eq. 6.30) results in a penta-diagonal matrix. For both flux (Fig. 6.12) and coefficient of LAEs (Fig. 6.15)-based methodology, note that the computational stencil for the neighboring grid points is obtained by shifting the stencil leftward by Δx for $T_{3,4}^n$, rightward by Δx for $T_{5,4}^n$, downward by Δy for $T_{4,3}^n$, and upward by Δy for $T_{4,5}^n$.

6.3.4 Solution Algorithm

For both the explicit and implicit methods and coefficient of LAEs-based solution methodology on a non-uniform grid, the algorithm for 2D unsteady-state heat advection is as follows:

1. User input:

    (a) Thermo-physical property of the fluid: enter density.
    (b) Geometrical parameters: enter the lengths L_1 and L_2 of the domain.
    (c) Grid: enter the maximum number of grid points $i$max and $j$max.
    (d) IC (initial condition): enter T_0.
    (e) BCs (Boundary conditions): enter the parameter corresponding to the BCs (such as T_{wb}, T_{sb} T_{eb}, and T_{nb}, for the example problem shown in Fig. 6.11).
    (f) Governing parameter: enter the velocity field in the domain.
    (g) Steady-state convergence tolerance: enter the value of ϵ_{st}. For the implicit method, also enter an iterative method convergence tolerance ϵ.
    (h) Select the advection scheme (FOU/SOU/QUICK) for the solution.

2. Non-uniform Cartesian grid generation: Determine the coordinates of vertices of the CVs, using an equation for the non-uniform grid generation. Thereafter, compute the non-uniform width of the CVs Δx_i and Δy_j, and the distance between the cell centers δx_i and δy_j. Compute the stability criterion-based time step Δt (using Eq. 6.29) for explicit method; the time step may result in unstable solution for QUICK scheme and needs to be decreased till the converged solution is obtained. For the implicit method, enter any value of Δt, usually much larger than the Δt used in the explicit method.

3. Using the non-uniform width of the CVs Δx_i and Δy_j, compute the weight $w_{e,i,k}^{\pm}$ at the east face center and $w_{n,j,k}^{\pm}$ at the north face center (Table 6.4). Note that they are computed at *internal* (not boundary) face centers

4. Apply IC at all the internal grid points and BCs at all the boundary grid points.

5. Calculate mass flux from the prescribed velocity field. Thereafter, compute $m_{x,i,j}^{\pm}$ and $m_{y,i,j}^{\pm}$ at the face centers (Eq. 6.8). Although the mass flux is same at all the face centers for the problem in Fig. 6.11, they are considered here as non-uniform for a generic case.

6. Compute the coefficient of LAEs: $a E_{i,j}$ and $a N_{i,j}$ at the face centers (Eq. 6.33), and $a P_{i,j}^{0}$ and $a P_{i,j}$ at the cell center (Eq. 6.31).

7. Apply BC for the boundary grid points which are subjected to a non-Dirichlet BC (FOU leading to this BC at the outlet boundary). Then, the temperature at present time instant is considered as the previous time instant temperature, $T_{old\,i,j} = T_{i,j}$ (for all the grid points), as the computations are to be continued further for the computation of the temperature of the present time instant $T_{i,j}$.

8. Computation of deferred temperature, deferred advection out-flux, and temperature:

   (a) For the iterative solution in the implicit method, the temperature at present iteration is considered as the previous iterative value, $T_{olditer\,i,j} = T_{i,j}$ (for all the grid points).

   (b) Compute the deferred temperature $T_{e,i,j}^{d,\pm}$ at the vertical and $T_{n,i,j}^{d,\pm}$ at the horizontal *face centers* in the domain (Eq. 6.34 and 6.35), using the cell-center temperature at previous time level n for the explicit method, and at previous iteration N for the implicit method.

   (c) Compute total heat out-flux/loss by deferred advection $Q_{adv,i,j}^{d,n}$ for the explicit method, and $Q_{adv,i,j}^{d,N}$ for the implicit method (Eq. 6.32), and then $b_{i,j}$ (Eq. 6.31). Finally, the temperature for the next time step is computed at the *internal grid point* (Eq. 6.31).

   (d) Go to step 9 below for the explicit method, and continue for the implicit method.

   (e) Calculate an iteration convergence criterion $\left| T_{i,j} - T_{olditer\,i,j} \right|_{max}$ for all(i,j) as the stopping parameter, called here as *error*. If the error is greater than convergence tolerance ϵ, then go to the step (a) above for the next iterative solution. Otherwise, the present iterative value becomes the transient temperature for the present time level $n + 1$.

9. Calculate a stopping parameter (steady-state convergence criterion; L.H.S of Eq. 6.13) called here as unsteadiness. If the unsteadiness is greater than steady-state tolerance ϵ_{st} (Eq. 6.13), then go to step 7, for the next time-instant solution.

10. Otherwise, stop and post-process the steady-state results for analysis of the advection problem. In an inherently unsteady problem, the condition in previous step will never be satisfied and a different stopping parameter is used.

A computer program for the coefficient of LAEs-based approach of CFD development is presented below in Prog. 6.2, considering an example problem for the 2D unsteady heat advection on a non-uniform grid.

Example 6.2: Coefficient of LAEs-based methodology of CFD development and testing for a 2D unsteady-state heat advection problem.

Consider the problem presented above in Example 6.1 (Fig. 6.10 and 6.11), with a *diagonally downward flow* here instead of upward flow above. This results in the north and east boundaries of the domain as the inlet (with an inflow velocity of $u = v = u_\infty = -5\,m/s$) and the inlet BCs as $T_{nb} = T_h = 100^\circ C$ and $T_{eb} = T_c = 0^\circ C$. Consider the procedure for the non-uniform grid generation presented in the previous chapter (Example 5.4).

(a) Using the coefficient of LAEs-based approach develop a *non-dimensional* computer program for *implicit method* on a non-uniform grid.

(b) Present a CFD application of the code for the FOU, SOU, and QUICK schemes, on the grid size as (i) 22×22 and (ii) 102×102 and the steady state as well as iteration convergence tolerance as $\epsilon_{st} = \epsilon = 10^{-4}$. For the various schemes, plot the steady-state *non-dimensional* temperature profiles along the vertical centerline and compare with the exact solution.

Solution:

(a) Using the coefficient of LAEs-based implementation for 2D unsteady heat advection, an implicit-method-based non-dimensional code for the present problem is presented below. The program uses a non-dimensional time step $\Delta\tau = 0.5$ for the implicit method, much larger than the time step needed for the explicit method.

Note that the Gauss-Seidel method is presented below by marching against and along the flow direction, with the respective sequence of point-by-point iteration as $j = 2$ to $(j\max - 1)$, $i = 2$ to $(i\max - 1)$, and $j = (j\max - 1)$ to 1, $i = (i\max - 1)$ to 1, for the inflow from the east and north boundaries of the domain. The iteration against the flow direction leads to a convergence problem — number blow up in the iterative solutions within a time step. However, this is taken care by using an under-relaxation factor ω which corresponds to a constant value less than one. It causes a constant reduction in an increase (or decrease) in the temperature after every iteration, at all the grid points. Thus, the present iterative value is modified for the next iteration using an equation given as

$$\left(T_{i,j}^{N+1}\right)_{modified} = T_{i,j}^N + \omega\left[\left(T_{i,j}^{N+1}\right)_{present} - T_{i,j}^N\right]$$

Listing 6.2 Scilab code for 2D unsteady state advection, using the implicit method and coefficient of LAEs based approach of CFD development on a non-uniform 2D Cartesian grid.

```
1   //STEP-1: User-Input
2   rho = 1000.0;
3   L1=1.0;  L2=1.0;  imax=22;  jmax=22;
4   T0=0.5; T_eb=0.0;T_nb=1.0;u=-1;v=-1;
5   epsilon_st=0.0001; Dt =0.5;
6   disp("SELECT THE ADVECTION SCHEME (1/2/3)");
7   printf("1. FOU \n"); printf("2. SOU \n"); printf(
        "3. QUICK\n");
8   scheme = input("ENTER THE ADVECTION SCHEME : ");
9   //STEP-2: Non-uniform Cartesian grid generation
10  Beta=1.2; xi=linspace(0,1,imax-1);
11  Beta_p1=Beta+1; Beta_m1=Beta-1;
12  Beta_p1_div_m1=(Beta_p1/Beta_m1)^(2*xi-1);
13  num=(Beta_p1*Beta_p1_div_m1)-Beta_m1;den=2*(1+
        Beta_p1_div_m1);
14  x=L1*num./den;  y=x;
15  xc(2:imax-1)=(x(2:imax-1)+x(1:imax-2))/2;
16  xc(1)=x(1);xc(imax)=x(imax-1);yc=xc;
17  Dx(2:imax-1)=x(2:imax-1)-x(1:imax-2);
18  Dx(1)=0;Dx(imax)=0; Dy=Dx;Dx=Dx';
19  // STEP-3: Computation of weights at  "internal"
        face-centers (TABLE 6.2)
20  //*** FUNCTION: weights for various Advection
        Schemes ***
21  function [w]= weights(scheme,Ds_D,Ds_U,Ds_UU)
22      if (scheme==1) then
23          w(1)=0;w(2)=1;w(3)=0;
24      elseif (scheme==2) then
25  w(1)=0;  w(2)=(2*Ds_U+Ds_UU)/(Ds_U+Ds_UU);w(3)=-
        Ds_U/(Ds_U+Ds_UU);
26      elseif (scheme==3) then
27  w(1)=Ds_U*(2*Ds_U+Ds_UU);
28  w(1)=w(1)/((Ds_D+Ds_U)*(Ds_D+2*Ds_U+Ds_UU));
29  w(2)=Ds_D*(2*Ds_U+Ds_UU);
30  w(2)=w(2)/((Ds_D+Ds_U)*(Ds_U+Ds_UU));
31  w(3)=-Ds_D*Ds_U;
32  w(3)=w(3)/((Ds_U+Ds_UU)*(Ds_D+2*Ds_U+Ds_UU))
33      else
34      end
35  endfunction
36  for i=2:imax-2
37  [we]=weights(scheme,Dx(i+1),Dx(i),Dx(i-1));
38  we_p(i,1:3)=we';
39  [we]=weights(scheme,Dx(i),Dx(i+1),Dx(i+2));
40  we_m(i,1:3)=we';
41  end
42  for j=2:jmax-2
43  [wn]=weights(scheme,Dy(j+1),Dy(j),Dy(j-1));
```

```
44   wn_p(j,1:3)=wn';
45   [wn]=weights(scheme,Dy(j),Dy(j+1),Dy(j+2));
46   wn_m(j,1:3)=wn';
47   end
48   //STEP-4: IC and BCs
49   T(2:imax-1,2:jmax-1)=T0;
50   T(imax,1:jmax)=T_eb; T(1:imax,jmax)=T_nb;
51   //STEP-5: Calculate mass flux
52   mx=rho*u;my=rho*v;
53   mx_p=max(mx,0); mx_m=min(mx,0);
54   my_p=max(my,0); my_m=min(my,0);
55   //STEP-6: Coefficient of implicit LAEs
56   for j=2:jmax-1
57       for i=1:imax-1
58           aE(i,j)=-mx_m*Dy(j);
59       end
60   end
61   for j=1:jmax-1
62       for i=2:imax-1
63           aN(i,j)=-my_m*Dx(i);
64       end
65   end
66   for j=2:jmax-1
67       for i=2:imax-1
68           aP0(i,j)=rho*Dx(i)*Dy(j)/Dt;
69       aP(i,j)=aP0(i,j)+aE(i,j)+aE(i-1,j)+mx*Dy(j)+
             aN(i,j)+aN(i,j-1)+my*Dx(i);
70       end
71   end
72   // Function to compute face-center temperature (
         TABLE 6.3)
73   function [Tfd1, Tfd2]=Temp_f_d(k,wf_p,wf_m,T1,T2,
         T3,T4)
74   Tfd1=wf_p(k,1)*T3+(wf_p(k,2)-1)*T2+wf_p(k,3)*T1;
75   Tfd2=wf_m(k,1)*T2+(wf_m(k,2)-1)*T3+wf_m(k,3)*T4;
76   endfunction
77   //Time-Marching for Implicit Unsteady State LAEs:
         START
78    unsteadiness_nd=1; n=0; Ntot=0;
79   while unsteadiness_nd>=epsilon_st
80       n=n+1;
81   // Step 7: Non-Dirichlet BC: FOU scheme leading
         to dT/Dx=0 for west and DT/dY for south
         outlet boundary
82   T(1,1:jmax)=T(2,1:jmax); T(1:imax,1)=T(1:imax,2);
83   T_old=T;
84   //STEP-8:Deferred temp., advection out-flux &
         temp
85   //Iterative soln. (by GS method) at each time
         step
86   epsilon=0.0001; N=0; Error=1;
```

```
87   while Error >= epsilon
88   T(1,1:jmax) = T(2,1:jmax); T(1:imax,1) = T(1:imax,2);
89       T_old_iter = T; N = N+1; // Step 8(a)
90   // Step 8(b): Deferred temp. at the face-centers
91       for j = 2:jmax-1
92           for i = 1:imax-1
93               if (i==1) | (i==imax-1) then   // FOU@bdy
                       .
94                   Ted_p(i,j) = 0; Ted_m(i,j) = 0;
95               else
96   [Tedp, Tedm] = Temp_f_d(i, we_p, we_m, T_old_iter(i-1,j
         ), T_old_iter(i,j), T_old_iter(i+1,j),
         T_old_iter(i+2,j));
97           Ted_p(i,j) = Tedp; Ted_m(i,j) = Tedm;
98               end
99           end
100      end
101      for j = 1:jmax-1
102          for i = 2:imax-1
103              if (j==1) | (j==jmax-1) then    //
                     FOU@bdy.
104                  Tnd_p(i,j) = 0; Tnd_m(i,j) = 0;
105              else
106  [Tndp, Tndm] = Temp_f_d(j, wn_p, wn_m, T_old_iter(i,j
         -1), T_old_iter(i,j), T_old_iter(i,j+1),
         T_old_iter(i,j+2));
107          Tnd_p(i,j) = Tndp; Tnd_m(i,j) = Tndm;
108              end
109          end
110      end
111  //Step 8(c): Deferred advection out-flux and temp
         .
112      for j = jmax-1:-1:2     // Does not require
             Under-Relaxation for GS
113          for i = imax-1:-1:2
114  //  for j = 2:jmax-1       // Requires Under-
             Relaxation For GS Method
115  //      for i = 2:imax-1
116  Qd_adv_old_iter = (mx_p * Ted_p(i,j) + mx_m * Ted_m(i,j))
         * Dy(j) - (mx_p * Ted_p(i-1,j) + mx_m * Ted_m(i-1,j)) *
         Dy(j);
117  Qd_adv_old_iter = Qd_adv_old_iter + (my_p * Tnd_p(i,j) +
         my_m * Tnd_m(i,j)) * Dx(i) - (my_p * Tnd_p(i,j-1) +
         my_m * Tnd_m(i,j-1)) * Dx(i);
118  b = aP0(i,j) * T_old(i,j) - Qd_adv_old_iter;
119  T(i,j) = aE(i,j) * T(i+1,j) + (aE(i-1,j) + mx * Dy(j)) * T(i
         -1,j) + aN(i,j) * T(i,j+1) + (aN(i,j-1) + my * Dx(i)) * T
         (i,j-1) + b;
120              T(i,j) = T(i,j) / aP(i,j);
121          end
122      end
```

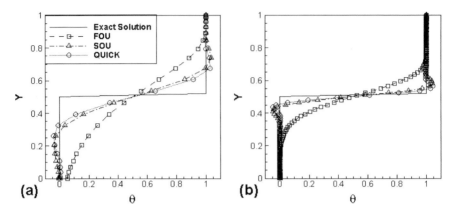

Fig. 6.16 CFD application of the *implicit-method*-based 2D unsteady advection code on *non-uniform* Cartesian grid: comparison of the present results (on various advection schemes) with the exact solution for steady-state *non-dimensional* temperature profiles along the vertical centerline, on a grid size of (**a**) 22 × 12 and (**b**) 102 × 102

```
123        Error=max(abs(T-T_old_iter)); // STEP 8(e)
124  //          printf("Iteration step no. %5d, Error
            = %8.4e\n", N , Error);
125  //omega=0.9; // omega=0.8 for SOU and 0.5 for
            QUICK for u=v=-1; refer the discussion below
            in (b)
126  //T=T_old_iter+omega*(T-T_old_iter)
127  end
128  Ntot=Ntot+N;
129  unsteadiness_nd=max(abs(T-T_old))/Dt; //STEP 9
130  printf("Time step no. %5d, unsteadiness_nd) = %8
            .4e\n", n , unsteadiness_nd);
131  end
```

(b) The program is applied for the present problem and the results are shown in Fig. 6.16. The nature of the result for the various schemes is similar to that discussed above in the previous example problem. *Using Gauss-Seidel point-by-point iteration against the flow stream* for the non-uniform grid size of 22 × 22, an under-relaxation factor ω of 0.8 and 0.5 is required for the SOU and QUICK schemes, respectively; however, ω is not required for the FOU scheme. Thus, as discussed above, the FOU as compared to the other scheme is robust. Whereas, on the finer grid size of 102 × 102, the SOU scheme requires *under-relaxation* $\omega = 0.9$ even while using the iteration *along the flow direction*.

Problems

6.1 Consider the example problem on 2D heat advection, *Example 6.1*, and modify the code (Prog. 6.1) to solve the problem on a non-uniform grid (Prog. 5.5). However, consider the *diagonally upward flow* such that the south and east boundaries of the domain are the inlet boundary, with the BCs as $T_{sb} = T_c = 0\,^\circ\text{C}$ and $T_{eb} = T_h = 100\,^\circ\text{C}$.

6.2 Consider the example problem on 2D heat advection, *Example 6.2*, and develop a code for the explicit method and the coefficient of LAEs-based solution methodology. However, consider the *diagonally downward flow* such that the west and north boundaries of the domain are the inlet boundary, with the BCs as $T_{wb} = T_c = 0\,^\circ\text{C}$ and $T_{nb} = T_h = 100\,^\circ\text{C}$. Present a CFD application of the code, with the results similar to the example problem.

References

Anderson, J. D. (1995). *Computational fluid dynamics: The basics with applications*. New York: McGraw Hill.

Ferziger, J. H., & Peric, M. (2002). *Computational methods for fluid dynamics* (3rd ed.). Berlin: Springer.

Khosla, P. K., & Rubin, S. G. (1974). A diagonally dominant second-order accurate implicit scheme. *Comput. Fluids, 2*, 207–218.

Leonard, B. P. (1979). A stable accurate convective modelling procedure based on quadratic upstream interpolation. *Comp. Methods Appl. Mech. Eng., 19*, 59–98.

Patankar, S. V. (1980). *Numerical heat transfer and fluid flow*. New York: Hemisphere Publishing Corporation.

Chapter 7
Computational Heat Convection

Convection is a combined advection and diffusion transport mechanism; the respective mechanism corresponds to the bulk motion and the random motion of the molecule. The diffusion, advection, and convection mechanisms are presented in Fig. 3.3, with an example for 1D heat transport. For the energy transport, the computation of non-uniform spatial distributions (in a computational domain) of conduction flux and enthalpy flux is presented separately in the last two chapters, and is considered in combination in this chapter. Thus, the numerical methodology for the present chapter is a combination of the methodology presented in the last two chapters. Similar to the previous chapter, the coupling of the energy transport with both mass and momentum transport is avoided in this chapter. This is done by considering a continuity satisfying steady-state velocity field, for the unsteady heat convection.

The mind map for this chapter is similar to the previous chapter, and is shown in Fig. 7.1. The figure shows that the present chapter starts with the physical law-based FVM, followed first by the flux-based solution methodology and then by the coefficient of LAEs-based methodology, presented in more detail for the former as compared to the latter methodology.

7.1 Physical Law-based Finite Volume Method

7.1.1 Energy Conservation Law for a Control Volume

Over a time interval Δt (encountered in CFD), the law presented in the previous chapters for conduction and advection is given here for a CV subjected to the heat convection (Fig. 7.2b) as

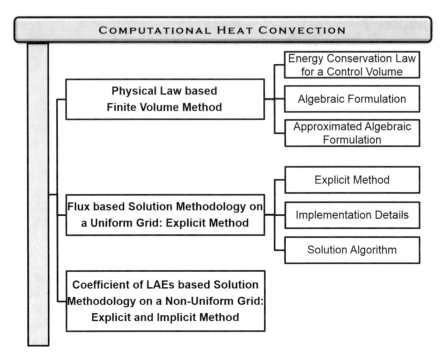

Fig. 7.1 Mind map for Chap. 7

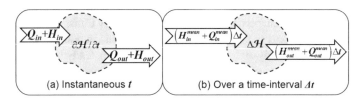

(a) Instantaneous t (b) Over a time-interval Δt

Fig. 7.2 Two different forms of conservation of energy for a CV, subjected to heat convection

Over a time interval Δt, the amount of convected (conducted and advected) thermal energy that enters a CV $(H_{in}^{mean} + Q_{in}^{mean})\Delta t$ minus the amount of convected thermal energy that leaves the CV $(H_{out}^{mean} + Q_{out}^{mean})\Delta t$ must equal the increase of enthalpy $\Delta\mathcal{H}$ stored within the CV.

$$\left(H_{in}^{mean} - H_{out}^{mean}\right)\Delta t + \left(Q_{in}^{mean} - Q_{out}^{mean}\right)\Delta t = \Delta\mathcal{H}$$

where $H^{mean} = \frac{1}{\Delta t}\int_{t}^{t+\Delta t} H\,dt$, $Q^{mean} = \frac{1}{\Delta t}\int_{t}^{t+\Delta t} Q\,dt$ (7.1)

$$\Delta\mathcal{H} = \int_{t}^{t+\Delta t} \frac{\partial\mathcal{H}}{\partial t}\,dt$$

where H is the enthalpy flow rate and Q is conduction heat transfer rate, and the superscript *mean* represents the time-averaged value over the time interval Δt. The above conservation statement, for the heat convection, corresponds to a law of conservation of energy for a fluid which is not experiencing a change in phase.

7.1.2   Algebraic Formulation

Similar to the last two chapters, accounting the spatial variation of the enthalpy $\mathcal{H}$ as the volumetric phenomenon, and the enthalpy flow rate H and the conduction heat transfer rate Q as the surface phenomenon, an algebraic formulation is finally obtained from Eq. 7.1 (for a 2D Cartesian CV in Fig. 7.3) as

$$\frac{\left(\overline{\mathcal{H}}_{av@vol}^{t+\Delta t} - \overline{\mathcal{H}}_{av@vol}^{t}\right)\Delta V_P}{\Delta t} + Q_{adv}^{mean} = Q_{cond}^{mean}$$

$$\text{where } Q_{adv}^{mean} = \left(h_{x,av@(x+\Delta x)}^{mean} - h_{x,av@x}^{mean}\right)\Delta y_P$$

$$+ \left(h_{y,av@(y+\Delta y)}^{mean} - h_{y,av@y}^{mean}\right)\Delta x_P \qquad (7.2)$$

$$Q_{cond}^{mean} = \left(q_{x,av@x}^{mean} - q_{x,av@(x+\Delta x)}^{mean}\right)\Delta y_P$$

$$+ \left(q_{y,av@y}^{mean} - q_{y,av@(y+\Delta y)}^{mean}\right)\Delta x_P$$

where the subscript av represents the *space-averaged* (within the volume/surface area) value, and the superscript *mean* represents the *time-averaged* (over the time interval Δt) value. Furthermore, the subscript $av@vol$ represents the average value of the enthalpy inside the volume, and the subscript $av@face$ represents the average value of the enthalpy flux and conduction flux at the various surfaces ($face = x, y,$ $x + \Delta x,$ and $y + \Delta y$), shown in Fig. 7.3.

7.1.3   Approximated Algebraic Formulation

The two approximations used to obtain linear algebraic equations are already discussed in the last two chapters, for the unsteady, conduction, and advection term. Since the combination of the three terms is considered for the unsteady-state convection, the approximations are not presented here for convection heat transfer; however, they are shown in Fig. 7.4.

Using the $I-$approximation for the volumetric term and the flux term (Eqs. 5.6, 5.8, and 6.3), the algebraic formulation in Eq. 7.2 is given as

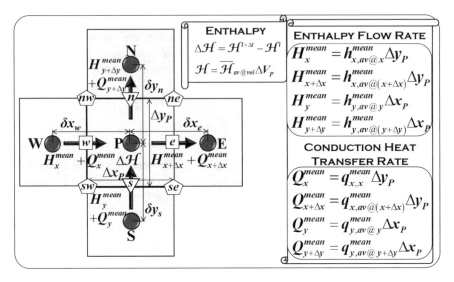

Fig. 7.3 Application of the time-interval-based law of conservation of energy (for heat convection) to a representative internal CV in a 2D Cartesian computational domain, with a non-uniform grid

$$\rho c_p \frac{T_P^{n+1} - T_P^n}{\Delta t} \Delta x_P \Delta y_P + Q_{adv}^\chi = Q_{cond}^\chi$$
$$\text{where } Q_{adv}^\chi = \left(h_{x,e}^\chi - h_{x,w}^\chi\right) \Delta y_P + \left(h_{y,n}^\chi - h_{y,s}^\chi\right) \Delta x_P \tag{7.3}$$
$$Q_{cond}^\chi = \left(q_{x,w}^\chi - q_{x,e}^\chi\right) \Delta y_P + \left(q_{y,s}^\chi - q_{y,n}^\chi\right) \Delta x_P$$

Using the $II-$approximation for the enthalpy flux (FOU scheme, Eq. 6.10) and conduction flux (CD scheme, Eq. 5.9) and substituting $\overline{\mathcal{H}} = \rho c_p T$, the above equation results in the final algebraic formulation as

$$\rho c_p \frac{T_P^{n+1} - T_P^n}{\Delta t} \Delta x_P \Delta y_P$$
$$+ \left\{\left[m_{x,e}^+ T_P^\chi + m_{x,e}^- T_E^\chi\right] - \left[m_{x,w}^+ T_W^\chi + m_{x,w}^- T_P^\chi\right]\right\} c_p \Delta y_P$$
$$+ \left\{\left[m_{y,n}^+ T_P^\chi + m_{y,n}^- T_N^\chi\right] - \left[m_{y,s}^+ T_S^\chi + m_{y,s}^- T_P^\chi\right]\right\} c_p \Delta x_P \tag{7.4}$$
$$= k \left(\frac{T_E^\chi - T_P^\chi}{\delta x_e} - \frac{T_P^\chi - T_W^\chi}{\delta x_w}\right) \Delta y_P + k \left(\frac{T_N^\chi - T_P^\chi}{\delta y_n} - \frac{T_P^\chi - T_S^\chi}{\delta y_s}\right) \Delta x_P$$

where $\chi = n, n+1$, and $n + 1/2$ for explicit, implicit, and Crank-Nicolson method, respectively. For simplicity, the above formulation is presented for the *FOU advection scheme,* as it remains same for both uniform and non-uniform grids.

Note that the above algebraic formulation for the temperature is also presented in two steps: first step, the LAE in terms of enthalpy flux and conduction flux (Eq. 7.3), and second step, final LAE for temperature (Eq. 7.4). This is analogous to the differential formulation shown in Fig. 7.5, consisting of $\nabla \cdot \overrightarrow{h}$ and $\nabla \cdot \overrightarrow{q}$ in the first step, and $\rho c_p \nabla \cdot T$ and $k \nabla^2 T$ in the second step.

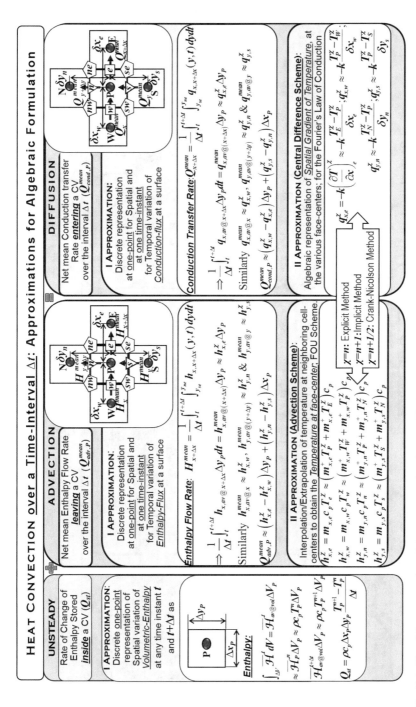

Fig. 7.4 Pictorial representation of the two approximations used in the physical law-based FVM for a 2D heat convection

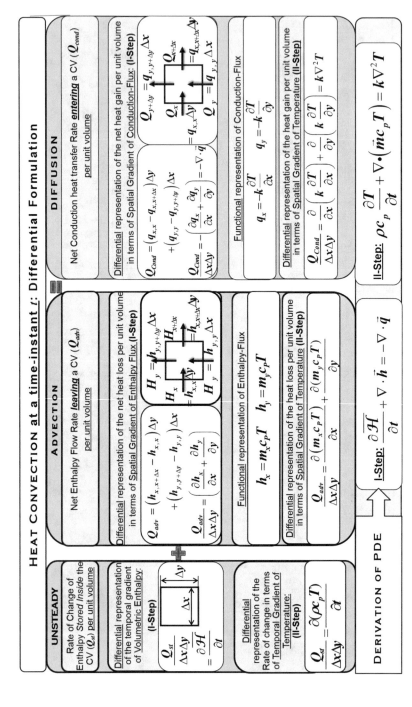

Fig. 7.5 Pictorial representation of physical law-based differential formulation for heat a 2D heat convection

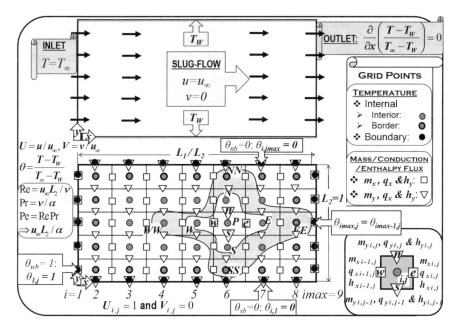

Fig. 7.6 Computational domain, discretized BCs, and different types of grid points for a 2D heat convection problem — slug flow in a plane channel

7.2 Flux-based Solution Methodology on a Uniform Grid: Explicit Method

Solution methodology for the 2D unsteady convection will be presented here for a prescribed continuity satisfying steady-state velocity field, similar to advection in the previous chapter. One such velocity field $u = u_\infty$ and $v = 0$ (spatially uniform) inside a plane channel leads to a heat convection, if there is a difference in the temperature of the incoming fluid and the channels walls. This problem is called as heat convection for the *slug flow* in a channel, shown in Fig. 7.6.

The figure shows a non-dimensional computational setup, with a non-dimensional representation of the domain size and discretized form of BCs. The characteristic scales are taken as the height of the channel for the length scale ($l_c = L_2$) and the streamwise velocity for the velocity scale ($u_c = u_\infty$). The slug flow is prescribed by non-dimensional velocity fields $U = 1$ and $V = 0$. The figure also shows the non-dimensional governing parameters as Reynolds number Re, Prandtl number Pr, and Peclet number Pe.

For the slug flow, note that the velocity field is not satisfying the no-slip BC on the stationary walls of the channel; thus, the problem is hypothetical. However, it has an analytical solution and acts as an excellent test problem. Similar to the last two chapters, the steady-state convergence criterion here also corresponds to the algebraic representation of the non-dimensional temporal gradient of the temperature, the criterion given in Eq. 6.13 for advection is considered here for the convection.

7.2.1 Explicit Method

Using the physical law-based FVM for 2D unsteady-state heat convection in a representative CV (Fig. 7.6), the explicit-method-based LAE for the conduction flux, enthalpy flux, and temperature (presented in the last two chapters) are given as follows:

$$q_{x,e}^n \approx -k\frac{T_E^n - T_P^n}{\delta x_e}, \; q_{x,w}^n \approx -k\frac{T_P^n - T_W^n}{\delta x_w}$$
$$q_{y,n}^n \approx -k\frac{T_N^n - T_P^n}{\delta y_n}, \; q_{y,s}^n \approx -k\frac{T_P^n - T_S^n}{\delta y_s} \tag{7.5}$$

$$h_{x/y,f}^n = c_p\left[m_{x/y,f}^+ T_f^+ + m_{x/y,f}^- T_f^-\right] \tag{7.6}$$

$$\text{where } T_f = w_1 T_D + w_2 T_U + w_3 T_{UU},$$
$$m_{x/y,f}^+ = \max(m_{x/y,f}, 0) \;\&\; m_f^- = \min(m_{x/y,f}, 0)$$
$$\therefore h_{x,e}^n = \left[m_{x,e}^+ \left(w_1 T_E^n + w_2 T_P^n + w_3 T_W^n\right) + m_{x,e}^- \left(w_1 T_P^n + w_2 T_E^n + w_3 T_{EE}^n\right)\right] c_p$$
$$h_{x,w}^n = \left[m_{x,w}^+ \left(w_1 T_P + w_2 T_W + w_3 T_{WW}\right) + m_{x,w}^- \left(w_1 T_W + w_2 T_P + w_3 T_E\right)\right] c_p$$
$$h_{y,n}^n = \left[m_{y,n}^+ \left(w_1 T_N + w_2 T_P + w_3 T_S\right) + m_{y,n}^- \left(w_1 T_P + w_2 T_N + w_3 T_{NN}\right)\right] c_p$$
$$h_{y,s}^n = \left[m_{y,s}^+ \left(w_1 T_P + w_2 T_S + w_3 T_{SS}\right) + m_{y,s}^- \left(w_1 T_S + w_2 T_P + w_3 T_N\right)\right] c_p$$

$$\rho c_p \frac{T_P^{n+1} - T_P^n}{\Delta t} \Delta x \Delta y = Q_{conv}^n$$
$$\Rightarrow T_P^{n+1} = T_P^n + \frac{\Delta t}{\rho c_p \Delta x \Delta y}\left(Q_{cond}^n - Q_{adv}^n\right) \tag{7.7}$$
$$\text{where } Q_{cond}^n = (q_{x,w}^n - q_{x,e}^n)\Delta y + (q_{y,s}^n - q_{y,n}^n)\Delta x$$
$$Q_{adv}^n = (h_{x,e}^n - h_{x,w}^n)\Delta y + (h_{y,n}^n - h_{y,s}^n)\Delta x$$

where Q_{cond} and Q_{conv} are the heat in-flux/gained and Q_{adv} is the heat out-flux/lost by the fluid in a CV. Furthermore, $m_{x,e} = \rho u_e, m_{x,w} = \rho u_w, m_{y,n} = \rho v_n$, and $m_{y,s} = \rho v_s$ are the mass fluxes at the various face centers of a 2D CV.

7.2.2 Implementation Details

Similar to the last two chapters, the flux-based implementation for 2D convection is also a two-step process. For the explicit method, the temperatures of the present time level $T_{i,j}^{n+1}$ are calculated in two steps as follows:

First step: For the internal as well as boundary grid points at the face centers (represented by square for vertical face centers and triangle for horizontal face centers, refer Fig. 7.6), calculate the conduction flux and enthalpy flux at the previous time level n as

$$
\left.
\begin{aligned}
q_{x\,i,j}^n &= -k\frac{T_{i+1,j}^n - T_{i,j}^n}{\Delta x/2} \\
h_{x\,i,j}^n &= c_p\left[m_{x,i,j}^+ T_{i,j}^n + m_{x,i,j}^- T_{i+1,j}^n\right] \text{ FOU Scheme}
\end{aligned}
\quad \right\} \text{for } i = 1\,\&\,imax - 1
$$

$$
\left.
\begin{aligned}
q_{x\,i,j}^n &= -k\frac{T_{i+1,j}^n - T_{i,j}^n}{\Delta x} \\
h_{x\,i,j}^n &= c_p\Big[m_{x,i,j}^+ (w_1 T_{i+1,j}^n + w_2 T_{i,j}^n + w_3 T_{i-1,j}^n) \\
&\quad +m_{x,i,j}^- (w_1 T_{i,j}^n + w_2 T_{i+1,j}^n + w_3 T_{i+2,j}^n)\Big]
\end{aligned}
\right\} \text{for } i = 2 \text{ to } imax - 2
$$

$$
\implies \text{for } j = 2 \text{ to } jmax - 1 \tag{7.8}
$$

$$
\left.
\begin{aligned}
q_{y\,i,j}^n &= -k\frac{T_{i,j+1}^n - T_{i,j}^n}{\Delta y/2} \\
h_{y\,i,j}^n &= c_p\left[m_{y,i,j}^+ T_{i,j}^n + m_{y,i,j}^- T_{i,j+1}^n\right] \text{ FOU Scheme}
\end{aligned}
\quad \right\} \text{for } j = 1\,\&\,jmax - 1
$$

$$
\left.
\begin{aligned}
q_{y\,i,j}^n &= -k\frac{T_{i,j+1}^n - T_{i,j}^n}{\Delta y} \\
h_{y\,i,j}^n &= c_p\Big[m_{y,i,j}^+ (w_1 T_{i,j+1}^n + w_2 T_{i,j}^n + w_3 T_{i,j-1}^n) \\
&\quad +m_{y,i,j}^- (w_1 T_{i,j}^n + w_2 T_{i,j+1}^n + w_3 T_{i,j+2}^n)\Big]
\end{aligned}
\right\} \text{for } j = 2\text{to } jmax - 2
$$

$$
\implies \text{for } i = 2 \text{ to } imax - 1 \tag{7.9}
$$

where the weights w_1, w_2, and w_3 are given in Table 6.1 and Fig. 6.7.

Second Step: Calculate the net mean convective heat transfer rate entering the internal CVs and then the temperature at the internal grid points (Eq. 7.7) as

$$
\left.
\begin{aligned}
Q_{cond\,i,j}^n &= (q_{x\,i-1,j}^n - q_{x\,i,j}^n)\Delta y + (q_{y\,i,j-1}^n - q_{y\,i,j}^n)\Delta x \\
Q_{adv\,i,j}^n &= (h_{x\,i,j}^n - h_{x\,i-1,j}^n)\Delta y + (h_{y\,i,j}^n - h_{y\,i,j-1}^n)\Delta x \\
Q_{conv\,i,j}^n &= Q_{cond\,i,j}^n - Q_{adv\,i,j}^n \\
T_{i,j}^{n+1} &= T_{i,j}^n + \frac{\Delta t}{\rho c_p \Delta x \Delta y} Q_{conv\,i,j}^n
\end{aligned}
\quad \right\}
\begin{aligned}
&\text{for } i = 2 \text{ to } imax - 1 \\
&\text{for } j = 2 \text{ to } jmax - 1
\end{aligned}
$$

$$
\tag{7.10}
$$

The computational stencil—corresponding to a geometrical representation of the grid points involved in Eqs. 7.8, 7.9, and 7.10—is shown in Fig. 7.7. The figure is

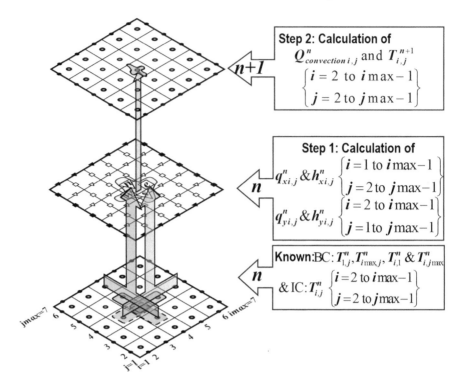

Fig. 7.7 Computational stencil or the computation of the convection flux in the first step and total heat in-flux/gained by convection in the second step of a flux-based 2D implementation details, for 2D unsteady-state convection, using explicit method

shown for the temperature at the maximum number of neighboring cell centers (two upstream and two downstream) needed for an advection scheme.

7.2.3 Solution Algorithm

For the flux and explicit-method-based solution methodology, the solution algorithm for the 2D unsteady-state convection on a uniform grid is very similar to that given in Sect. 6.2.3 for 2D advection, with the computation of the fluxes and the heat transfer by convection instead of advection in step 6.

Example 7.1: Flux-based solution methodology of CFD development and code verification for a 2D unsteady-state heat convection problem.
Consider convection for a slug flow in a plane channel of *non-dimensional* length $L_1/L_2 = 20$ and height $L_2 = 1$, shown in Fig. 7.6. The figure shows the

non-dimensional computational setup for the problem, with prescribed steady-state velocity field ($U = 1$ and $V = 0$) and the temperature BCs; $\theta = 1$ at the west boundary, $\theta = 0$ at the north/south boundary, and $\partial\theta/\partial X = 0$ at the east boundary. The initial condition for the temperature of the fluid is $\theta = 0$.

(a) Using the flux-based approach of CFD development presented above, develop a *non-dimensional* computer program for the explicit method on a uniform grid.

(b) For a Reynolds number $Re \equiv u_\infty L_2/\nu = 100$ and a Prandtl number $Pr = 0.7$, present a CFD application of the code for the FOU, SOU, and QUICK schemes on the grid size of 22×12 and steady-state convergence tolerance as $\epsilon_{st} = 10^{-4}$. For the various schemes, plot the axial variation of the temperature at the horizontal centerline of the channel. Also plot the steady-state temperature profiles $\theta(Y)$ at the various axial locations $X = 1, 2, 3$, and 5. Compare the profiles for the various advection schemes with the analytical solution (Burmeister 1993) given as

$$\theta(x, y) = \frac{4}{\pi} \sum_{n=0}^{\infty} \frac{(-1)^n}{2n + 1} \cos\,[\eta\,(Y - 0.5)]\,exp\left[-\eta^2 \frac{X}{Re\,Pr}\right]$$
$$\text{where } \eta = (2n + 1)\,\pi$$

Solution:

(a) The program on the flux-based implementation for 2D unsteady heat advection, presented in the previous chapter in Program 6.1, is extended here for the present problem on convection. Thus, the additions (in between certain line numbers in Program 6.1) and modifications for the present problem are presented in Program 7.1 . It uses the maximum non-dimensional time step for a stable solution in the explicit-method-based 2D convection problem, presented in Eq. 7.17. The setup results in a non-dimensional grid size and time step, with the non-dimensional form of steady-state convergence criterion for $\partial\theta/\partial\tau$.

Listing 7.1 Modification as well as additions in the Scilab code for 2D unsteady-state advection in Program 6.1, for its extension to the present problem on convection, using the explicit method and flux-based approach of CFD development on a uniform 2D Cartesian grid.

```
//STEP-1: MODIFICATION in User-Input(line no.
   2-6)
Re=100; Pr=0.7;
rho=1;  cp=1;  k=1/(Re*Pr);
L1=20.0;  L2=1.0;  imax=22; jmax=12;
T0=0;  T_wb=1.0;  T_sb=0.0;  T_nb=0.0;
```

```
u=1; v=0; epsilon_st=0.0001;
//STEP-2: MODIFICATION in line no. 30-37
Dx = L1/(imax-2);Dy = L2/(jmax-2);
alpha=k/(rho*cp);
t1=(abs(u)/Dx)+(abs(v)/Dx); t2=2*alpha*((1/Dx^2)
   +(1/Dy^2));
if (scheme==1) then
       Dt=(t1+t2)^-1;   // FOU scheme
     elseif (scheme==2) then
         Dt=((3/2)*t1+t2)^-1;// SOU scheme
     else
        Dt=((9/4)*t1+t2)^-1;//QUICK scheme
end
//STEP-3: MODIFICATIONS in Line No. 39-40
T(2:imax-1,2:jmax-1)=T0; T(1,1:jmax)=T_wb;
T(1:imax,1)=T_sb;T(1:imax,jmax)=T_nb;
// STEP 5: MODIFICATIONS in Line No. 55
T(imax,1:jmax)=T(imax-1,1:jmax);
// STEP 6: ADDITION between Line No. 61 & 62   as
qx_old(i,j)=-k*(T_old(i+1,j)-T_old(i,j))/(Dx
    /2.0);
// ADDITION between Line No. 63 & 64 as
qx_old(i,j)=-k*(T_old(i+1,j)-T_old(i,j))/Dx;
//ADDITION between Line No. 72 & 73
qy_old(i,j)=-k*(T_old(i,j+1)-T_old(i,j))/(Dy
    /2.0);
// ADDITION between Line No. 74 & 75 as
qy_old(i,j)=-k*(T_old(i,j+1)-T_old(i,j))/Dy;
// ADDITION between Line No. 82 & 83 as
Q_cond_old(i,j)=((qx_old(i-1,j)-qx_old(i,j))*Dy)
    +((qy_old(i,j-1)-qy_old(i,j))*Dx);
// ADDITION after Line No. 83
Q_conv_old(i,j)=Q_cond_old(i,j)-Q_adv_old(i,j);
// MODIFICATION in Line No. 84
T(i,j)=T_old(i,j)+(Dt/(rho*cp*Dx*Dy))*Q_conv_old
    (i,j);
```

(b) The program is applied for the present problem and the results are shown in Fig. 7.8. Comparing the present computational with the analytical results, Fig. 7.8a shows that the maximum temperature is slightly larger near the inlet and smaller near the outlet. Further comparison in Fig. 7.8b–c shows a much better agreement of the SOU and QUICK as compared to the FOU scheme for the various temperature profiles. Both the figures show almost same results by the SOU and QUICK schemes, with a slightly better result by the QUICK scheme for the temperature profile.

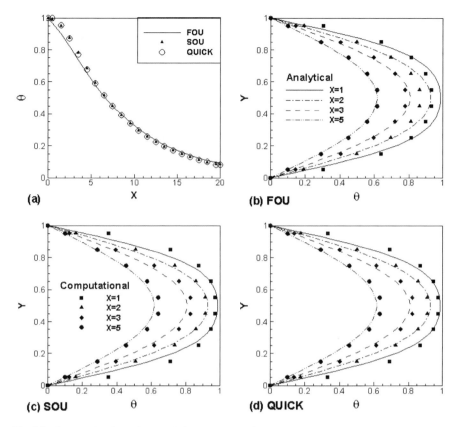

Fig. 7.8 CFD application of the explicit-method-based unsteady convection code, for the slug flow in a channel at $Re = 100$ and $Pr = 0.7$: (**a**) axial variation of the maximum in the temperature profile; and (**b**)–(**d**) comparison of the present computational with the analytical solution for steady-state *non-dimensional* temperature profiles along the transverse direction, at different axial locations, using the various advection schemes

7.3   Coefficients of LAEs-based Solution Methodology on a Non-Uniform Grid: Explicit and Implicit Method

The methodology is presented here for all the solution methods (explicit, implicit, and Crank-Nicolson) on a non-uniform grid. For the 2D unsteady state convection, the FVM-based discretized equation (Eq. 7.4) is expressed in a general form as

$$\rho c_p \frac{T_P^{n+1} - T_P^n}{\Delta t} \Delta x_P \Delta y_P = \lambda Q_{conv,P}^{n+1} + (1 - \lambda) Q_{conv,P}^n$$

where $Q_{conv,P} = Q_{cond,P} - Q_{adv,P}$

$$Q_{cond,P} = k \left\{ \left(\frac{T_E - T_P}{\delta x_e} - \frac{T_P - T_W}{\delta x_w} \right) \Delta y_P + \left(\frac{T_N - T_P}{\delta y_n} - \frac{T_P - T_S}{\delta y_s} \right) \Delta x_P \right\}$$

$$Q_{adv,P} = c_p \left\{ \left(m_{x,e} T_e - m_{x,w} T_w \right) \Delta y_P + \left(m_{y,n} T_n - m_{y,s} T_s \right) \Delta x_P \right\}$$

$$(7.11)$$

where $\lambda = 0, 1$, and 0.5 correspond to explicit, implicit, and Crank-Nicolson method, respectively. Furthermore, Q_{cond} and Q_{conv} are the heat in-flux/gained, and Q_{adv} is the heat out-flux/lost by the fluid in a CV. Finally, $m_e = \rho u_e, m_w = \rho u_w, m_n = \rho v_n$, and $m_s = \rho v_s$ are the mass fluxes at the various face centers of a 2D CV.

Applying the various advection schemes to the above equation, the temperature at a face center is expressed in terms of the neighboring cell-center temperature. As discussed in the previous chapter for the *implicit* and *Crank-Nicolson* method, a *deferred correction approach* is used for all the advection schemes except the FOU scheme. This is to ensure diagonal dominance in the system of LAEs, and thereby ensure convergence by an iterative method. The approach involves implicit consideration of only the nearest cell-center values (T_E, T_W, T_N, and T_S) and the temperature corresponding to the other cell centers (T_{WW}, T_{EE}, T_{NN}, and T_{SS}) are taken from the previous iteration N at each time step. Although *the deferred correction approach is not relevant for the explicit method and the implicit method with the FOU scheme*, a general formulation (for all the solution method and advection schemes) of the final LAE is given as

$$aP\, T_P^{n+1} = aE\left[\lambda\, T_E^{n+1} + (1-\lambda)\, T_E^n\right] + aW\left[\lambda\, T_W^{n+1} + (1-\lambda)\, T_W^n\right]$$
$$+aN\left[\lambda\, T_N^{n+1} + (1-\lambda)\, T_N^n\right] + aS\left[\lambda\, T_S^{n+1} + (1-\lambda)\, T_S^n\right] + b$$
$$(7.12)$$

where the coefficient of the above LAE is given in Table 7.1. Substituting λ (in the table and the above equation) as 0 for the explicit method and 1 for the implicit method, the final LAE form for the two methods is given as

Explicit : $\begin{cases} aP\, T_P^{n+1} = aE\, T_E^n + aW\, T_W^n + aN\, T_N^n + aS\, T_S^n + b \\ \text{where } aP = aP^0 \text{ and } b = \left(aP^0 - \sum_{NB} aNB\right) T_P^n - Q_{adv,P}^{d,n} \end{cases}$
$$(7.13)$$

Implicit : $\begin{cases} aP\, T_P^{n+1} = aE\, T_E^{n+1} + aW\, T_W^{n+1} + aN\, T_N^{n+1} + aS\, T_S^{n+1} + b \\ \text{where } aP = \sum_{NB} aNB + aP^0 \text{ and } b = aP^0 T_P^n - Q_{adv,P}^{d,N} \end{cases}$
$$(7.14)$$

where aE, aW, aN, aS, aP^0, and $Q_{adv,P}^d$ are given in Table 7.1 for both the methods.

For the derivation of the stability criterion for the explicit method, Eq. 7.13 is expressed in terms of all the neighboring coefficients and the weights as

$$aP\, T_P^{n+1} = aEE\, T_{EE}^n + aE\, T_E^n + aW\, T_W^n + aWW\, T_{WW}^n$$
$$+aNN\, T_{NN}^n + aN\, T_N^n + aS\, T_S^n + aSS\, T_{SS}^n + b$$
$$\text{where } aE = \left(-w_{e,1}m_{x,e}^+ - w_{e,2}m_{x,e}^- + w_{w,3}m_{x,w}^- + k/\left(c_p\delta x_e\right)\right)\Delta y_P$$
$$aW = \left(w_{w,1}m_{x,w}^- + w_{w,2}m_{x,w}^+ - w_{e,3}m_{x,e}^+ + k/\left(c_p\delta x_w\right)\right)\Delta y_P$$
$$aN = \left(-w_{n,1}m_{y,n}^+ - w_{n,2}m_{y,n}^- + w_{s,3}m_{y,s}^- + k/\left(c_p\delta y_n\right)\right)\Delta x_P \qquad (7.15)$$
$$aS = \left(w_{s,1}m_{y,s}^- + w_{s,2}m_{y,s}^+ - w_{n,3}m_{y,n}^+ + k/\left(c_p\delta y_s\right)\right)\Delta x_P$$
$$aEE = -w_{e,3}m_{x,e}^-\Delta y_P,\ aWW = w_{w,3}m_{x,w}^+\Delta y_P,$$
$$aNN = -w_{n,3}m_{y,n}^-\Delta x_P,\ aSS = w_{s,3}m_{y,s}^+\Delta x_P,$$
$$b = \left(aP^0 - \sum_{NB} aNB\right) T_P^n\ \&\ aP = aP^0 = \rho\Delta x_P\Delta y_P/\Delta t$$

Table 7.1 Coefficient of the LAE given in Eq. 7.12, for 2D unsteady heat convection

$aE = \left(-m_{x,e}^- + \dfrac{k}{c_p \delta x_e}\right) \Delta y_P$	$aW = \left(m_{x,w}^+ + \dfrac{k}{c_p \delta x_w}\right) \Delta y_P$
$aN = \left(-m_{y,n}^- + \dfrac{k}{c_p \delta y_n}\right) \Delta x_P$	$aS = \left(m_{y,s}^+ + \dfrac{k}{c_p \delta y_s}\right) \Delta x_P$

$$aP = \lambda \left(\sum_{NB} aNB + S_{m,P}\right) + aP^0, \quad \sum_{NB} aNB = aE + aW + aN + aS$$

$$b = \left[aP^0 - (1-\lambda)\left(\sum_{NB} aNB + S_{m,P}\right)\right] T_P^n - \lambda Q_{adv,P}^{d,N} - (1-\lambda) Q_{adv,P}^{d,n}$$

$$aP^0 = \rho \Delta x_P \Delta y_P / \Delta t, \quad S_{m,P} = \left(m_{x,e} - m_{x,w}\right)\Delta y_P + \left(m_{y,n} - m_{y,s}\right)\Delta x_P \approx 0$$

$$Q_{adv,P}^d = \left[\left(m_{x,e}^+ T_e^{d,+} + m_{x,e}^- T_e^{d,-}\right) - \left(m_{x,w}^+ T_w^{d,+} + m_{x,w}^- T_w^{d,-}\right)\right]\Delta y_P$$
$$+ \left[\left(m_{y,n}^+ T_n^{d,+} + m_{y,n}^- T_n^{d,-}\right) - \left(m_{y,s}^+ T_s^{d,+} + m_{y,s}^- T_s^{d,-}\right)\right]\Delta x_P$$

where $T_{f=e,w,n,s}^{d,\pm} = T_f^{HOS,\pm} - T_f^{FOU,\pm} = T_f^{\pm} - T_U$

As discussed in the last two chapters, for stability criterion in an explicit method, the coefficients of all the temperatures of the previous time step in the LAE should be positive to correctly capture the heat transfer phenomenon. Similar to the discussion in the previous chapter, the positive coefficient for all the neighboring temperatures is ensured only by the FOU scheme. Thus, for the FOU scheme, with $w_{f,1} = w_{f,3} = 0$, $w_{f,2} = 1$, and $aEE = aWW = aNN = aSS = 0$, the condition for the coefficient of T_P^n (seen in the expression for b in Eq. 7.15) to be positive is given as

$$aE + aW + aN + aS \qquad\qquad \le aP^0$$

$$\left\{\left(-m_{x,e}^- + m_{x,w}^+\right)\frac{\Delta t}{\rho \Delta x_P} + \frac{\alpha \Delta t}{\Delta x_P}\left(\frac{\delta x_e + \delta x_w}{\delta x_e \, \delta x_w}\right)\right. $$
$$\left. + \left(-m_{y,n}^- + m_{y,s}^+\right)\frac{\Delta t}{\rho \Delta y_P} + \frac{\alpha \Delta t}{\Delta y_P}\left(\frac{\delta y_n + \delta y_s}{\delta y_n \, \delta y_s}\right)\right\} \le 1$$

For the uniform grid, substituting $\delta x_e = \delta x_w = \Delta x_P = \Delta x$ and $\delta y_n = \delta y_s = \Delta y_P = \Delta y$ in the above equation, the stability criterion is given finally as

$$\left(\frac{|u|\,\Delta t}{\Delta x} + \frac{|v|\,\Delta t}{\Delta y}\right) + 2\alpha \Delta t \left(\frac{1}{\Delta x^2} + \frac{1}{\Delta y^2}\right) \le 1 \qquad (7.16)$$

Thus, the limiting/stable time step is *proposed* here as

$$\Delta t_{stab} = \left[\left(\frac{|u|\,\Delta t}{\Delta x} + \frac{|v|\,\Delta t}{\Delta y}\right) + 2\alpha \Delta t\left(\frac{1}{\Delta x^2} + \frac{1}{\Delta y^2}\right)\right]^{-1} \qquad (7.17)$$

The application of above time step is presented above in Program 7.1 for the uniform grid, with $\Delta t = \Delta t_{stab}$.

Note that the final LAE for convection (Eq. 7.12 and Table 7.1) as compared to that for the advection (Eq. 6.23) consists of an extra term (for conduction) in aE, aW, aN, and aS. Thus, the implementation details and solution algorithm presented for advection in the previous chapter can be easily extended to the convection, and therefore not presented here.

Example 7.2: Coefficient of LAEs-based methodology of CFD development and code verification for a 2D unsteady-state heat convection problem.
Solve the problem presented above in Example 7.1 (Fig. 7.6), using the coefficient of LAEs-based approach of CFD development for implicit method on a non-uniform grid. Consider the procedure for the *non-uniform* grid generation presented in Example 5.4.

(a) The physics of the present problem is such that the temperature profile $\theta(Y)$ is symmetric about the axis of the channel and there is a larger temperature gradient near both the walls and near the inlet (which reduces gradually along the length of the channel), shown in Fig. 7.8. Thus, for the accurate computation of the temperature gradients, the equal clustering of grid at both the ends of the domain (using Eq. 5.36) is justified for the non-uniform grid distribution in the $y-$direction, presented in previous chapter in Example 5.4. However, for the distribution in $x-$direction, finest grid near the inlet which gradually becomes coarser toward the outlet is desirable for the accurate results. This is achieved by an algebraic method, using an equation given (Hoffmann and Chiang 2000) as

$$x = L_1 \frac{(\beta + 1) - (\beta - 1) \times [(\beta + 1) / (\beta - 1)]^{(1-\xi)}}{1 + [(\beta + 1) / (\beta - 1)]^{(1-\xi)}} \qquad (7.18)$$

Generate a non-uniform 2D Cartesian grid, using Eq. 5.36 for the equal clustering at $Y = 0$ and $Y = 1$; and using Eq. 7.18 for clustering at $X = 0$, along with a gradual stretching. For both the types of clustering, consider $\beta = 1.2$ and a grid size of 22×12. Show the non-uniform grid in the computational domain.

(b) Using the coefficient of LAEs-based approach, develop a *non-dimensional* computer program for the implicit method on the non-uniform grid.

(c) For a Reynolds number $Re \equiv u_\infty L_2 / \nu = 100$ and a Prandtl number $Pr = 0.7$, present a CFD application of the code for the FOU, SOU, and QUICK schemes on the non-uniform grid size of 22×12; using the steady state as well as iteration convergence tolerance as $\epsilon_{st} = \epsilon = 10^{-4}$. Plot the steady-state temperature profiles $\theta(Y)$ at various axial locations $X = 1, 2, 3$, and 5. Compare the profiles with the exact solution (Burmeister 1993).

Solution:

(a) Using the solution algorithm, a computer program for the non-uniform grid generation is presented in Program 7.2.

Listing 7.2 Scilab code for non-uniform 2D Cartesian grid generation, using an algebraic method, for slug flow in a plane channel.

```
1   L1=20.0;L2=1.0;  imax=22;jmax=12;  Beta=1.2;  //
       STEP-1
2   xi=linspace(0,1,imax-1);zeta=linspace(0,1,jmax
       -1)  // STEP-2
3   Beta_p1=Beta+1;Beta_m1=Beta-1;Beta_p1_div_m1=
       Beta_p1/Beta_m1;
4   temp_x=Beta_p1_div_m1^(1-xi);
5   num_x=Beta_p1-(Beta_m1*temp_x);den_x=1+temp_x;
6   temp_y=Beta_p1_div_m1^(2*zeta-1);
7   num_y=(Beta_p1*temp_y)-Beta_m1;den_y=2*(1+temp_y
       );
8   x=L1*num_x./den_x;y=L2*num_y./den_y;  // STEP-3 [
       X
9   Y]=meshgrid(x,y);Z=zeros(jmax-1,imax-1);
10  xc(2:imax-1)=(x(2:imax-1)+x(1:imax-2))/2;  //STEP
       -4
11  yc(2:jmax-1)=(y(2:jmax-1)+y(1:jmax-2))/2;  xc(1)
       =0;xc(imax)=L1;
12  yc(1)=0;yc(jmax)=L2;  [Xc Yc]=meshgrid(xc,yc);Zc=
       zeros(jmax,imax);
13  xset('window',1);  surf(X,Y,Z);  plot(Xc,Yc,'bo')
       // plotting
14  dx(1:imax-1)=xc(2:imax)-xc(1:imax-1);  // STEP-5
15  dy(1:jmax-1)=yc(2:jmax)-yc(1:jmax-1);
16  Dx(2:imax-1)=x(2:imax-1)-x(1:imax-2);Dx(1)=0;Dx(
       imax)=0;
17  Dy(2:jmax-1)=y(2:jmax-1)-y(1:jmax-2);Dy(1)=0;Dy(
       jmax)=0;
```

(b) The program on coefficient of LAEs-based implementation for 2D unsteady heat advection, presented in the previous chapter in Program 6.2, is extended here for the present problem on convection. The modification needed for the *user input and the BCs* is already presented in Program 7.1, as step-1 and step-3, respectively. Consider a constant non-dimensional *time step* $\Delta \tau = 2$ for the implicit method, much larger than the time step needed for the explicit method. Further modification is for *grid*; replace line no. 9–18 in Program 6.2, with Program 7.2. The modifications (in Program 6.2) also correspond to line no. 58 as "aE(i,j)=(-mx_m+k/(cp*dx(i)))*Dy(j)", line no. 63 as "aN(i,j)=(-my_m + k/(cp*dy(j)))*Dx(i)", and the *outlet BC* (Line No. 82 and 88) as "T(imax,1:jmax)=T(imax-1,1:jmax)".

(c) The program is applied for the present problem and the non-uniform grid as well as the velocity profile is shown in Fig. 7.9. Figure 7.9a shows the non-uniform grid, with clustering of grids near the walls and near the inlet of the channel. Comparing the velocity profile for the non-uniform grid (Fig. 7.9b–d) and the uniform grid (Fig. 7.8b–d), a much better agreement between the present computational and exact solution is seen for the non-uniform grid here as compared to the earlier uniform grid (Fig. 7.8).

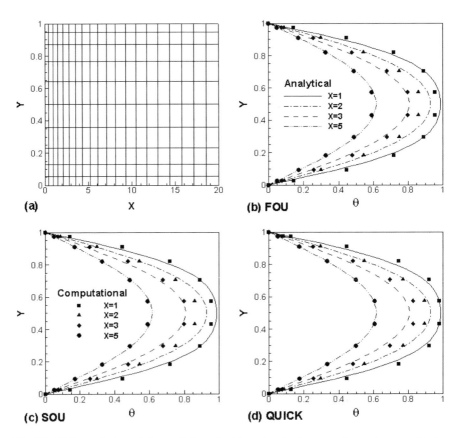

Fig. 7.9 CFD application of the implicit-method-based unsteady convection code on a grid size of 22 × 12, for the slug flow in a channel at $Re = 100$ and $Pr = 0.7$: (**a**) non-uniform grid in the domain, and (**b**)–(**d**) comparison of the present computational results with the analytical solution for steady-state *non-dimensional* temperature profiles along the vertical centerline, at different axial locations, using the various advection schemes

Problems

7.1 Consider the example problem on 2D heat convection, *Example* 7.1, and modify the code (Program 7.1) to solve the problem on a non-uniform grid (Program 7.2).

7.2 Consider the example problem on 2D heat convection, *Example* 7.2, and develop a code for the explicit method and the coefficient of LAEs-based solution methodology. Present a CFD application of the code, with the results similar to the example problem.

7.3 Solve the problems, (a) *Example* 7.1 and (b) *Example* 7.2, using a fully developed flow field ($U = 6Y(1 - Y)$ and $V = 0$) instead of the slug flow in the plane channel. The problem corresponds to hydro-dynamically fully developed and thermally developing flow, with the channel maintained at the <u>C</u>onstant <u>W</u>all <u>T</u>emperature (CWT). Derive the expression for the *local Nusselt number* as $Nu_{L,sb} = \partial\theta/\partial Y$ at the south wall, and $Nu_{L,nb} = -\partial\theta/\partial Y$ at the north wall of the channel. Thereafter, incorporate the implementation of the second-order numerical differentiation (Sect. 4.3) in the code to compute the variation of the local Nusselt number at the south and north boundary grid points. Plot as well as discuss the *steady-state* Nu_L at the various walls of the channel. Also incorporate the trapezoidal rule (Sect. 4.4) in the code and compute the space-averaged Nusselt number at each of the walls.

7.4 Repeat the previous problem for the <u>U</u>niform <u>H</u>eat <u>F</u>lux (UHF) BC at the channel walls, with the non-dimensional temperature defined as $\theta = (T - T_\infty)/(q_W L_2/k)$. Thus, as compared to the thermal BCs for the CWT case in Fig. 7.6, the non-dimensional BCs for the UHF case is $\theta = 0$ at the west/inlet boundary, $\partial\theta/\partial Y = -1$ at the south boundary, $\partial\theta/\partial Y = 1$ at the north boundary, and $\partial\theta/\partial X = 0$ at the east/outlet boundary of the domain. Derive the expression for the BCs at the north and south boundaries ($\partial\theta/\partial Y = \pm 1$). Also derive the local Nusselt number as $Nu_{L,sb} = 1/\theta_{sb}$ at the south boundary, and $Nu_{L,nb} = 1/\theta_{nb}$ at the north boundary, for the computation of the local and average Nusselt number.

References

1. Hoffmann, K. A. and Chiang, S. T. (2000). *Computational Fluid Dynamics* (Vols. 1, 2 and 3). Kansas: Engineering Education System.
2. Burmeister, L. C. (1993). *Convective heat transfer* (2nd ed.). New York: Wiley.

Chapter 8
Computational Fluid Dynamics: Physical Law-Based Finite Volume Method

After the $\underline{C}$omputational $\underline{H}$eat $\underline{T}$ransfer (CHT) in the last three chapters, computational fluid dynamics is presented in this chapter and the next two chapters; for a Cartesian geometry problem. Note that the *CFD* here includes heat transfer along with the fluid dynamics, i.e., combined *mass, momentum and energy transport phenomenon*. Thus, as compared to the last two chapters on the unsteady heat advection/-convection with a known steady state velocity field, the unsteady velocity (along with the temperature) field is considered as unknown for this as well as later chapters on CFD. Furthermore, as compared to only energy conservation law for the CHT in the last three chapters, the mass and momentum conservation law are also considered for the CFD. Since the underlying molecular motion-based diffusion mechanism as well as bulk motion-based advection mechanism governs the momentum transport along with the energy transport, the earlier presented CHT development methodology for the heat transfer is generalized and extended here to include fluid dynamics.

The mind map for this chapter is shown in Fig. 8.1. The figure shows that this chapter starts with a discussion on the generalized variables for the combined heat and fluid flow followed by the presentation of the physical law-based FVM for CFD (consisting of conservation laws, algebraic formulation, approximations, and approximated algebraic formulation), and ending with a discussion.

8.1 Generalized Variables for the Combined Heat and Fluid Flow

The various volumetric and flux terms, encountered during the physical law-based algebraic formulation, was presented earlier for the energy transport. Now, for the combined momentum and energy transport here, the conduction heat transfer rate Q is generalized as diffusion D_ϕ (with $D_T = Q$), and the enthalpy flow rate H is

© The Author(s), under exclusive license to Springer Nature Switzerland AG 2022 221
A. Sharma, *Introduction to Computational Fluid Dynamics*,
https://doi.org/10.1007/978-3-030-72884-7_8

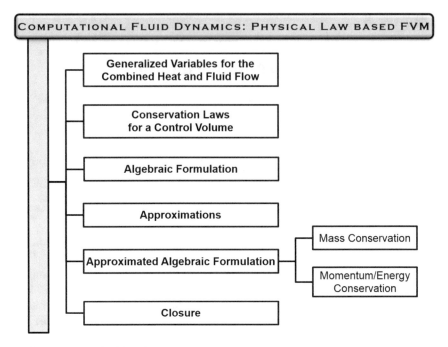

Fig. 8.1 Mind map for Chap. 8

generalized as advection A_ϕ (with $A_T = H$). Similarly, the conduction flux $\overrightarrow{q}$ is generalized as diffusion flux $\overrightarrow{d}_\phi$ (with $\overrightarrow{d}_T = -\overrightarrow{q}$), and the enthalpy flux $\overrightarrow{h}$ as advection flux $\overrightarrow{a}_\phi$ (with $\overrightarrow{a}_T = \overrightarrow{h}$). Other than the flux term, the volumetric term enthalpy H for the heat transfer is generalized here as V_ϕ (with $V_T = H$).

Note that the purpose of the generalization is to create a common framework for the algebraic formulation as well as CFD development, from the energy and momentum conservation law; as both the laws are driven by the diffusion and advection transport process. Furthermore, the purpose is to use the unsteady code developed, for the *net* mean (over the time interval Δt) conduction heat transfer rate *entering* a CV $Q^{mean}_{cond,P}$, as a generic subroutine to compute $D^{mean}_{\phi,P}$ where $D^{mean}_{T,P} = Q^{mean}_{cond,P}$. The subroutine is also used to compute a fluid flow variable which is also based on the diffusion transport mechanism; such as, *net* mean viscous force—$D^{mean}_{u,P}$ in the $x-direction$ and $D^{mean}_{v,P}$ $y-direction$ for a CV. Similarly, the purpose is to create a subroutine from the already developed code on the *net* mean enthalpy flow rate *leaving* a CV $Q^{mean}_{adv,P}$, and compute a fluid flow variable which is also based on the advection transport mechanism; such as, *net* mean $x-$momentum flow rate $A^{mean}_{u,P}$ and $y-$momentum flow rate $A^{mean}_{v,P}$ *leaving* a CV.

Since the generalization now encounters momentum which is a vector (as compared to the energy as the scalar), the flux terms are second order tensors for the momentum conservation as compared to the fluxes as the vectors in the energy conservation. However, the second-order tensors/fluxes, corresponding to momentum flux $[\vec{m}\,\vec{u}\,]$ and viscous stress $[\sigma]$, are converted into vector by taking their dot product with the unit vector $\hat{i}$ and $\hat{j}$; as there is a componentwise (in the $x-$direction and $y-$direction for a 2D case) application of the momentum conservation law for a CV. Thus, for the $x-$momentum conservation, the advection flux is given as $\vec{a}_u = [\vec{m}\,\vec{u}\,].\hat{i}$ and the diffusion flux is given as $\vec{d}_u = [\sigma].\hat{i}$. The respective flux (multiplied by the associated surface area) results in $x-$momentum flow rate A_u and viscous force in the $x-$direction D_u; at the various faces of a CV. Similarly, for the $y-$momentum conservation law, the fluxes $\vec{a}_v = [\vec{m}\,\vec{u}\,].\hat{j}$ results in $y-$momentum flow rate A_v and $\vec{d}_v = [\sigma].\hat{j}$ results in viscous force in the $y-$direction D_v.

Thus, the generalization results in the advection A_ϕ and diffusion D_ϕ at the various faces of a CV, with the respective flux term as $\vec{a}_\phi$ and $\vec{d}_\phi$, where $\phi = u$, v, and T for the application of $x-$momentum, $y-$momentum, and energy conservation law on a CV, respectively. The generalization of the variables from the CHT to CFD is shown in Table 8.1. Furthermore, a generic diffusion coefficient Γ_ϕ and a constant C are defined, with $\Gamma_\phi = \mu$ and $C = 1$ for the momentum conservation, and $\Gamma_\phi = k$ and $C = c_p$ for the energy conservation.

The generic representation ensures that the CFD development methodology in Chap. 5 for heat conduction is also applicable to the computation of net viscous force on a CV. Furthermore, the methodology for heat advection (corresponding to the net enthalpy flow rate leaving a CV) in Chap. 6 is also applicable to computation of net x/y momentum flow rate out of the CV. Thus, the developed and well-tested code of conduction and advection heat transfer—presented in the previous chapters—can be converted into generic subroutines for the heat and fluid flow. During the development of a CFD software as a product, the subroutines are the various components. The modular/component-by-component development as well as testing leads to the successful development of our product—the CFD software.

The already presented physical law, algebraic formulation, two approximations, and the approximated algebraic formulation for the energy transport (in the previous chapter) is generalized here for a coupled mass, momentum and energy transport process; presented below in separate sections.

8.2 Conservation Laws for a Control Volume

Since the velocity and temperature field ϕ_P^t (at a certain time instant t) are used to determine the flow properties after a certain time interval Δt in CFD $\phi_P^{t+\Delta t}$, the algebraic formulation in the FVM starts with the application of the physical laws on a CV over a time interval Δt; not the laws at certain time instant t. Figure 8.2 presents the two forms of the physical laws: instantaneous t and over a time interval Δt; the former used for the differential formulation in a course in fluid dynamics, and the later used for the algebraic formulation here in CFD. The law over a time

Table 8.1 Volumetric and flux terms for a generalized representation in a combined heat transfer and fluid dynamics problem. Here, M is the mass flow rate, $\vec{m}$ is the mass flux, $\mathcal{M}$ is the mass, and $\mathcal{H}$ is the enthalpy

A_ϕ	Advection flow rate at the various faces	D_ϕ	Diffusion at the various faces
$A_T = H = Mc_p T$	Enthalpy flow rate	$D_T = Q$	Conduction heat transfer rate
$A_u = Mu$	$x-$momentum flow rate	D_u	Viscous force in the $x-$direction
$A_v = Mv$	$y-$momentum flow rate	D_v	Viscous force in the $y-$direction
$\vec{a}_\phi$	Advection Flux at the various faces	$\vec{d}_\phi$	Diffusion Flux at the various faces
$\vec{a}_T = \vec{h}$	Enthalpy flux	$\vec{d}_T = \vec{q}$	Conduction flux
$\vec{a}_u = [\vec{m}\,\vec{u}] \cdot \hat{i}$	$x-$momentum flux	$\vec{d}_u = [\sigma] \cdot \hat{i}$	Viscous stress in the $x-$direction
$\vec{a}_v = [\vec{m}\,\vec{u}] \cdot \hat{j}$	$y-$momentum flux	$\vec{d}_v = [\sigma] \cdot \hat{j}$	Viscous stress in the $y-$direction
$A_{\phi,P}$	Net Advection for a representative CV P	$D_{\phi,P}$	Net Diffusion for a representative CV P
$A_{T,P} = Q_{adv,P}$	Net enthalpy flow rate *leaving* a CV	$D_{T,P} = Q_{cond,P}$	Net conduction rate *entering* a CV
$A_{u,P}$	Net $x-$momentum flow rate *leaving* a CV	$D_{u,P}$	Net viscous force in the $x-$direction
$A_{v,P}$	Net $y-$momentum flow rate *leaving* a CV	$D_{v,P}$	Net viscous force in the $y-$direction
$\overline{\mathcal{V}}_\phi$	Volumetric Term	S_ϕ	Net Source Term
$\overline{\mathcal{V}}_T = \overline{\mathcal{H}} = \overline{\mathcal{M}}c_p T$	Volumetric enthalpy	$S_T = Q_{gen}$	Total heat generation
$\overline{\mathcal{V}}_u = \overline{\mathcal{M}}u = \rho u$	Volumetric $x-$momentum	S_u	Net pressure force in the $x-$direction
$\overline{\mathcal{V}}_v = \overline{\mathcal{M}}v = \rho v$	Volumetric $y-$momentum	S_v	Net pressure force in the $y-$direction

interval Δt, presented in the previous chapter on heat convection, is generalized here for a CV subjected to the advection and diffusion transport phenomenon (Fig. 8.2b); presented separately for the momentum/vector (componentwise) and energy/scalar conservation laws as

---

x- **Momentum Conservation Law for a CV, over a time interval** Δt:

The amount of increase of $x-$momentum $\Delta(\mathcal{M}u)$ stored within the CV, plus the net amount of advected $x-$momentum that leaves the CV $\left(A_{u,out}^{mean} - A_{u,in}^{mean}\right)\Delta t$ must equal the net *impulse* acting in the $x-$direction; given as

---

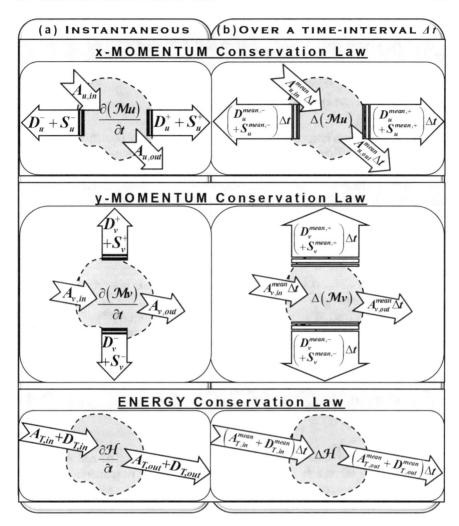

Fig. 8.2 Two different forms of conservation of x-momentum, y-momentum, and energy for a CV, subjected to molecular motion-based diffusion and bulk motion-based advection phenomenon. A corresponds to the advection flow rate—A_u, A_v, and A_T as the x−momentum, y−momentum, and enthalpy flow rate, respectively. D and S correspond to forces—D_u and D_v as the net viscous force, and S_u and S_v as the net pressure force, in the $x−$ and $y−$direction, respectively. D_T corresponds to net conduction heat transfer rate. Figures in the first column, on instantaneous law, lead to a differential formulation; and in the second column, on the laws over a discrete time interval Δt, leads to an algebraic formulation

$$\Delta(\mathcal{M}u) + \left(A_{u,out}^{mean} - A_{u,in}^{mean}\right)\Delta t = \left(D_u^{mean,+} - D_u^{mean,-}\right)\Delta t$$
$$+ \left(S_u^{mean,+} - S_u^{mean,-}\right)\Delta t$$
$$\text{where } (A/D/S)_u^{mean} = \frac{1}{\Delta t}\int_t^{t+\Delta t}(A/D/S)_u\,dt, \tag{8.1}$$
$$\Delta(\mathcal{M}u) = \int_t^{t+\Delta t}\frac{\partial(\mathcal{M}u)}{\partial t}dt$$

where, over the time interval Δt, A_u^{mean} is the mean $x-$momentum flow rate; and $A_{u,out}^{mean}\Delta t$ and $A_{u,in}^{mean}\Delta t$ are the $x-$momentum leaving and entering the CV over the time interval, respectively. Furthermore, D_u^{mean} and S_u^{mean} are the mean viscous force and pressure force in the $x-$direction, respectively. Impulse (defined as integral of a force over a time interval) corresponding to the respective forces are $D_u^{mean,+}\Delta t$ and $S_u^{mean,+}\Delta t$ in the positive $x-$direction, and $D_u^{mean,-}\Delta t$ and $S_u^{mean,-}\Delta t$ in the negative $x-$direction.

y- **Momentum Conservation Law for a CV, over a time interval** Δt: The amount of increase of $y-$momentum $\Delta(\mathcal{M}v)$ stored within the CV, plus the net amount of advected $y-$momentum that leaves the CV $\left(A_{v,out}^{mean} - A_{v,in}^{mean}\right)\Delta t$ must equal the net *impulse* acting in the $y-$direction; given as

$$\Delta(\mathcal{M}v) + \left(A_{v,out}^{mean} - A_{v,in}^{mean}\right)\Delta t = \left(D_v^{mean,+} - D_v^{mean,-}\right)\Delta t$$
$$+ \left(S_v^{mean,+} - S_v^{mean,-}\right)\Delta t$$
$$\text{where } (A/D/S)_v^{mean} = \frac{1}{\Delta t}\int_t^{t+\Delta t}(A/D/S)_v\,dt, \tag{8.2}$$
$$\Delta(\mathcal{M}v) = \int_t^{t+\Delta t}\frac{\partial(\mathcal{M}v)}{\partial t}dt$$

where A_v^{mean} is the mean $y-$momentum flow rate; and $A_{v,out}^{mean}\Delta t$ and $A_{v,in}^{mean}\Delta t$ are the $y-$momentum leaving and entering the CV over the time interval, respectively. Furthermore, over the time interval, $D_v^{mean,+}\Delta t$ is the viscous impulse and $S_v^{mean,+}\Delta t$ is pressure impulse in the positive $y-$direction; and $D_v^{mean,-}\Delta t$ and $S_v^{mean,-}\Delta t$ are the respective impulses in the negative $y-$direction.

Energy Conservation Law for a CV, over a time interval Δt: The amount of increase of enthalpy $\Delta\mathcal{H}$ stored within the CV plus the net amount of advected enthalpy that leaves the CV $\left(A_{T,out}^{mean} - A_{T,in}^{mean}\right)\Delta t$, must equal the net amount of conducted thermal energy that enters the CV $\left(D_{T,in}^{mean} - D_{T,out}^{mean}\right)\Delta t$ plus the amount of thermal energy that is generated within the CV $S_T^{mean}\Delta t$; given as

$$\Delta\mathcal{H} + \left(A_{T,out}^{mean} - A_{T,in}^{mean}\right)\Delta t = \left(D_{T,in}^{mean} - D_{T,out}^{mean}\right)\Delta t + S_T^{mean}\Delta t$$
$$\text{where } (A/D/S)_T^{mean} = \frac{1}{\Delta t}\int_t^{t+\Delta t}(A/D/S)_T\,dt, \tag{8.3}$$
$$\Delta\mathcal{H} = \int_t^{t+\Delta t}\frac{\partial\mathcal{H}}{\partial t}dt$$

where A_T^{mean} is the mean enthalpy flow rate, and D_T^{mean} is the mean conduction heat transfer rate. Furthermore, over the time interval, $A_{T,out}^{mean} \Delta t$ is the enthalpy and $D_{T,out}^{mean} \Delta t$ is the conduction heat transfer leaving the CV; and $A_{T,in}^{mean} \Delta t$ and $D_{T,in}^{mean} \Delta t$ are the respective thermal energy entering a CV.

where the superscript *mean* represents the time-averaged value over the time interval Δt; + and − represents the viscous/pressure force in the positive and negative x/y direction, respectively. The above conservation statements correspond to negligible body force and negligible viscous dissipation.

Finally, the mass-conservation law for a CV is given as

Mass-Conservation Law for a CV, over a time interval Δt: For an incompressible flow over the time interval, the net mass that leaves the CV $\left(M_{out}^{mean} - M_{in}^{mean}\right) \Delta t$ must be equal zero; given as

$$S_{m,P}^{mean} \Delta t = \left(M_{out}^{mean} - M_{in}^{mean}\right) \Delta t = 0 \quad \left\{ M^{mean} = \frac{1}{\Delta t} \int_t^{t+\Delta t} M dt \right\} \quad (8.4)$$

where $S_{m,P}^{mean} \Delta t$ is the mass source. Furthermore, over the Δt, M^{mean} is the mean mass flow rate; and $M_{out}^{mean} \Delta t$ is the mass leaving and $M_{in}^{mean} \Delta t$ is the mass entering the CV.

Since the *present book is on incompressible CFD*, note that the above law is presented for an incompressible flow.

8.3 Algebraic Formulation

Similar to the last three chapters, the spatial variation of the general volumetric term $\mathcal{V}_\phi$ is considered as the volumetric phenomenon, and advection term A_ϕ and diffusion term D_ϕ (at the various surfaces) as the surface phenomenon. For a 2D Cartesian CV, the advection and diffusion term are represented in terms of fluxes, shown in Fig. 8.3. The figure shows that the direction of arrows—at the surfaces of a CV—is normal to the surfaces. The sense of the arrows are seen as outward for a positive (north/east) surfaces, and inward for a negative (south/west) surfaces. The sense of the arrows correspond to the balance statement for all the fluxes encountered in the scalar (energy and mass) conservation law. Whereas, for the fluxes (momentum flux and viscous stress) in the Cartesian componentwise application of the vector/momentum conservation law, the direction of the arrows on the faces of the CV are all horizontal for the x−momentum conservation, and are all vertical for the y−momentum conservation law; not shown here and will be shown below.

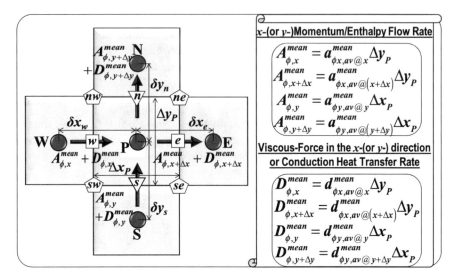

Fig. 8.3 Application of law of conservation of momentum/energy, over a time interval Δt, for a representative internal CV in a 2D Cartesian computational domain; with a non-uniform grid

Considering the sense of the arrows (discussed above), the time interval-based momentum and energy conservation laws for a CV (Eqs. 8.1–8.3), and the generalization of the variables encountered for CHT to CFD (Table 8.1), a generic algebraic formulation is finally obtained for a 2D Cartesian CV (Fig. 8.3) as

$$\frac{\left(\overline{\mathcal{V}}^{t+\Delta t}_{\phi,av@vol} - \overline{\mathcal{V}}^{t}_{\phi,av@vol}\right)\Delta x_P \Delta y_P}{\Delta t} + A^{mean}_{\phi,net} = D^{mean}_{\phi,net} + S^{mean}_{\phi}$$

$$\text{where } A^{mean}_{\phi,net} = \left(a^{mean}_{\phi x,av@(x+\Delta x)} - a^{mean}_{\phi x,av@x}\right)\Delta y_P$$

$$+ \left(a^{mean}_{\phi y,av@(y+\Delta y)} - a^{mean}_{\phi y,av@y}\right)\Delta x_P$$

$$D^{mean}_{\phi,net} = \left(d^{mean}_{\phi x,av@(x+\Delta x)} - d^{mean}_{\phi x,av@x}\right)\Delta y_P + \left(d^{mean}_{\phi y,av@(y+\Delta y)} - d^{mean}_{\phi y,av@y}\right)\Delta x_P$$

$$S^{mean}_{u} = \left(p^{mean}_{av@x} - p^{mean}_{av@(x+\Delta x)}\right)\Delta y_P, \; S^{mean}_{v} = \left(p^{mean}_{av@y} - p^{mean}_{av@(y+\Delta y)}\right)\Delta x_P,$$

$$S^{mean}_{T} = \overline{Q}_{gen,av@vol}\Delta x_P \Delta y_P, \; \left(\overline{\mathcal{V}/Q}_{gen}\right)_{av} = \frac{1}{\Delta V_P}\int_{\Delta V_P}\left(\overline{\mathcal{V}/Q}_{gen}\right)dV$$

$$\text{and } f^{mean}_{x/y,av@face} = \frac{1}{\Delta S}\int_{\Delta S} f^{mean}_{x/y,face}dS$$

$$(8.5)$$

where the subscript av represents the space-averaged (within the volume/surface area) value, and the superscript *mean* represents the time-averaged (over the time interval Δt) value. Furthermore, the subscript $av@face$ represents the average value of advection and diffusion fluxes at the various surfaces ($face = x$, y, $x + \Delta x$ and $y + \Delta y$), shown in Fig. 8.3; and $av@vol$ represents the average value of the volumetric term $\overline{\mathcal{V}}_{\phi}$ inside the volume of the CV.

Similarly, the time interval-based mass-conservation law for a CV (Eq. 8.4) results in an algebraic formulation as

$$S_{m,P}^{mean} = \left(m_{x,av@(x+\Delta x)}^{mean} - m_{x,av@x}^{mean}\right)\Delta y_P + \left(m_{y,av@(y+\Delta y)}^{mean} - m_{y,av@y}^{mean}\right)\Delta x_P = 0$$
(8.6)

where m^{mean} is the mean mass flux, over the time interval Δt, at the various surfaces of a 2D Cartesian CV.

8.4 Approximations

The two approximations presented in previous chapter for volumetric (enthalpy) and flux (enthalpy flux and conduction flux) terms in the energy conservation is also applicable for the momentum conservation law. For a 2D Cartesian CFD problem, there are four volumetric terms $\overline{\mathcal{V}}_\phi$ (Table 8.1): first, volumetric enthalpy $\overline{\mathcal{H}}$; second, volumetric x–momentum $\overline{\mathcal{M}u}$; third, volumetric y–momentum $\overline{\mathcal{M}v}$; and fourth, volumetric heat generation $\overrightarrow{\mathcal{Q}}_{gen}$. The fluxes (represented by $\overrightarrow{f}$) encountered in the mass, momentum and energy transport are also of four different types: first, mass flux $\overrightarrow{m}$; second, advection flux $\overrightarrow{a}_\phi$; third, diffusion flux $\overrightarrow{d}_\phi$ and fourth, pressure p. Thus, for the volumetric and the flux term, the generalization of the I–approximation (presented in the previous chapters for CHT) is given for CFD here as

$$\begin{aligned} \text{Volumetric term}: \quad & \overline{\mathcal{V}}_{\phi,av@vol} \approx \overline{\mathcal{V}}_{\phi,P} \\ \text{Flux Term}: \quad & f_{x/y,av@face}^{mean} \approx f_{x/y,f}^{\chi} \end{aligned}$$
(8.7)

where $face = x, y, x + \Delta x$, and $y + \Delta y$ and the centroid of the respective faces are $f = e, w, n$, and s; and $\chi = n, n + 1$ and $n + 1/2$ for the explicit, implicit and Crank-Nicolson method, respectively. The I–approximation for the four volumetric and the four flux terms is presented in Table 8.2. For the temporal variation over the time interval Δt, the I–approximation for the *mean* value is seen in the table as *implicit* for the mass flux $m_{x/y,f}^{n+1}$ and pressure p_f^{n+1}, and either explicit, implicit or Crank-Nicolson method for the advection flux $a_{\phi x/y,f}^{\chi}$ and diffusion flux $d_{\phi x/y,f}^{\chi}$. However, note that both implicit and explicit approximation of the mass flux are used; $m_{x/y,f}^{n+1}$ for the mass conservation and $m_{x/y,f}^{n}$ for the advection term $a_{\phi x/y,f}^{\chi}$ (seen in the Table 8.2). The explicit treatment for the mass flux is to avoid the non-linearity (for $a_{\phi x/y,f}^{\chi}$, due to the product of two unknown quantities: $m_{x/y,f}^{n+1}$ and ϕ_f^{n+1}) in the algebraic formulation for the momentum and energy equations.

Similarly, for the II–approximation, the central difference scheme (earlier for the conduction flux) is generalized here for the diffusion flux; and the advection scheme (for the enthalpy flux) is generalized for the advection flux. The generalized II–approximations are given as

Table 8.2 The approximations for the various terms in the momentum conservation law as a generalization of the similar approximation for the energy conservation law (presented in last three chapters). The volumetric term $\overline{\mathcal{V}}_u$ is $x-$momentum, $\overline{\mathcal{V}}_v$ is $y-$momentum and $\overline{\mathcal{V}}_T$ is enthalpy per unit volume

Terms ↓			First and Second Approximation
Vol.	x-Mom.	I	$\overline{\mathcal{V}}_{u,av@vol} = \left(\mathcal{M}u\right)_{av@vol} \approx \left(\mathcal{M}u\right)_P$
	y-Mom.		$\overline{\mathcal{V}}_{v,av@vol} = \left(\mathcal{M}v\right)_{av@vol} \approx \left(\mathcal{M}v\right)_P$
	Energy		$\overline{\mathcal{V}}_{T,av@vol} = \overline{\mathcal{H}}_{av@vol} \approx \overline{\mathcal{H}}_P$
Mass Flux		I	$m^{mean}_{x,av@(x+\Delta x)} \approx m^{n+1}_{x,e} \approx$ ρu^{n+1}_e, $m^{mean}_{x,av@x} \approx \rho u^{n+1}_w$
			$m^{mean}_{y,av@(y+\Delta y)} \approx m^{n+1}_{y,n} \approx$ ρv^{n+1}_n, $m^{mean}_{y,av@y} \approx \rho v^{n+1}_s$
		II	$u^{n+1}_e \approx \overline{u^{n+1}_E}, u^{n+1}_P, u^{n+1}_w \approx \overline{u^{n+1}_P}, u^{n+1}_W$
			$v^{n+1}_n \approx \overline{v^{n+1}_N}, v^{n+1}_P, v^{n+1}_s \approx \overline{v^{n+1}_P}, v^{n+1}_S$
Advection Flux		I	$a^{mean}_{\phi x,av@(x+\Delta x)} \approx a^{\chi}_{\phi x,e} \approx$ $m^n_{x,e} C\phi^{\chi}_e, a^{mean}_{\phi x,av@x} \approx m^n_{x,w} C\phi^{\chi}_w$
			$a^{mean}_{\phi y,av@(y+\Delta y)} \approx a^{\chi}_{\phi y,n} \approx$ $m^n_{y,n} C\phi^{\chi}_n, a^{mean}_{\phi y,av@y} \approx m^n_{y,s} C\phi^{\chi}_s$
		II	$a^{\chi}_{\phi x,e} \approx C\left(m^{+,n}_{x,e}\phi^{+,\chi}_e + m^{-,n}_{x,e}\phi^{-,\chi}_e\right)$
			$a^{\chi}_{\phi x,w} \approx C\left(m^{+,n}_{x,w}\phi^{+,\chi}_w + m^{-,n}_{x,w}\phi^{-,\chi}_w\right)$
			$a^{\chi}_{\phi y,n} \approx C\left(m^{+,n}_{y,n}\phi^{+,\chi}_n + m^{-,n}_{y,n}\phi^{-,\chi}_n\right)$
			$a^{\chi}_{\phi y,s} \approx C\left(m^{+,n}_{y,s}\phi^{+,\chi}_s + m^{-,n}_{y,s}\phi^{-,\chi}_s\right)$
Diffusion Flux		I	$d^{mean}_{\phi x,av@(x+\Delta x)} \approx d^{\chi}_{\phi x,e}, d^{mean}_{\phi x,av@x} \approx d^{\chi}_{\phi x,w}$
			$d^{mean}_{\phi y,av@(y+\Delta y)} \approx d^{\chi}_{\phi y,n}, d^{mean}_{\phi y,av@y} \approx d^{\chi}_{\phi x,s}$
		II	$d^{\chi}_{\phi x,e} \approx \Gamma_\phi\left(\phi^{\chi}_E - \phi^{\chi}_P\right)/\delta x_e, d^{\chi}_{\phi x,w} \approx$ $\Gamma_\phi\left(\phi^{\chi}_P - \phi^{\chi}_W\right)/\delta x_w,$
			$d^{\chi}_{\phi y,n} \approx \Gamma_\phi\left(\phi^{\chi}_N - \phi^{\chi}_P\right)/\delta y_n, d^{\chi}_{\phi y,s} \approx$ $\Gamma_\phi\left(\phi^{\chi}_P - \phi^{\chi}_S\right)/\delta y_s$
Source	x-Mom.	I	$p^{mean}_{av@(x+\Delta x)} \approx p^{n+1}_e, p^{mean}_{av@x} \approx p^{n+1}_w$
	(Flux)	II	$p^{n+1}_e \approx \overline{p^{n+1}_E}, p^{n+1}_P, p^{n+1}_w \approx \overline{p^{n+1}_P}, p^{n+1}_W$
	y-Mom.	I	$p^{mean}_{av@(y+\Delta y)} \approx p^{n+1}_n \Delta x_P, p^{mean}_{av@y} \approx p^{n+1}_s$
	(Flux)	II	$p^{n+1}_n \approx \overline{p^{n+1}_N}, p^{n+1}_P, p^{n+1}_s \approx \overline{p^{n+1}_P}, p^{n+1}_S$
	Energy	I	$\overline{Q}_{gen,av@vol} \approx \overline{Q}_{gen,P}$
	(Vol.)		

$$d_{\phi x,e}^\chi \approx \Gamma_\phi \frac{\phi_E^\chi - \phi_P^\chi}{\delta x_e}, \; d_{\phi x,w}^\chi \approx \Gamma_\phi \frac{\phi_P^\chi - \phi_W^\chi}{\delta x_w}$$
$$d_{\phi y,n}^\chi \approx \Gamma_\phi \frac{\phi_N^\chi - \phi_P^\chi}{\delta y_n}, \; d_{\phi y,s}^\chi \approx \Gamma_\phi \frac{\phi_P^\chi - \phi_S^\chi}{\delta y_s} \tag{8.8}$$

$$a_{\phi x,e}^\chi = m_{x,e}^n C\phi_e^\chi, \; a_{\phi x,w}^\chi = m_{x,w}^n C\phi_w^\chi, \; a_{\phi y,n}^\chi = m_{y,n}^n C\phi_n^\chi \text{ and } a_{\phi y,s}^\chi = m_{y,s}^n C\phi_s^\chi$$

$$\text{where } m_{x/y,f}\phi_f = m_{x/y,f}^+\phi_f^+ + m_{x/y,f}^-\phi_f^- \left.\begin{array}{l} f = e, w, n, s \end{array}\right.$$
$$\phi_f^\pm = w_{f,1}^\pm\phi_{f,D}^\pm + w_{f,2}^\pm\phi_{f,U}^\pm + w_{f,3}^\pm\phi_{f,UU}^\pm$$
$$\left.\begin{array}{l} m_{x/y,f}^+ \equiv \max(m_{x/y,f}, 0) = +\left|m_{x/y,f}\right| \text{ if } m_{x/y,f} > 0 \\ m_{x/y,f}^- \equiv \min(m_{x/y,f}, 0) = -\left|m_{x/y,f}\right| \text{ if } m_{x/y,f} < 0 \end{array}\right\} \text{otherwise 0}$$
$$\tag{8.9}$$

where the transport variable $\phi = u$, v, and T for $x-$momentum, $y-$momentum, and energy conservation, respectively. Furthermore, $\phi_f^\pm$ is an advected variable at a face center f, presented in Table 6.3; replace the temperature T with the generalized transport variable ϕ. Also, $w_{f,1}^\pm$, $w_{f,2}^\pm$, and $w_{f,3}^\pm$ are the weights of the downstream, upstream, and upstream-of-upstream neighbor of the a face center f, respectively; presented in Table 6.2.

Other than the advection and diffusion flux, there are two additional fluxes in CFD as compared to CHT: mass flux and pressure. The $II-$approximation for the respective flux term, corresponding to computation of normal velocity and pressure at the various face centers, is a linear interpolation method. The approximation is given for $m_e(= \rho u_e)$ and p_e, at the east face center, as

$$u_e \approx \overline{u_E, u_P}, \text{ and } p_e \approx \overline{p_E, p_P}$$

where the over-bar operator above represents linear interpolation, given as $p_e \approx (\Delta x_P \Delta p_E + \Delta x_E \Delta p_P)/(\Delta x_E + \Delta x_P)$. The $II-$approximation for the four flux terms is also presented in Table 8.2.

8.5 Approximated Algebraic Formulation

8.5.1 Mass Conservation

Using the $I-$approximation for the mass flux, by substituting $f = m$ and $\chi = n + 1$ (implicit method) in Eq. 8.7, Eq. 8.6 for the algebraic formulation (from the time interval-based mass-conservation law) is approximated as

$$S_{m,P}^{mean} \approx \left(m_{x,e}^{n+1} - m_{x,w}^{n+1}\right) \Delta y_P + \left(m_{y,n}^{n+1} - m_{y,s}^{n+1}\right) \Delta x_P = 0$$
$$\Rightarrow \rho\left[\left(u_e^{n+1} - u_w^{n+1}\right) \Delta y_P + \left(v_n^{n+1} - v_s^{n+1}\right) \Delta x_P\right] = 0 \tag{8.10}$$

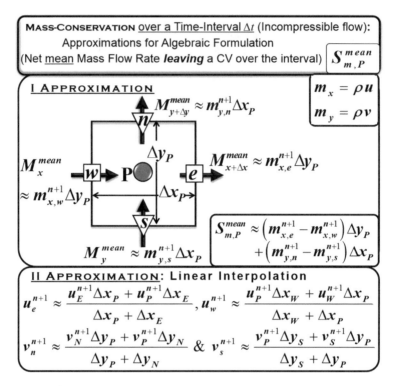

Fig. 8.4 A representative fluid CV to demonstrate two approximations for algebraic representation of the balance statement in the mass-conservation law. Here, $\vec{m}$ is the mass flux, M is the mass flow rate, and $S_{m,P}$ is the mass source

where $n+1$ represents the present time level. Furthermore, the $II-$approximation is needed to compute the normal velocity at a face center; obtained by an *interpolation procedure,* using the velocity of the first neighboring cell center on the either side of the face. The I and II approximation, presented earlier in Table 8.2, is shown in Fig. 8.4. For a *uniform* grid, $\Delta x_P = \Delta x_E = \Delta x_W$ and $\Delta y_P = \Delta y_N = \Delta y_S$, the $II-$approximation is given as

$$u_e^{n+1} = (u_E^{n+1} + u_P^{n+1})/2; \ u_w^{n+1} = (u_W^{n+1} + u_P^{n+1})/2$$
$$v_n^{n+1} = (v_N^{n+1} + v_P^{n+1})/2; \ v_s^{n+1} = (v_S^{n+1} + v_P^{n+1})/2$$

Substituting the above equation for the velocities to Eq. 8.10, the final LAE for the mass conservation on a *uniform* grid is given as

$$\left(u_E^{n+1} - u_W^{n+1}\right)\Delta y + \left(v_N^{n+1} - v_S^{n+1}\right)\Delta x = 0 \tag{8.11}$$

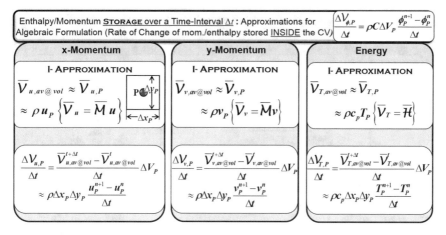

Fig. 8.5 Pictorial representation of the two approximations used in the derivation of governing LAE, for the unsteady term of the law of conservation of momentum and energy

8.5.2 Momentum/Energy Conservation

For the various generic volumetric and flux terms (Table 8.1), encountered during the application of the time interval-based momentum and energy conservation law for a CV, the approximations (presented in Table 8.2) is shown separately for the unsteady, advection, diffusion, and source terms in Figs. 8.5, 8.6 and 8.7; and together for all the terms in Fig. 8.8.

For the momentum conservation, a Cartesian componentwise application of the approximations are shown in Fig. 8.5 for the volumetric term, Fig. 8.6 for the advection flux and diffusion flux term, and Fig. 8.7 for the pressure as the source term; using the equation for the volumetric and the flux terms presented in Table 8.1.

During the application of momentum conservation law to a CV, note that the sense of all the arrows are horizontal for the x−momentum and vertical for the y−momentum conservation. The zig-zag arrows are shown in Fig. 8.6a, for the momentum flux $a_{uy} = m_y u$ and $a_{vx} = m_x v$, where the direction of the normal velocity v and u (respective velocity corresponding to the direction of m_y and m_x) is perpendicular the direction of the advected velocity u and v, respectively. Also note that the expression for the diffusion flux $\vec{d}_u$ and $\vec{d}_v$ in Fig. 8.6b, corresponds to viscous stress for the *incompressible* flow; as discussed in Chap. 3 (Fig. 3.7 and Eq. 3.9).

During the application of energy conservation law for a CV, note that the $Q^{mean}_{cond,P}$ is presented as net in-flux (Chap. 5) and $Q^{mean}_{adv,P}$ as the net out-flux (Chap. 6). However, since the generalized formulation here involves the net out-flux for advection $A^{mean}_{\phi,P}$ as well as diffusion $D^{mean}_{\phi,P}$ in Fig. 8.6, note from the figure $\vec{d}_T \equiv -\vec{q}$ to ensure $D_{T,P} = Q_{cond,P}$; thus, here for the energy transport, *the out-flux of diffusion is equal to in-flux of conduction.*

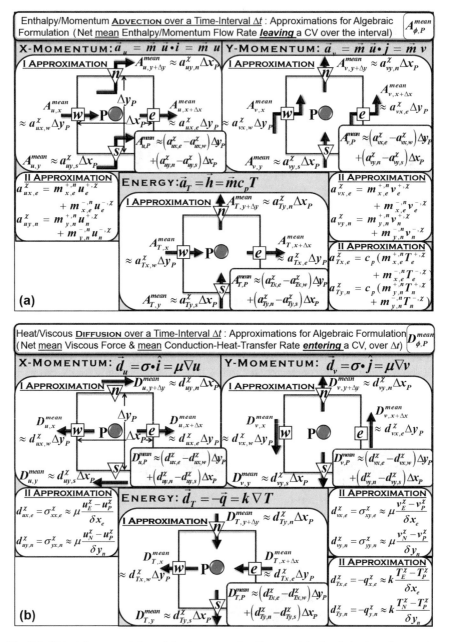

Fig. 8.6 Pictorial representation of the two approximations used in the derivation of governing LAE for (**a**) advection and (**b**) diffusion term of the law of conservation of momentum and energy. Here, the transported variable $\phi = u$ for $x-$momentum, $\phi = v$ for $y-$momentum, and $\phi = T$ for energy conservation

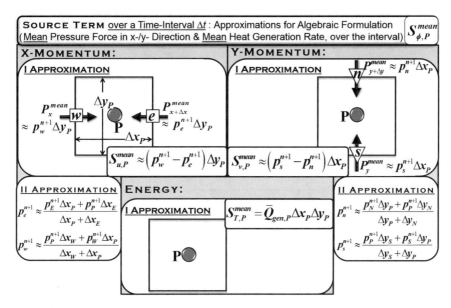

Fig. 8.7 Pictorial representation of the two approximations used in the derivation of governing LAE for source term in law of conservation of momentum and energy. Here, S_u and S_v are the net pressure force in the $x-$ and y-direction, respectively; and S_T is heat gain by volumetric heat generation

The conservation laws for momentum and energy transport are expressed as unsteady advection-diffusion equation, with the remaining term called as source term. The source term corresponds to pressure force in $x-$ $(y-)$ direction for the $x-$ $(y-)$ momentum conservation, and volumetric heat generation is the source term in the energy conservation law. The approximations for the source term, presented earlier in Table 8.2, is shown in Fig. 8.7. Note from the table that the pressure is a flux and the heat generation is a volumetric term. Thus, during the I approximation, the former term is shown at the face center and the latter term at the cell center. Although pressure is a flux term, it is not used to compute the net out-flux as in case of mass/advection/diffusion flux. Instead, the pressure at the vertical and horizontal faces are used to calculate the pressure force in $x-$ and y-direction, for the $x-$momentum and $y-$momentum conservation, respectively. This is shown in Fig. 8.7, with linear interpolation as the $II-$approximation for the pressure. For a uniform grid, substituting $\Delta x_E = \Delta x_P = \Delta x_W$ and $\Delta y_N = \Delta y_P = \Delta y_S$ for the $II-$approximation shown in the figure, the source term is finally given as

$$S_{u,P}^{mean} \approx (p_W^{n+1} - p_E^{n+1})\Delta y \text{ and } S_{v,P}^{mean} \approx (p_S^{n+1} - p_N^{n+1})\Delta y \quad (8.12)$$

Using the physical law-based FVM for the unsteady, advection, diffusion, and source term, the various terms are shown together in Fig. 8.8; and the final discretized equations are given (for 2D unsteady-state momentum and energy transport) as

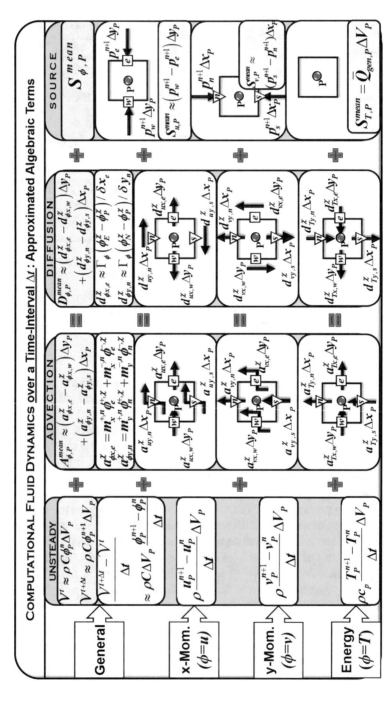

Fig. 8.8 A representative 2D CV to demonstrate an approximate form of the balance statement for the momentum and energy conservation laws, over a time interval Δt

$$\frac{\Delta \mathcal{V}_{\phi,P}}{\Delta t} + A_{\phi,P}^{mean} = D_{\phi,P}^{mean} + S_{\phi,P}^{mean}$$

$$\rho \frac{u_P^{n+1} - u_P^n}{\Delta t} \Delta V_P + A_{u,P}^{\chi} = D_{u,P}^{\chi} + \left(p_w^{n+1} - p_e^{n+1}\right) \Delta y_P$$

$$\rho \frac{v_P^{n+1} - v_P^n}{\Delta t} \Delta V_P + A_{v,P}^{\chi} = D_{v,P}^{\chi} + \left(p_s^{n+1} - p_n^{n+1}\right) \Delta x_P$$

$$\rho c_p \frac{T_P^{n+1} - T_P^n}{\Delta t} \Delta V_P + A_{T,P}^{\chi} = D_{T,P}^{\chi} + \overline{Q}_{gen,P} \Delta V_P$$

where

$$A_{\phi,P}^{\chi} = \left[\left(m_{x,e}^{+,n}\phi_e^{+,\chi} + m_{x,e}^{-,n}\phi_e^{-,\chi}\right)\Delta y_P - \left(m_{x,w}^{+,n}\phi_w^{+,\chi} + m_{x,w}^{-,n}\phi_w^{-,\chi}\right)\Delta y_P\right]C$$
$$+ \left[\left(m_{y,n}^{+,n}\phi_n^{+,\chi} + m_{y,n}^{-,n}\phi_n^{-,\chi}\right)\Delta x_P - \left(m_{y,s}^{+,n}\phi_s^{+,\chi} + m_{y,s}^{-,n}\phi_s^{-,\chi}\right)\Delta x_P\right]C$$

$$D_{\phi,P}^{\chi} = \Gamma_\phi \left(\frac{\phi_E^{\chi} - \phi_P^{\chi}}{\delta x_e}\Delta y_P - \frac{\phi_P^{\chi} - \phi_W^{\chi}}{\delta x_w}\Delta y_P + \frac{\phi_N^{\chi} - \phi_P^{\chi}}{\delta y_n}\Delta x_P - \frac{\phi_P^{\chi} - \phi_S^{\chi}}{\delta y_s}\Delta x_P\right)$$

$$p_w^{n+1} - p_e^{n+1} = \frac{p_W^{n+1}\Delta x_P + p_P^{n+1}\Delta x_W}{\Delta x_P + \Delta x_W} - \frac{p_P^{n+1}\Delta x_E + p_E^{n+1}\Delta x_P}{\Delta x_E + \Delta x_P}$$

$$p_s^{n+1} - p_n^{n+1} = \frac{p_S^{n+1}\Delta y_P + p_P^{n+1}\Delta y_S}{\Delta y_P + \Delta y_S} - \frac{p_P^{n+1}\Delta y_N + p_N^{n+1}\Delta y_P}{\Delta y_N + \Delta y_P}$$

$$(8.13)$$

where the transport variable at a face center $\phi_{f=e,w,n,s}^{\pm}$ (for $A_{\phi,P}^{\chi}$) is presented in Table 6.3 for $T_{f=e,w,n,s}^{\pm}$.

8.6 Closure

It is important to notice the difference in the time level for the various terms in LAE equation considered in the FVM above. The normal velocity at the face center (corresponding to mass flux) is encountered in both mass and momentum (driving the advection process) conservation; it is considered implicit for the mass conservation (Fig. 8.4 and Eq. 8.10) and explicit (to avoid non-linearity) for the advection term (Fig. 8.6a) in the momentum conservation. For the momentum conservation law, pressure is always considered implicit (Fig. 8.7); whereas, advection and diffusion term are considered explicit and implicit (Fig. 8.6) in a semi-explicit and semi-implicit method, respectively. The two types of solution method are presented in the following chapters.

Chapter 9
Computational Fluid Dynamics on a Staggered Grid

The solution methodology for heat transfer in the previous chapters is extended to the fluid flow in this chapter. However, in Chap. 7, the energy conservation-based heat convection was presented for a given steady-state velocity field, more for the convenience and commonly not encountered in the real-world situation. Thus, the practically encountered unknown unsteady velocity field is computed from the momentum conservation-based algebraic formulation. The solution for the velocity field introduces a non-linearity in advection term of the energy equation, and also a coupling between the energy and the momentum equation.

The previous chapter presents the physical law-based algebraic formulation for the momentum equations as a specific case of the advection and diffusion mechanism-based transport equation with a source term, generalized from the energy conservation-based algebraic formulations and approximations in the earlier chapters. However, there is a substantial increase in the numerical complexity to model the source term for the momentum conservation as compared to that for the energy conservation. This is because the source term consists of a dependent variable (pressure) for the momentum equation, as compared to mostly an independent variable (volumetric heat generation) for the energy equation. Thus, the solution for the pressure leads to the mass conservation law while extending the methodology from the energy to the momentum conservation law. The solution of the non-linear and coupled system of governing equations poses numerous challenges in a CFD development.

The mind map for the present chapter is shown in Fig. 9.1. The figure shows that the chapter starts with a discussion on the non-linearity, equation for pressure, and pressure-velocity decoupling, as the challenges in the development of the numerical formulation for CFD. Thereafter, it can be seen that a staggered grid and the associated physical law-based FVM are presented to clearly demonstrate one of the first remedies to probably the biggest challenge in the CFD development—a pressure-velocity decoupling. This is followed by a flux-based solution methodology which is presented on a uniform grid. Figure 9.1 shows that the methodology starts with a philosophy, followed by a mathematical formulation for a pressure-correction-based

© The Author(s), under exclusive license to Springer Nature Switzerland AG 2022   239
A. Sharma, *Introduction to Computational Fluid Dynamics*,
https://doi.org/10.1007/978-3-030-72884-7_9

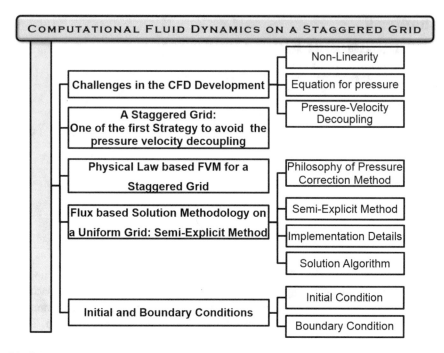

Fig. 9.1 Mind map for Chap. 9

semi-explicit method, and ends with the implementation details and solution algorithm. Finally, the figure shows that this chapter ends with the initial and boundary conditions encountered in the various CFD problems. The solution methodology for the CFD development is introduced in this chapter using the flux-based methodology on a uniform grid, and will be continued for the coefficient of LAEs-based methodology on a non-uniform grid in the next chapter.

9.1 Challenges in the CFD Development

As compared to the methodology for heat transfer presented in the previous chapters, there are various types of challenges involved in the development of a numerical methodology for fluid dynamics. They are presented below in separate subsections.

9.1.1 Non-Linearity

The non-linearity in the transport equation corresponds to the advection term in the general transport equation. The non-linearity arises due to both mass flux and advected variable as unknown in the advection term of the transport equations. This is handled using a lagged (previous time step n) velocity field for the mass flux, $\vec{m}\,^n_f$, to obtain present time-instant field of the advected variable ϕ^{n+1}_P—velocity $\vec{u}^{\,n+1}_P$ from the momentum equation and temperature T^{n+1}_P from the energy equation (Eq. 8.13). The explicit consideration of the mass flux $\vec{m}\,^n_f$ can also be seen in equations for the II-approximation, shown in Fig. 8.6a. However, the implicit consideration of the mass flux $\vec{m}\,^{n+1}_f$, during the mass conservation, can be seen in Fig. 8.4 and Eq. 8.10.

9.1.2 Equation for Pressure

During the solution for the flow properties, a much bigger challenge is to algebraically formulate the equation for the unsteady pressure field. The unsteady velocity field is obtained from the momentum equation, and the unsteady temperature field from the energy equation (Eq. 8.13). This is obvious because the conservation law for momentum and energy consists of rate of change of velocity and temperature (unsteady term), respectively (Fig. 8.5). However, for pressure, we are left with mass conservation law which does not consist of rate of change of pressure; moreover, its unsteady term becomes zero for incompressible flow (compare Eqs. 3.5 and 3.6). Also the mass conservation equation does not contain any pressure term. Thus, there is no such obvious unsteady conservation law for pressure, and the challenge is to convert the *steady-state* mass conservation law as an algebraic equation for the *unsteady* pressure field.

9.1.3 Pressure-Velocity Decoupling

The FVM presented in the previous chapter shows that the II-approximation corresponds to the linear interpolation for the computation of the normal velocity (Fig. 8.4) and the pressure (Fig. 8.7) at the various face centers. For a *uniform grid*, the respective terms result in the LAE for the mass and the momentum conservation equation which involves difference of velocity (Eq. 8.11) and that of pressure (Eq. 8.12) at two *alternate* cell center values (E and W, and N and S); not the *adjacent* ones. Thus, the alternate cell center pressure-based LAE for momentum conservation is used to compute the velocity, and Eq. 8.13 is given for a uniform grid as

$$u_P^{n+1} = u_P^n + \frac{\Delta t}{\rho \Delta V_P} \left[\left(D_{u,P}^\chi - A_{u,P}^\chi \right) + \left(p_W^{n+1} - p_E^{n+1} \right) \Delta y \right] \qquad (9.1)$$

$$v_P^{n+1} = v_P^n + \frac{\Delta t}{\rho \Delta V_P} \left[\left(D_{v,P}^\chi - A_{v,P}^\chi \right) + \left(p_S^{n+1} - p_N^{n+1} \right) \Delta x \right]$$

Since the above equation does not contain p_P, the II-approximation for the pressure eliminates the usage of the pressure p_P in the LAE, for the computation of the velocity u_P^{n+1} and v_P^{n+1}. The resulting breakage in communication between the pressure and velocity at the *same* grid point P is called as *pressure-velocity decoupling*. The decoupling leads to a *superposition* of the correct field with a physically unrealistic oscillatory type of velocity and pressure field.

The oscillatory field was called as checker-board distribution of velocity/pressure by Patankar (1980), an example shown in Fig. 9.2. For any grid point P, it can be seen that the oscillatory velocity field obeys the alternate grid-point-based LAE for mass conservation (Eq. 8.11). Whereas the oscillatory pressure field results in a zero pressure force (in the x as well as y direction) for the alternate grid point pressure-based LAE for momentum conservation (Eq. 9.1). Thus, for any problem of interest, such type of checker-board distribution *over and above* the correct velocity/pressure distribution will also obey the mass and momentum conservation. The superposition leads to a physically unrealistic *oscillatory* velocity and pressure field as the solution of the LAEs. The FVM, presented in the previous chapter, corresponds to the grid points for all the flow properties *located* at the same discrete locations—called as a *co-located* grid. Although the pressure-velocity decoupling-based unrealistic solution is discussed here for the uniform grid, they are also applicable to a non-uniform grid; however, they appear in a less clear form (Patankar 1980) and is not presented here. Note that the II approximation for the mass flux in the mass conservation, and for the pressure in the momentum conservation, leads to the pressure-velocity decoupling. Furthermore, the II approximation for the advection flux leads to the various types of advection schemes. These II-approximations (for the various flux terms) correspond to the value of a flow property at a face center, whereas the II-approximation for the gradient of a flow property (in a diffusion flux) corresponds to the central difference scheme only. Thus, for the II approximations in the physical law-based FVM, there is more difficulty in the numerical formulation for the value as compared to first derivative of a flow property.

9.2 A Staggered Grid: One of the First Strategies to Avoid the Pressure-Velocity Decoupling

One of the first strategies was proposed to *stagger* the CVs/grid points for velocity and pressure (instead of using the same ones for all the flow property, presented in the previous chapter) such that the *II approximation for the mass flux and the pressure is completely avoided*. The II approximation for mass flux (Eq. 8.4) is

Fig. 9.2 An example of a checker-board velocity/pressure field inside a computational domain, for the solution of the LAEs on a co-located grid

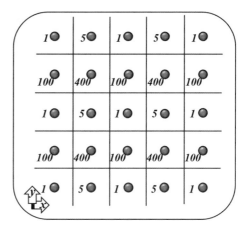

avoided by defining the grid points for the u velocity at the east and west face centers, and the points for v-velocity at the north and south face centers, of the mass conservation obeying CVs. Whereas the II approximation for the pressure (Eq. 8.7) is avoided by defining the pressure grid points appropriately; at the east and west face centers of x-momentum conservation obeying CVs, and north and south face centers of y-momentum conservation obeying CVs. The spatially staggered CVs for the application of mass, x-momentum, and y-momentum conservation law are called as pressure p-CV, u-velocity CV, and v-velocity CV, respectively.

The II approximation avoiding three different types of CVs, *each for the application of a different conservation laws*, is shown spatially staggered in Fig. 9.3. The figure shows the staggered grid points for the various flow properties by different symbols: square for u, triangle for v, and circle for p. A representative CV for a flow property is also shown separately in the figure, along with the various grid points of the other flow properties encountered in the CV. Since the pressure is obtained in CFD from mass conservation, note that the CV for the application of the mass conservation law is considered as the CV for the pressure. Furthermore, the CV for the pressure is more generic and is called as scalar or main CV. The pressure CV is applicable to any scalar variable such as temperature; thus, the energy conservation law is applied to the scalar CV. The staggered grid was first used by Harlow and Welch (1965) and later in a popular procedure (called as SIMPLE algorithm) by Patankar and Spalding (1972).

Note from the p-CV (Fig. 9.3) that the staggered grid points are such that location of the u-velocity grid points here is the same as presented earlier for q_x and h_x, and that for v-velocity grid points is the same as presented for q_y and h_y (Fig. 7.6). For an efficient numerical implementation, it was discussed that only one subscript (and running indices) is used for both temperature at cell center P and heat flux at the face centers. A convention was proposed where the subscript P for temperature corresponds to that for the heat flux at the positive (east and north) faces. Using this convention for normal velocity here, Fig. 9.3 shows u_P at the east, v_P at the north,

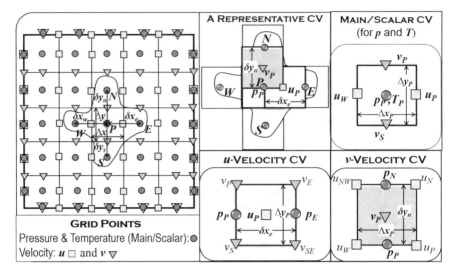

Fig. 9.3 Staggered grid points in a computational domain, and a representative CV for the scalar and vector flow properties.

u_W (east face of grid point W) at the west, and v_S (north face of grid point S) at the south face centers of the main/scalar CV.

9.3   Physical Law-Based FVM for a Staggered Grid

The physical law-based FVM presented in the previous chapter for the co-located grid is also applicable here for the staggered grid. For the co-located grid, the two approximations were presented in Fig. 8.4 for the mass conservation, and Figs. 8.5–8.5 for the momentum conservation. The respective figure for the staggered grid is shown in Figs. 9.4, 9.5, 9.6, and 9.7.

However, the *I I* approximations which leads to the pressure-velocity decoupling are completely avoided with a staggered grid. This is shown in Figs. 9.4 and 9.7, with only *I* approximation for the mass flux and pressure at the face centers, respectively. Thus, the final discretized form of the law of conservation of mass (mass source S_{Pm}) and the pressure force in the x-direction and y-direction (pressure as the source terms S_{Pu} and S_{Pv}) are given as

$$S_{m,P}^{n+1} \approx \rho \left[\left(u_P^{n+1} - u_W^{n+1} \right) \Delta y + \left(v_P^{n+1} - v_S^{n+1} \right) \Delta x \right] = 0 \qquad (9.2)$$

$$S_{u,P}^{n+1} = \left(p_P^{n+1} - p_E^{n+1} \right) \Delta y; \ S_{v,P} = \left(p_P^{n+1} - p_N^{n+1} \right) \Delta x$$

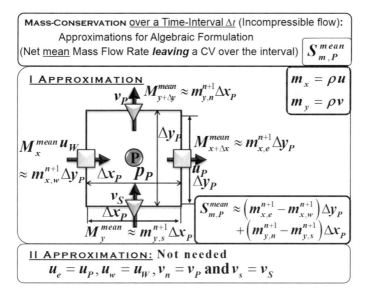

Fig. 9.4 Application of mass conservation law on a representative pressure CV to demonstrate the two approximations for algebraic formulation on a staggered grid

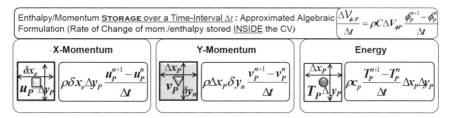

Fig. 9.5 Pictorial representation of the approximated algebraic formulation on a staggered grid for the unsteady term of the law of conservation of momentum and energy.

Note that the above equations for the staggered grid involve velocity/pressure at adjacent as compared to that at alternate grid points (Eqs. 8.11 and 8.12) for the co-located grid.

Finally, for all the terms, the approximations (presented earlier in Fig. 8.8 for the co-located grid) are shown in Fig. 9.8 for the staggered grid.

Since the mass flux is involved in the mass as well as momentum/energy (advection term) conservation laws, and the associated CVs are staggered, note that the *II* approximation for mass flux is avoided only for the mass and the energy conservation obeying main/scalar CV (Figs. 9.4 and 9.6a for $T-$CV). Whereas, for the velocity CVs, the *II* approximation for the mass flux (at the associated face center) corresponds to a *linear interpolation* of the velocity of the cell centers adjacent to the face center. The mass flux for the velocity and temperature (and pressure) CVs is given for the *uniform* grid (refer the *various* CVs in Fig. 9.3) as

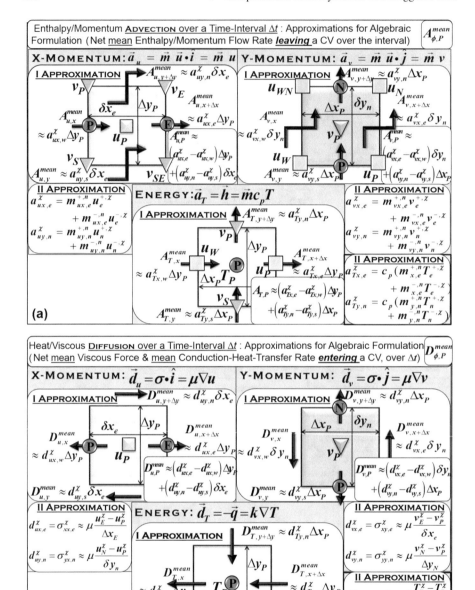

Fig. 9.6 Application of the momentum and the energy conservation law on a representative velocity and temperature CV, respectively. This is shown separately for (**a**) advection and (**b**) diffusion term to demonstrate the two approximations on a staggered grid.

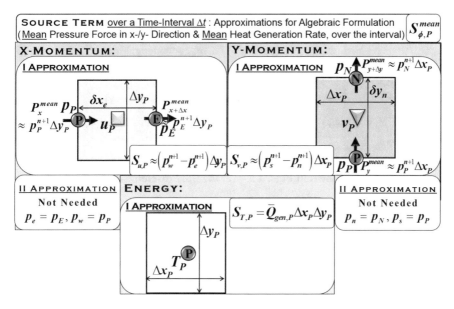

Fig. 9.7 Pictorial representation of the two approximations on a representative velocity and temperature CV during the FVM for the source term of the momentum and energy conservation law, respectively.

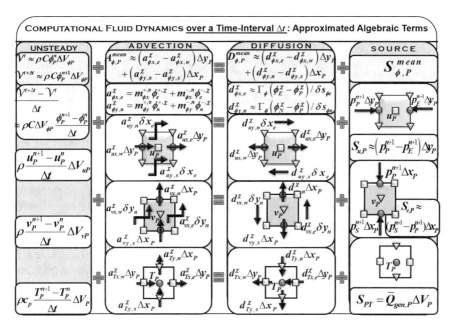

Fig. 9.8 A representative 2D CV to demonstrate an approximate algebraic representation of the various terms for the momentum and energy conservation laws, on a staggered grid.

$$\text{For} u - CV \begin{cases} m_{ux,e} = \rho(u_E + u_P)/2, \ m_{ux,w} = \rho(u_P + u_W)/2 \\ m_{uy,n} = \rho(v_E + v_P)/2, \ m_{uy,s} = \rho(v_S + v_{SE})/2 \end{cases}$$

$$\text{For} v - CV \begin{cases} m_{vx,e} = \rho(u_N + u_P)/2, \ m_{vx,w} = \rho(u_{WN} + u_W)/2 \quad (9.3) \\ m_{vy,n} = \rho(v_N + v_P)/2, \ m_{vy,s} = \rho(v_P + v_S)/2 \end{cases}$$

$$\text{For} T - CV \ m_{x,e} = \rho u_P, \ m_{x,w} = \rho u_W, \ m_{y,n} = \rho v_P, \ m_{y,s} = \rho v_S$$

Using the physical law-based FVM for the various terms of the 2D unsteady-state transport equation, shown in Figs. 9.5–9.7, the final discretized equations are given as

$$\frac{\Delta \mathcal{V}_{\phi,P}}{\Delta t} + A^{\chi}_{\phi,P} = D^{\chi}_{\phi,P} + S^{n+1}_{\phi,P}$$

$$\rho \frac{u^{n+1}_P - u^n_P}{\Delta t} \Delta V_P + A^{\chi}_{u,P} = D^{\chi}_{u,P} + \left(p^{n+1}_P - p^{n+1}_E\right) \Delta y_P$$

$$\rho \frac{v^{n+1}_P - v^n_P}{\Delta t} \Delta V_P + A^{\chi}_{v,P} = D^{\chi}_{v,P} + \left(p^{n+1}_P - p^{n+1}_N\right) \Delta x_P$$

$$\rho c_p \frac{T^{n+1}_P - T^n_P}{\Delta t} \Delta V_P + A^{\chi}_{T,P} = D^{\chi}_{T,P} + \overline{Q}_{gen,P} \Delta V_P$$

where $A^{\chi}_{\phi,P} = \left[\left(m^{+,n}_{\phi x,e}\phi^{+,\chi}_e + m^{-,n}_{\phi x,e}\phi^{-,\chi}_e\right) \Delta S_{\phi,e}\right.$

$$- \left(m^{+,n}_{\phi x,w}\phi^{+,\chi}_w + m^{-,n}_{\phi x,w}\phi^{-,\chi}_w\right) \Delta S_{\phi,w} \Big] C + \left[\left(m^{+,n}_{\phi y,n}\phi^{+,\chi}_n\right)\right. \quad (9.4)$$

$$\left. + m^{-,n}_{\phi y,n}\phi^{-,\chi}_n\right) \Delta S_{\phi,n} - \left(m^{+,n}_{\phi y,s}\phi^{+,\chi}_s + m^{-,n}_{\phi y,s}\phi^{-,\chi}_s\right) \Delta S_{\phi,s}\Big] C$$

$$D^{\chi}_{\phi,P} = \Gamma_\phi \left(\frac{\phi^{\chi}_E - \phi^{\chi}_P}{\delta s_{\phi,e}} \Delta S_{\phi,e} - \frac{\phi^{\chi}_P - \phi^{\chi}_W}{\delta s_{\phi,w}} \Delta S_{\phi,w} \right.$$

$$\left. + \frac{\phi^{\chi}_N - \phi^{\chi}_P}{\delta s_{\phi,n}} \Delta S_{\phi,n} - \frac{\phi^{\chi}_P - \phi^{\chi}_S}{\delta s_{\phi,s}} \Delta S_{\phi,s} \right)$$

where the transport variable $\phi = u, v$, and T for the x-momentum, y-momentum, and energy equation, respectively, and the diffusion coefficient Γ_ϕ is μ for the momentum and k for the energy equation. Furthermore, ΔV is the volume and ΔS is the surface area of the *staggered* CVs, and δs is the distance between a cell center for $u_P/v_P/p_P/T_P$ and a neighboring cell center. For the various CVs, the equation for volumes $\Delta V_{\phi,P}$ is shown in Fig. 9.5; and the surface area $\Delta S_{\phi,f}$ and the distances $\delta s_{\phi,f}$ are seen in Fig. 9.6b. For the advected velocity at a face center $\phi^{\pm}_{f=e,w,n,s}$ in the above equation, an advection scheme-based equation is presented in Table 6.3. The weights $w^{\pm}_{f,i=1-3}$ in the equations are given in Table 6.1 for a uniform grid, and Table 6.2 for a non-uniform grid.

Note that the FVM for obtaining the LAEs is presented above for the velocity but not for the pressure. The LAE for the pressure is presented below as a part of the solution methodology, as it involves a coupled and iterative solution of the velocity and the pressure.

9.4 Flux-based Solution Methodology on a Uniform Grid: Semi-Explicit Method

Figure 9.9 shows the staggered CVs separately for the velocity and pressure (and temperature), for a uniform Cartesian grid.

The figure shows that the *shaded* CVs for the velocity and all the CVs for temperature (and pressure) are of same volume $\Delta V_{\phi,P} = \Delta x \Delta y$ and surface area $\Delta S_{\phi,e} = \Delta S_{\phi,w} = \Delta y$ and $\Delta S_{\phi,n} = \Delta S_{\phi,s} = \Delta x$. The volume is reduced by half

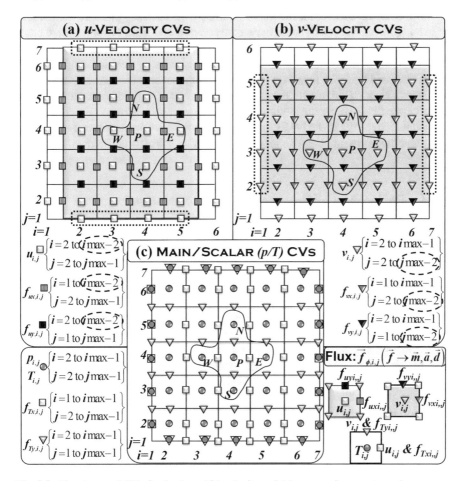

Fig. 9.9 The staggered CVs for the (**a** and **b**) velocity and (**c**) pressure/temperature, shown separately in the computational domain, on a uniform grid with $imax = jmax = 7$. The shaded region in the velocity CVs shows the internal CVs, obeying the momentum conservation law. The loops to compute the various flow properties ϕ (u, v, p, and T at the internal grid points) and fluxes $\vec{f}$ (mass $\vec{m}$, advection $\vec{a}$, and diffusion $\vec{d}$ at the vertical and horizontal face centers) for each of the three CVs are also shown. The *dashed* encircled region in the loops shows that the total number of CVs as well as the face centers in the x-direction are one less for the u-CVs as compared to that for the p/T CVs, similarly in the y-direction for the v-CVs.

for the unshaded CVs, seen near the east and west boundaries for the u-CVs in Fig. 9.9a, and near the north and south boundaries for the v-CVs in Fig. 9.9b. Since these *half-sized CVs* correspond to the *boundary* grid points, the velocity for these boundary CVs/grid points is obtained from the *BCs*. These grid points for the velocity CVs, represented geometrically by square and triangle, coincide with the circles for the boundary cell centers (for the scalar CVs) in Fig. 9.9c. Since the half-CVs are not involved in the present flux-based methodology for the *momentum conservation law*, the FVM is applied to the *shaded CVs* only in Fig. 9.9a,b. Whereas the mass and energy conservation law obeying FVM is applied to all the CVs in Fig. 9.9c.

As discussed above, for the incompressible flow, the unsteady pressure field should be obtained from the steady-state mass conservation law. This becomes more challenging as there is no explicit relationship between the pressure and the mass conservation law. In contrast, the solution for the unsteady velocity field is straightforward as it is obtained from the unsteady-state momentum conservation law. Thus, at each time step, a *coupled solution* of the pressure and the velocity field is established by an *iterative predictor-corrector* method—called as a *pressure-correction method*. This method which results in the LAEs for the pressure field is first introduced *philosophically* and then presented *mathematically* in separate subsections below.

9.4.1 Philosophy of Pressure-Correction Method

Since we do not know the pressure p^{n+1} at the present time level $n+1$, the pressure p^n of previous time level n is used in Eq. 9.4 (substitute $p_P^{n+1} = p_P^n$) to *predict* an *approximate "*" value* of the velocity for the present time level as $u_P^{n+1} \approx u_P^*$ and $v_P^{n+1} \approx v_P^*$; the superscript $*$ represents the predicted value. The predicted velocity results in a mass flux at the various face centers $m_{x/y,f}^*$ of a representative pressure CV (for mass conservation), shown as an example in Fig. 9.10a. Considering the surface area of the CV as $\Delta x = \Delta y = 1\,m$, the representative figure shows a total mass flow rate $M_{out}^* = 4\,kg/s$ leaving and $M_{in}^* = 2\,kg/s$ entering the CV (obtained from the u_P^*, u_W^*, v_P^*, and v_S^*) with a net out-flux/*mass source* prediction (Eq. 9.2) as $S_{m,P}^* = M_{out}^* - M_{in}^* = 2\,kg/s$. Since the velocity field is not conserving mass with $S_{m,P} \approx 0$, the predicted velocity needs a correction until the mass conservation is almost obeyed in the CV. This led to an *iterative predictor-corrector method*, with the convergence criterion for the mass source as $S_{m,P}^* \le \epsilon$; ϵ is the convergence tolerance, which is a practically zero. The predicted mass flux $m_{x/y,f}^*$ as well as the pressure p^* (equal to p^n for the first iteration) is updated by adding a *correction* value *iteratively*, $m_{x/y,f}^{*,N+1} = m_{x/y,f}^{*,N} + m_{x/y,f}'$ and $p^{*,N+1} = p^{*,N} + p'$ (superscript $'$ represents a correction value and N is the iteration number), until the $S_{m,P} \le \epsilon$ for all the CVs in the domain. Then, the resulting velocity, mass flux, and pressure become the correct value for the *present* time instant, i.e., $\vec{u}_P^{*,N+1} = \vec{u}_P^{n+1}$, $\vec{m}_f^{*,N+1} = \vec{m}_f^{n+1}$, and $p_P^{*,N+1} = p_P^{n+1}$. The above discussion refers to the FVM-based Eq. 9.4 for the predictor step but not for the corrector step. The equations for $m_{x/y,f}'$ and p_P'

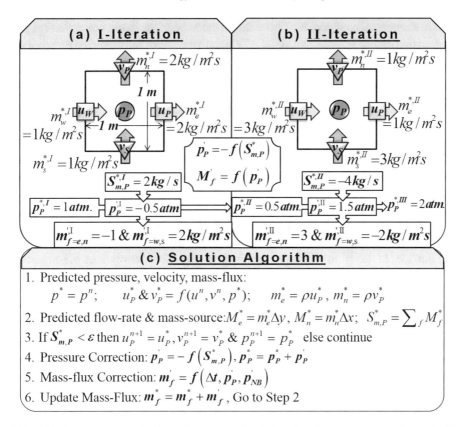

Fig. 9.10 A representative 2D CV to demonstrate the philosophy of a pressure-correction method

are presented philosophically in this section and derived mathematically in the next section.

For the I-iteration of mass conservation at a particular time step, Fig. 9.10a shows a CV with the pressure at the cell center as $1\,atm$. This is the pressure from the previous time instant which is considered as the predicted pressure for the I-iteration $p_P^{*,I} = p_P^n = 1\,atm$. The figure also shows the resulting predicted mass flux $m_{x/y,f}^{*,I}$ (at the various face centers) and the mass source $S_{m,P}^{*,I} = 2\,kg/s$, shown as *step-1* and *step-2* in Fig. 9.10c, respectively. A corrective method for the mass conservation corresponds to tuning/correcting the pressure inside the CV. Using our physical understanding that the pressure should be *decreased* inside the CV to induce more inflow and counter-balance the positive mass source $S_{m,P}^{*,I} = 2\,kg/s$, hoping that it becomes almost zero. Any change in the pressure is called as *pressure correction*; negative for the decrease and positive for the increase in pressure. Thus, the *positive mass source leads to a negative pressure correction*. Here, the negative correlation is discussed philosophically and represented symbolically as $p' = -f(S_{m,P})$ as *step 4* in Fig. 9.10c, and will be presented mathematically in the next subsection. Using the mathematical equation, the figure shows the resulting pressure correction for the I-iteration as $p_P^{\prime,I} = -0.5$ atm and the II iterative value of the predicted pressure of the CV as $p_P^{*,II} = p_P^{*,I} + p_P^{\prime,I} = 0.5$ atm.

Furthermore, as $p_P^{\prime,I} = -0.5$ atm in the I iteration, the reduction in the pressure inside the CV should lead to an increase in the incoming flow for the next iteration. Thus, the pressure correction should lead to a mass flux correction at the various faces of the CV. This is represented symbolically as $m_{x/y,f}^{\prime} = f(p^{\prime})$, shown as *step 5* in Fig. 9.10c, and will be derived mathematically in the next subsection. Using the mathematical equation, the figure shows the resulting mass-flux correction of $m_{x,w}^{\prime,I} = m_{y,s}^{\prime,I} = 2\,kg/m^2s$ at the west (and south) faces, and $m_{y,n}^{\prime,I} = m_{x,e}^{\prime,I} = -1\,kg/m^2s$ at the north (and east) faces of the CV. The resulting corrected mass flux at the various faces $(m_f^{*,II} = m_f^{*,I} + m_f^{\prime,I})$ is shown in Fig. 9.10b. The figure shows a total flow rate of $2\,kg/s$ going out and $6\,kg/s$ coming into the CV. Since the mass source $S_{m,P}^{*,II} = -4\,kg/s$ is non-zero, we need to continue with the iterations.

For the II iteration of mass conservation, Fig. 9.10b shows that the negative mass source $S_{m,P}^{*,II} = -4\,kg/s$ results in positive pressure correction $p^{\prime,II} = 1.5\,atm$ (as $p^{\prime} = -f(S_{m,P})$) and a corrected pressure of $p_P^{III} = p_P^{II} + p^{\prime,II} = 2\,atm$. The increase in the pressure results in a corrected mass flux for the III−iteration $m_f^{*,III}$, not shown here. This is the *correct* mass flow rate and pressure for the present time instant $(m_{x/y,f}^{n+1} = m_{x/y,f}^{*,III}$ and $p_P^{n+1} = p_P^{III})$ if $S_{m,P}^{*,III} \leq \epsilon$, otherwise the iterations are continued till we reach an iteration where the condition for the mass conservation is obeyed; *step 3* in Fig. 9.10c. Thus, other than the outer timewise loop for the steady-state convergence, the pressure-correction method results in an inner iteration loop to obtain both mass and momentum conservation law obeying velocity and the pressure field at each time step. The inner loop is also needed during the unsteady solution of the momentum equation by an implicit method.

Note that the above non-mass-conserving mass flow rates are physically impossible. The iterative values in the figure show a numerical artifact of the flow which fortunately approaches to the mass-conserving physical flow field. Furthermore, the tuning of pressure is presented above for one of the CV; however, the mass flow rate at the faces of a CV depends on the pressure of this CV as well as the adjoining CV (not shown in the figure for clarity). Thus, it is a coupled system which is solved by the pressure-correction (iterative predictor-corrector) method to establish a relationship and obtain the unsteady pressure field from the steady-state mass conservation law. Note that the continuity equation, having no explicit link to the pressure, is just an additional constraint on the velocity field that must be satisfied together with the momentum equations, and it is an appropriate manipulation of that constraint that leads to an equation for the pressure correction (Patankar 1980).

9.4.2 Semi-Explicit Method

Substituting the time level $\chi = n$ for the advection-diffusion term in Eq. 9.4, the discretized form of momentum equations is given as

$$u_P^{n+1} = u_P^n + \frac{\Delta t}{\rho \Delta V_{u,P}} \left[\left(D_{u,P}^n - A_{u,P}^n \right) + \left(p_P^{n+1} - p_E^{n+1} \right) \Delta y_P \right]$$

$$v_P^{n+1} = v_P^n + \frac{\Delta t}{\rho \Delta V_{v,P}} \left[\left(D_{v,P}^n - A_{v,P}^n \right) + \left(p_P^{n+1} - p_N^{n+1} \right) \Delta x_P \right] \qquad (9.5)$$

where n is the previous time level and $n + 1$ is the present time level. Since all the terms are considered explicit except the pressure, this is called as the semi-explicit method. The *fully explicit* method with pressure at time level n is not considered, as the velocity resulting from the momentum equation does not satisfy the mass conservation. Moreover, it will *not* lead to the pressure-correction method-based algebraic formulation for the solution of the unsteady pressure field.

9.4.2.1 Predictor Step

For the *original proposition* of the semi-explicit method in Eq. 9.5, there is a starting trouble as we do not know the pressure at the present time level $n + 1$. Thus, using its value from the previous time level n, the *predictor step* involves the prediction of the velocity, mass flux, and mass source as follows.

Velocity and Mass Flux Prediction:

For the semi-explicit method, this involves first the *explicit* computation of the *mass flux* (Eq. 9.3) given (for the uniform grid) as

$$x - \text{Momentum: } m_{ux,e}^n = \rho(u_E^n + u_P^n)/2, \; m_{ux,w} = \rho(u_P^n + u_W^n)/2,$$
$$m_{uy,n}^n = \rho(v_E^n + v_P^n)/2, \; m_{uy,s} = \rho(v_S^n + v_{SE}^n)/2$$
$$y - \text{Momentum: } m_{vx,e}^n = \rho(u_N^n + u_P^n)/2, \; m_{vx,w} = \rho(u_{WN}^n + u_W^n)/2,$$
$$m_{vy,n}^n = \rho(v_N^n + v_P^n)/2, \; m_{vy,s} = \rho(v_P^n + v_S^n)/2 \qquad (9.6)$$

where linear interpolation is used as the II approximation, for the computation of mass fluxes at the face centers of the staggered velocity CVs. Then, the explicit computation of advection and diffusion flux (Fig. 9.6, with $\chi = n$ and Table 6.1 for $\phi_{f=e,w,n,s}^{\pm}$), is given as

$$a_{\phi x,e}^n = \left[m_{\phi x,e}^{+,n} \left(w_1 \phi_E^n + w_2 \phi_P^n + w_3 \phi_W^n \right) + m_{\phi x,e}^{-,n} \left(w_1 \phi_P^n + w_2 \phi_E^n + w_3 \phi_{EE}^n \right) \right]$$
$$a_{\phi x,w}^n = \left[m_{\phi x,w}^{+,n} \left(w_1 \phi_P + w_2 \phi_W + w_3 \phi_{WW} \right) + m_{\phi x,w}^{-,n} \left(w_1 \phi_W + w_2 \phi_P + w_3 \phi_E \right) \right]$$
$$a_{\phi y,n}^n = \left[m_{\phi y,n}^{+,n} \left(w_1 \phi_N + w_2 \phi_P + w_3 \phi_S \right) + m_{\phi y,n}^{-,n} \left(w_1 \phi_P + w_2 \phi_N + w_3 \phi_{NN} \right) \right]$$
$$a_{\phi y,s}^n = \left[m_{\phi y,s}^{+,n} \left(w_1 \phi_P + w_2 \phi_S + w_3 \phi_{SS} \right) + m_{\phi y,s}^{-,n} \left(w_1 \phi_S + w_2 \phi_P + w_3 \phi_N \right) \right] \quad (9.7)$$

$$d_{\phi x,e}^n = \mu \frac{\phi_E^n - \phi_P^n}{\delta s_{\phi,e}}, \ d_{\phi x,w}^n = \mu \frac{\phi_P^n - \phi_W^n}{\delta s_{\phi,w}}$$

$$d_{\phi y,n}^n = \mu \frac{\phi_N^n - \phi_P^n}{\delta s_{\phi,n}} \ d_{\phi y,s}^n = \mu \frac{\phi_P^n - \phi_S^n}{\delta s_{\phi,s}} \tag{9.8}$$

where, for the advection flux, the weights $w_{i=1-3}$ are given in Table 6.1 for the uniform grid. Furthermore, for the uniform grid, the distance between the cell centers $\delta s_{\phi,e} = \delta s_{\phi,w} = \Delta x$ and $\delta s_{\phi,n} = \delta s_{\phi,s} = \Delta y$ for all the face centers of the *shaded* CVs in Fig. 9.9a,b except the *boundary face centers* of the velocity CVs, shown by the *dotted* encircled region at the north and south boundaries of the domain in Fig. 9.9a and at the east and west boundaries in Fig. 9.9b. For the boundary face centers, the distance is reduced to half; $\delta s_{\phi,w} = \Delta x/2$ at the west, $\delta s_{\phi,e} = \Delta x/2$ at the east, $\delta s_{\phi,n} = \Delta y/2$ at the north, and $\delta s_{\phi,s} = \Delta y/2$ at the south boundary face centers.

After the explicit computation of the various fluxes above, p^n instead of p^{n+1} is considered in Eq. 9.4 to predict the velocity and the mass fluxes (for the p-CV), given as

$$u_P^* = u_P^n + \frac{\Delta t}{\rho \Delta V_{u,P}} \left[\left(D_{u,P}^n - A_{u,P}^n \right) + \left(p_P^n - p_E^n \right) \Delta y \right]$$

$$v_P^* = v_P^n + \frac{\Delta t}{\rho \Delta V_{v,P}} \left[\left(D_{v,P}^n - A_{v,P}^n \right) + \left(p_P^n - p_N^n \right) \Delta y \right] \tag{9.9}$$

$$\text{where } \ D_{\phi,P}^n = (d_{\phi x,e}^n - d_{\phi x,w}^n)\Delta y + (d_{\phi y,n}^n - d_{\phi y,s}^n)\Delta x$$

$$A_{\phi,P}^n = (a_{\phi x,e}^n - a_{\phi x,w}^n)\Delta y + (a_{\phi y,n}^n - a_{\phi y,s}^n)\Delta x$$

$$m_{x,e}^* = \rho u_P^*, \ m_{x,w}^* = \rho u_W^*, \ m_{y,n}^* = \rho v_P^*, \ \text{and} \ m_{y,s}^* = \rho v_S^* \tag{9.10}$$

Mass Source Prediction:
Using the predicted mass fluxes at the various faces of the p-CV, the predicted *mass source* is given as

$$S_{m,P}^* = \left(m_{x,e}^* - m_{x,w}^* \right) \Delta y + \left(m_{y,n}^* - m_{y,s}^* \right) \Delta x \tag{9.11}$$

9.4.2.2 Corrector Step

Velocity and Mass-Flux Correction: Since the original proposition is considered as the sum of the predictor and the corrector step, the equations for the *corrector step* are obtained by subtracting Eq. 9.9 from Eq. 9.5, and substituting $p^{n+1} - p^n = p'$, $\Delta V_{u,P} = \delta x_e \Delta y_P$, and $\Delta V_{v,P} = \Delta x_P \delta y_n$. The resulting velocity-correction equations are given as

$$u_P' = \frac{\Delta t}{\rho} \frac{(p_P' - p_E')}{\delta x_e} \ \text{and} \ v_P' = \frac{\Delta t}{\rho} \frac{(p_P' - p_N')}{\delta y_n} \tag{9.12}$$

Thereafter, the mass-flux correction equations at the various faces of a p-CV are given as

$$m'_{x,e} = \rho u'_P = -\Delta t \frac{(p'_E - p'_P)}{\delta x_e} \; ; \; m'_{x,w} = \rho u'_W = -\Delta t \frac{(p'_P - p'_W)}{\delta x_w}$$

$$m'_{y,n} = \rho v'_P = -\Delta t \frac{(p'_N - p'_P)}{\delta y_n} \; ; \; m'_{y,s} = \rho v'_S = -\Delta t \frac{(p'_P - p'_S)}{\delta y_s} \quad (9.13)$$

With reference to the already discussed philosophy of the pressure-correction method, the above equation is for the mass-flux correction as a function of pressure-correction $m'_f = f(\Delta t, p'_p, p'_{NB})$, seen as *step*-5 in Fig. 9.10c. It is interesting to note that the equation for the mass-flux correction is analogous to algebraic representation of the subsidiary law for the diffusion flux (Eq. 8.8), such as Fourier's law in case of conduction flux. Thus, the mass-flux correction acts like a diffusion flux, with pressure correction as the diffused variable and $-\Delta t$ as the diffusion coefficient (analogous to $-k$ in heat transfer and μ in fluid flow). Thus, the algebraic formulation in Eq. 9.13 for mass flux correction can be expressed in a differential form as

$$\overrightarrow{m}' = -\Delta t \nabla p' \quad (9.14)$$

Pressure Correction: The discretized form of mass conservation (Eq. 9.2 in terms of mass flux) is given as

$$S_{m,P} = \left(m^{n+1}_{x,e} - m^{n+1}_{x,w}\right) \Delta y_P + \left(m^{n+1}_{y,n} - m^{n+1}_{y,s}\right) \Delta x_P = 0$$

Substituting $m^{n+1}_{x/y,f} = m^*_{x/y,f} + m'_{x/y,f}$ in the above equation, we get

$$\left(m'_{x,e} - m'_{x,w}\right) \Delta y_P + \left(m'_{y,n} - m'_{y,s}\right) \Delta x_P$$
$$= -\left[\left(m^*_{x,e} - m^*_{x,w}\right) \Delta y_P + \left(m^*_{y,n} - m^*_{y,s}\right) \Delta x_P\right]$$
$$\Rightarrow S'_{m,P} = -S^*_{m,P} \quad (9.15)$$

where $S^*_{m,P}$ and $S'_{m,P}$ are the mass source for the CV, considering the predicted mass flux $\overrightarrow{m}^*_f$ and the mass-flux correction $\overrightarrow{m}'_f$, respectively. The above algebraic formulation can be expressed in differential form as

$$\nabla \cdot \overrightarrow{m}' = -\nabla \cdot \overrightarrow{m}^* = -\rho \nabla \cdot \overrightarrow{u}^*$$

Substituting $\overrightarrow{m}' = -\Delta t \nabla p'$ (from Eq. 9.14) in the above equation, we get the differential formulation of the pressure-correction equation as

$$-\Delta t \nabla^2 p' = -\rho \nabla \cdot \overrightarrow{u}^*$$

This is a Poisson equation and is similar to steady-state heat conduction equation with a source term; the source term is heat generation in the heat transfer and the predicted mass source term here.

For the various faces of a CV, substituting the mass-flux correction from Eq. 9.13, Eq. 9.15 is given as

$$
-\Delta t \left\{ \left[\frac{(p'_E - p'_P)}{\delta x_e} - \frac{(p'_P - p'_W)}{\delta x_w} \right] \Delta y_P \right.
$$
$$
\left. + \left[\frac{(p'_N - p'_P)}{\delta y_n} - \frac{(p'_P - p'_S)}{\delta y_s} \right] \Delta x_P \right\} = -S^*_{m,P} \qquad (9.16)
$$

With reference to the already discussed philosophy of the pressure-correction method in Fig. 9.10, the above equation is for the pressure correction as a function of the mass source $p'_P = -f(S^*_{m,P})$, seen as *step*-4 in Fig. 9.10c.

9.4.2.3 Iterative Solution of the Pressure-Correction Equation

The steady-state mass conservation equation for the pressure correction (Eq. 9.16) can be solved *iteratively* by *two methods*. *First*, the updated values are obtained for both $p'_P{}^{,N+1}$ and $S^{*,N+1}_{m,p}$ (considering $N+1$ as the present and N as the previous iterative value) in each iteration, and *second*, $S^*_{m,P}$ (Eq. 9.11) is kept fixed (during all the iterations) and the updated values are obtained only for $p'_P{}^{,N+1}$. Considering ϵ as the convergence tolerance, the convergence criterion is $\left| S^{*,N+1}_{m,i,j} \right|_{max} \leq \epsilon$ for the first method, and $\left| p'^{,N+1}_{i,j} - p'^{,N}_{i,j} \right|_{max} \leq \epsilon$ for the second method. However, the converged solution automatically results in $\left| p'^{,N+1}_{i,j} - p'^{,N}_{i,j} \right|_{max} \leq \epsilon/a_P$ in the first method, and $\left| S^{*,n+1}_{m,i,j} \right|_{max} \leq a_P \epsilon$ in the second method; the a_P is presented in Eq. 9.17.

The mass flux is corrected after every iteration in the first method, and after the last iteration (converged solution) of pressure correction in the second method. Thus, the first method strictly follows the iterations seen in Fig. 9.10, for the philosophy of the pressure-correction method. Whereas the second method involves finding the correct pressure correction iteratively such that the converged pressure-correction-based mass-flux corrections (Eq. 9.13) when added to the predicted mass flux (Eq. 9.10, shown in Fig. 9.10a for the *I*-iteration) obey the mass conservation. The solution methodology is presented here for the first method and the second method will be presented in the next chapter.

Considering Gauss-Seidel method for the iterative solution, Eq. 9.16 is given as

$$-\Delta t \left\{ \left[\frac{(p_E^{',N} - p_P^{',N+1})}{\delta x_e} - \frac{(p_P^{',N+1} - p_W^{',N+1})}{\delta x_w} \right] \Delta y_P \right.$$
$$\left. + \left[\frac{(p_N^{',N} - p_P^{',N+1})}{\delta y_n} - \frac{(p_P^{',N+1} - p_S^{',N+1})}{\delta y_s} \right] \Delta x_P \right\} = -S_{m,P}^{*,N}$$

where the value of west and south neighbors will be a present iterative value $(N+1)$ if the loop for the Gauss-Seidel iteration is $i = 2$ to $imax - 1$ and $j = 2$ to $jmax - 1$. Since the predicted mass source gets updated due to pressure correction in each iteration, the RHS of the above equation also corresponds to the previous iterative value N. For each of the four terms in the numerator of LHS of the above equation, adding and subtracting $p_P^{',N}$ result in a residual form of the above equation as

$$p_P^{',N+1} = p_P^{',N} - \frac{1}{a_P}\left(S_{m,P}^{*,N} + S_{m,P}'\right) \tag{9.17}$$

where $S_{m,P}^* = \left[\left(m_{x,e}^* - m_{x,w}^*\right)\Delta y + \left(m_{y,n}^* - m_{y,s}^*\right)\Delta x\right]$

$$S_{m,P}' = -\Delta t \left[\left(\frac{p_E^{',N} - p_P^{',N}}{\delta x_e}\right) - \left(\frac{p_P^{',N} - p_W^{',N+1}}{\delta x_w}\right)\right]\Delta y_P$$
$$-\Delta t \left[\left(\frac{p_N^{',N} - p_P^{',N}}{\delta y_n}\right) - \left(\frac{p_P^{',N} - p_S^{',N+1}}{\delta y_s}\right)\right]\Delta x_P$$

$$a_P = \Delta t \left[\left(\frac{1}{\delta x_e} + \frac{1}{\delta x_w}\right)\Delta y_P + \left(\frac{1}{\delta y_n} + \frac{1}{\delta y_s}\right)\Delta x_P\right]$$

In the above equation, for the first iteration $(N + 1 = 1)$, note that the previous iterative value of the predicted mass source $S_{m,P}^{*,0}$ is calculated from the predicted mass flux (Eq. 9.10), and the mass source correction $S_{m,P}'^{,0}$ is obtained from the initial condition for the pressure correction.

After each iteration of the above equation, the pressure correction $p'^{,N+1}$ is substituted in Eq. 9.13 to obtain $m_f'^{,N+1}$. The mass-flux correction is used to obtain the updated value of the predicted mass flux and the predicted velocity as

$$m_{x/y,f=e,w,n,s}^{*,N+1} = m_{x/y,f}^{*,N} + m_{x/y,f}'^{,N+1}$$

$$u_P^{*,N+1} = \frac{m_{x,e}^{*,N+1}}{\rho}; \ u_W^* = \frac{m_{x,w}^{*,N+1}}{\rho}; \ v_P^{*,N+1} = \frac{m_{y,n}^{*,N+1}}{\rho}; \ v_S^{*,N+1} = \frac{m_{y,s}^{*,N+1}}{\rho} \tag{9.18}$$

Furthermore, the updated (present iterative) values of the mass source and pressure are given as

$$S_{m,P}^{*,N+1} = (m_{x,e}^{*,N+1} - m_{x,w}^{*,N+1})\Delta y_P + (m_{x,s}^{*,N+1} - m_{x,n}^{*,N+1})\Delta x_P$$
$$p_P^{*,N+1} = p_P^{*,N} + p_P'^{,N+1} \tag{9.19}$$

The above mass source is checked for convergence $S_{m,P}^{*,N+1} \leq \epsilon$. If not converged, Eqs. 9.17–9.19 are solved sequentially, shown as *step* 3 − 6 in Fig. 9.10c. Once converged, then the *updated* value of pressure, velocity, and mass flux becomes the *correct* value for the present time level, *i.e.*,

$$p_P^{n+1} = p_P^*; \ u_P^{n+1} = u_P^* \text{ and } v_P^{n+1} = v_P^*$$

$$m_{x,e}^{n+1} = m_{x,e}^*; \ m_{x,w}^{n+1} = m_{x,w}^*; \ m_{y,n}^{n+1} = m_{y,n}^* \text{ and } m_{y,s}^{n+1} = m_{y,s}^*$$

The above philosophy-based iterative solution methodology for the pressure-correction equation is proposed here similar to the flux-based solution methodology for the unsteady heat conduction in Chap. 5. However, since the pressure correction equation is obtained from a steady-state (due to the incompressible flow) mass conservation law, the iteration number N (for the steady state) here is taken analogous to time level n for the unsteady heat transfer. Thus, Eq. 9.17 can be interpreted as the pressure correction for the present iteration is equal to that of previous iteration value plus an additional term which depends on the net mass *entering* by a CV, *i.e.*, $-\left(S_{m,P}^{*,N} + S_{m,P}^{'}\right)$. Thus, the resulting mass gain by the CV is the sum of negative value of both predicted mass source $S_{m,P}^{*,N}$ (updated after every iteration) and mass source correction $S_{m,P}^{*,'}$. This is analogous to the temperature for the present time level is equal to that of previous time level plus an additional term which is obtained from total heat gained by a CV due to conduction flux and heat source, presented in Chap. 5. The conduction heat gain Q_{cond}^n earlier is analogous to the mass source correction $S_{m,P}^{*,'}$ here, and the heat source Q_{gen} is analogous to the mass source $S_{m,P}^{*,N}$.

9.4.3 Implementation Details

For the semi-explicit method presented above, the implementation details correspond to substituting the subscript (corresponding to the cell centers and face centers for the three types of CVs) in the above equations with appropriate running indices. It also uses the convention presented in the previous chapters to represent a single running indices (i, j) for the variables at the cell and the face centers. The details are presented separately for the predictor and corrector step below and for the *FOU as the advection scheme*.

9.4.3.1 Predictor Step

The u and v velocities are predicted by applying the already presented *two-step* flux-based implementation (for the heat convection) to 2D momentum equation here. The *first step* involves the computation of the various fluxes (mass flux, momentum flux as advection flux, and viscous stress as diffusion flux) at the various face centers of the CVs. The *second step* corresponds to computation of net out-flux of advection and diffusion, and then the prediction of the velocity at the next time step.

u-Velocity and Mass Flux (in the x-direction) Prediction:

First step: For the x-momentum conservation, the previous time level values of mass, advection, and diffusion fluxes are given in Eqs. 9.6–9.8; substitute $\phi = u$. At the face centers of the u-CVs in Fig. 9.9a, the grid points for the components of the flux vector $\overrightarrow{f}_u$ (mass flux $\overrightarrow{m}_u$, x-momentum flux $\overrightarrow{a}_u$, and viscous stress in the x-direction $\overrightarrow{d}_u$; refer Fig. 9.6) are represented as gray-filled square for f_{ux} and black-filled squares for f_{uy}. The figure also shows the range of the running indices (i, j) for the fluxes $f_{ux,i,j}$ and $f_{uy,i,j}$. Thus, the equations and the loops used for the prediction of the various fluxes for the u-CVs are given as

$$
\left.
\begin{aligned}
m^n_{ux\,i,j} &= \rho(u^n_{i+1,j} + u^n_{i,j})/2 \\
a^n_{ux\,i,j} &= m^{+,n}_{ux,i,j} u^n_{i,j} + m^{-,n}_{ux,i,j} u^n_{i+1,j} \\
d^n_{ux\,i,j} &= \mu \frac{u^n_{i+1,j} - u^n_{i,j}}{\Delta x}
\end{aligned}
\right\}
\quad
\begin{aligned}
&\text{for } i = 1 \text{ to } imax - 2 \\
&\text{and } j = 2 \text{ to } jmax - 1
\end{aligned}
\tag{9.20}
$$

$$
\left.
\begin{aligned}
m^n_{uy\,i,j} &= \rho(v^n_{i+1,j} + v^n_{i,j})/2 \\
a^n_{uy\,i,j} &= m^{+,n}_{uy,i,j} u^n_{i,j} + m^{-,n}_{uy,i,j} u^n_{i,j+1}
\end{aligned}
\right\}
\quad
\begin{aligned}
&\text{for } i = 2 \text{ to } imax - 2 \\
&\text{and } j = 1 \text{ to } jmax - 1
\end{aligned}
$$

$$
d^n_{uy\,i,j} =
\begin{cases}
\mu \dfrac{u^n_{i,j+1} - u^n_{i,j}}{\Delta y/2} & \text{for } j = 1 \text{ and } jmax - 1 \\[2ex]
\mu \dfrac{u^n_{i,j+1} - u^n_{i,j}}{\Delta y} & \text{for } j = 2 \text{ to } jmax - 2
\end{cases}
\tag{9.21}
$$

$$
\implies \text{for } i = 2 \text{ to } imax - 2
$$

The above equation uses the fact that the vertical distance between the two adjoining u-velocity grid point at the north and south boundaries (of the domain) is half of the distance Δy between the internal grid points, shown in Fig. 9.9a.

Second Step: Calculate the total advection, diffusion, and source term in the x-momentum transport, and then the u-velocity (Eq. 9.9) at the internal grid points of the u-CVs (Fig. 9.9a) and the corresponding mass flux in x-direction (9.10) as

$$
\begin{aligned}
A^n_{ux\,i,j} &= (a^n_{ux\,i,j} - a^n_{ux\,i-1,j})\Delta y + (a^n_{uy\,i,j} - a^n_{uy\,i,j-1})\Delta x \\
D^n_{ux\,i,j} &= (d^n_{ux\,i,j} - d^n_{ux\,i-1,j})\Delta y + (d^n_{uy\,i,j} - d^n_{uy\,i,j-1})\Delta x \\
S^n_{u\,i,j} &= (p^n_{i,j} - p^n_{i+1,j})\Delta y \\
u^*_{i,j} &= u^n_{i,j} + \frac{\Delta t}{\rho c_p \Delta x \Delta y}\left(D^n_{ux\,i,j} - A^n_{ux\,i,j} + S^n_{u\,i,j}\right) \\
m^*_{x\,i,j} &= \rho u^*_{i,j}
\end{aligned}
\tag{9.22}
$$

$$
\implies \text{for } i = 2 \text{ to } imax - 2 \text{ and } j = 2 \text{ to } jmax - 1
$$

where the above mass fluxes are at the centroid of the shaded u-CVs (Fig. 9.9a), and at the east and west face centers of the p-CVs (Fig. 9.9c). The mass fluxes corresponding to the u-velocity at the west and east boundaries of the domain are obtained from the boundary condition (for the u-velocity) as

$$m^*_{x,1,j} = \rho u_{1,j} \text{ and } m^*_{x,imax-1,j} = \rho u_{imax-1,j} \tag{9.23}$$

v-**Velocity and Mass Flux (in the y-direction) Prediction:**

First step: Similar to the implementation details discussed above, the various fluxes (Eqs. 9.6–9.8; substitute $\phi = v$) at the face centers of the v-CV (Fig. 9.9b) are computed as

$$
\left.
\begin{aligned}
m_{vx\,i,j} &= \rho(u^n_{i,j+1} + u^n_{i,j})/2 \\
a_{vx\,i,j} &= m^{+,n}_{vx,i,j} v^n_{i,j} + m^{-,n}_{vx,i,j} v^n_{i+1,j}
\end{aligned}
\right\}
\quad
\begin{aligned}
&\text{for } i = 1 \text{ to } imax - 1 \\
&\text{for } j = 2 \text{ to } jmax - 2
\end{aligned}
$$

$$
d^n_{vx\,i,j} =
\begin{cases}
\mu \dfrac{v^n_{i+1,j} - v^n_{i,j}}{\Delta x/2} & \text{for } i = 1 \text{ and } imax - 1 \\[3mm]
\mu \dfrac{v^n_{i+1,j} - v^n_{i,j}}{\Delta x} & \text{for } i = 2 \text{ to } imax - 2
\end{cases}
\tag{9.24}
$$

$$\Longrightarrow \text{ for } j = 2 \text{ to } jmax - 2$$

$$
\left.
\begin{aligned}
m_{vy\,i,j} &= \rho(v^n_{i,j+1} + v^n_{i,j})/2 \\
a_{vy\,i,j} &= m^{+,n}_{vy,i,j} v^n_{i,j} + m^{-,n}_{vy,i,j} v^n_{i,j+1} \\
d_{vy\,i,j} &= \mu \dfrac{v^n_{i,j+1} - v^n_{i,j}}{\Delta y}
\end{aligned}
\right\}
\quad
\begin{aligned}
&\text{for } i = 2 \text{ to } imax - 1 \\
&\text{for } j = 1 \text{ to } jmax - 2
\end{aligned}
\tag{9.25}
$$

Second Step: Calculate the total advection, diffusion, and source term in the y-momentum transport, and then the v-velocity (Eq. 9.9) at the internal grid points of the v-CVs (Fig. 9.9b) and the corresponding mass flux in y-direction (Eq. 9.10) as

$$
\begin{aligned}
A^n_{vx\,i,j} &= (a^n_{vx\,i,j} - a^n_{vx\,i-1,j})\Delta y + (a^n_{vy\,i,j} - a^n_{vy\,i,j-1})\Delta x \\
D^n_{vx\,i,j} &= (d^n_{vx\,i,j} - d^n_{vx\,i-1,j})\Delta y + (d^n_{vy\,i,j} - d^n_{vy\,i,j-1})\Delta x \\
S^n_{v\,i,j} &= (p^n_{i,j} - p^n_{i,j+1})\Delta x \\
v^*_{i,j} &= v^n_{i,j} + \frac{\Delta t}{\rho c_p \Delta x \Delta y}\left(D^n_{vx\,i,j} - A^n_{vx\,i,j} + S^n_{v\,i,j}\right) \\
m^*_{y\,i,j} &= \rho v^*_{i,j} \\
&\Longrightarrow \text{for } i = 2 \text{ to } imax - 1 \text{ and } j = 2 \text{ to } jmax - 2
\end{aligned}
\tag{9.26}
$$

The above mass flux is at the centroid of the shaded v-CVs shown in Fig. 9.9b, and at the north and south face centers of the p-CVs (Fig. 9.9c). The mass flux corresponding to the v-velocity at the south and north boundaries of the domain is obtained from the boundary condition as

$$m^*_{y\,i,1} = \rho v_{i,1} \text{ and } m^*_{y,i,jmax-1} = \rho v_{i,jmax-1} \tag{9.27}$$

Mass Source Prediction:
Using the above predicted mass flux, calculate the mass source (Eq. 9.11) at the internal grid points of the p-CVs (Fig. 9.9c) as

$$S^*_{m\,i,j} = (m^*_{x\,i,j} - m^*_{x\,i-1,j})\Delta y + (m^*_{y\,i,j} - m^*_{y\,i,j-1})\Delta x$$
$$\Longrightarrow \text{for } i = 2 \text{ to } imax - 1 \text{ and } j = 2\text{to}jmax - 1 \tag{9.28}$$

9.4.3.2 Corrector Step

Using the Gauss-Seidel method, compute the pressure correction (Eq. 9.17) at the internal grid points of the p-CVs (Fig. 9.9c) as

$$
\left.
\begin{aligned}
p'^{,N+1}_{i,j} &= p'^{,N}_{i,j} - \frac{1}{a_p}\left(S^{*,N}_{m\,i,j} + S'_m\right) \\
S'_m &= -\Delta t \left[\left(\frac{p'^{,N}_{i+1,j} - p'^{,N}_{i,j}}{\delta x_i}\right) - \left(\frac{p'^{,N}_{i,j} - p'^{,N+1}_{i-1,j}}{\delta x_{i-1}}\right)\right]\Delta y \\
&\quad -\Delta t \left[\left(\frac{p'^{,N}_{i,j+1} - p'^{,N}_{i,j}}{\delta y_j}\right) - \left(\frac{p'^{,N}_{i,j} - p'^{,N+1}_{i,j-1}}{\delta y_{j-1}}\right)\right]\Delta x \\
a_p &= \Delta t \left[\left(\frac{1}{\delta x_i} + \frac{1}{\delta x_{i-1}}\right)\Delta y_i + \left(\frac{1}{\delta y_j} + \frac{1}{\delta y_{j-1}}\right)\Delta x_i\right]
\end{aligned}
\right\}
\begin{aligned}
&\text{for } i = 2 \text{ to } imax - 1 \\
&\text{for } j = 2 \text{ to } jmax - 1
\end{aligned}
$$
$$\tag{9.29}$$

After an iteration $N + 1$ for the pressure-correction, update the value of the mass-flux correction (Eq. 9.13) and the predicted mass flux as well as velocity (Eq. 9.18) as

$$
\begin{aligned}
m'^{,N+1}_{x\,i,j} &= -\Delta t\, \frac{p'^{N+1}_{i+1,j} - p'^{,N+1}_{i,j}}{\delta x_i}; \; m'^{,N+1}_{y\,i,j} = -\Delta t\, \frac{p'^{N+1}_{i,j+1} - p'^{,N+1}_{i,j}}{\delta y_j}; \\
m^{*,N+1}_{x\,i,j} &= m^{*,N}_{x\,i,j} + m'^{,N+1}_{x\,i,j}; \; m^{*,N+1}_{y\,i,j} = m^{*,N}_{y\,i,j} + m'^{,N+1}_{y\,i,j} \\
u^{*,N+1}_{i,j} &= m^{*,N+1}_{x\,i,j}/\rho; \; v^{*,N+1}_{i,j} = m^{*,N+1}_{y\,i,j}/\rho \\
&\Longrightarrow \text{for } i = i1\text{to}imax - 1 \text{ and } j = j1\text{to}jmax - 1
\end{aligned}
\tag{9.30}
$$

where $i1 = 1$ and $j1 = 2$ for m_x and vice versa for m_y; computed at the face centers shown by square and triangle symbols of the p-CVs in Fig. 9.9c. Furthermore, the new iterative value of the mass source and pressure (Eq. 9.19) is given as

$$
\begin{aligned}
S^{*,N+1}_{m\,i,j} &= (m^{*,N+1}_{x\,i,j} - m^{*,N+1}_{x\,i-1,j})\Delta y + (m^{*,N+1}_{y\,i,j} - m^{*,N+1}_{y\,i,j-1})\Delta x \\
p^{*,N+1}_{i,j} &= p^{*,N}_{i,j} + p'^{,N+1}_{i,j} \\
&\Longrightarrow \text{for } i = 2\text{to}imax - 1 \text{ and } j = 2\text{to}jmax - 1
\end{aligned}
\tag{9.31}
$$

For the solution of Eq. 9.29 for the pressure-correction, note that the mass source $S_{m\,i,j}^{*,N}$ is the predicted value (Eq. 9.28) for the first iteration and the corrected/updated value (Eq. 9.31) for the subsequent iterations. Once converged with $S_{m,p}^{*} \leq 0$, the pressure, velocity, and mass flux for the next time step are given as

$$p_{i,j}^{n+1} = p_{i,j}^{*,N+1}, \quad u_{i,j}^{n+1} = u_{i,j}^{*,N+1}, \quad v_{i,j}^{n+1} = v_{i,j}^{*,N+1},$$
$$m_{x\,i,j}^{n+1} = m_{x\,i,j}^{*,N+1} \,\&\, m_{y\,i,j}^{n+1} = m_{y\,i,j}^{*,N+1} \tag{9.32}$$

9.4.3.3 Boundary Conditions for the Pressure-Correction

The commonly encountered BCs in the fluid mechanics correspond to a Dirichlet BCs for either pressure or normal velocity (and mass flux). Thus, at a boundary of the flow domain, the pressure-correction BC is given as

$$\text{Dirichlet BC} \begin{cases} \text{Pressure } p : & p' = 0 \\ \text{Normal Velocity } u_n : & \partial p'/\partial \eta = 0 \end{cases} \tag{9.33}$$

where η is the coordinate normal to the boundary.

For the prescribed pressure BC, *no correction* is needed for the pressure; thus, the above equation shows zero pressure-correction. Whereas, for the prescribed BC for the normal velocity u_n, the mass flux (ρu_n) is also known at the boundary; thus, the mass-flux correction should be zero which leads to the above BC $\partial p'/\partial \eta = 0$ (the gradient of the pressure-correction normal to the boundary as zero) from Eq. 9.13. Since the mass-flux correction as a function of pressure-correction is similar to the Fourier law of heat conduction, the *zero mass flux correction*-based BC $(\partial p'/\partial \eta = 0)$ is analogous to the *zero conduction flux* $(\partial T/\partial \eta = 0)$-based insulated BC in the heat transfer. Note from Eq. 9.33 that the pressure-correction BC is either the Dirichlet or Neumann type of BC.

9.4.3.4 Stability and Steady-State Convergence Criterion

For the semi-explicit method, there is a time-step restriction for the timewise convergence of the flow field; otherwise, the method is numerically unstable. This was presented as a stability criterion for the explicit method in the previous chapters on the heat transfer. An exact stability analysis is available for a diffusion equation and an advection equation, but not for a Navier-Stokes equation. Thus, the stability criterion presented earlier separately for the diffusion and advection phenomenon corresponds here as the *necessary criterion* for convergence; however, they are not the *sufficient criterion* for convergence. The stability criterion commonly considered in the CFD is as follows:

- **CFL (Courant-Friedrichs-Lewy) Criterion:** This is obtained for a 2D advection with the FOU scheme (Eq. 6.28) as

$$\Delta t_{adv} \left(\frac{|u|}{\Delta x} + \frac{|v|}{\Delta y} \right) \leq 1 \tag{9.34}$$

- **Grid Fourier Number Criterion:** This is obtained for a 2D diffusion (Eq. 5.39, with v instead of α) as

$$v \Delta t_{diff} \left(\frac{1}{\Delta x^2} + \frac{1}{\Delta y^2} \right) \leq \frac{1}{2} \tag{9.35}$$

Here, the time step is computed individually from both the above criteria, and the smaller time step is used in the CFD development, given as

$$\Delta t_{stab} = \min(\Delta t_{adv}, \Delta t_{diff}) \tag{9.36}$$

However, since the time step Δt_{stab} is necessary but not sufficient for convergence, a slightly smaller value of the time step is used until the steady-state convergence is achieved.

For those CFD problems which reaches to the steady state, one of the commonly used convergence criteria during the unsteady CFD solution is given as

$$max \left(\left| \frac{U_{i,j}^{n+1} - U_{i,j}^n}{\Delta \tau} \right|_{max} \text{for all } (i,j), \left| \frac{V_{i,j}^{n+1} - V_{i,j}^n}{\Delta \tau} \right|_{max} \text{for all } (i,j) \right) \leq \epsilon_{st} \tag{9.37}$$

where the non-dimensional velocity $\vec{U} = \vec{u}/u_c$ and time $\tau = u_c t/l_c$. The above equation represents the stopping criterion as the maximum value of the non-dimensional velocity gradient (among all the grid points) should be less than the steady-state convergence tolerance ϵ_{st}.

9.4.4 Solution Algorithm

For the explicit method and the flux-based solution methodology on a uniform grid, presented above with the FOU scheme, the algorithm for 2D unsteady-state CFD is as follows:

1. User input:

   (a) Thermo-physical property of the fluid: enter density and viscosity.
   (b) Geometrical parameters: enter the length of the domain L_1 and L_2.
   (c) Grid: enter the maximum number of grid points $imax$ and $jmax$.
   (d) IC (initial condition) for the velocity and pressure.
   (e) BCs (Boundary conditions): enter the parameter corresponding to the BCs (such as lid velocity for the example problem shown in Fig. 9.11).

(f) Governing parameter: enter the parameter in a non-dimensional form (such as Reynolds number in the example problem below).

(g) Steady-state convergence tolerance: enter the value of ϵ_{st} (practically zero defined by the user). For the iterative solution of the pressure-correction equation, also enter the convergence tolerance ϵ for the predicted mass source $S^*_{m,P}$.

2. Calculate geometric parameters: $\Delta x = L_1/(imax - 2)$ and $\Delta y = L_2/(jmax - 2)$.

3. Apply the IC for u, v and p, and also for the predicted velocity with $u^* = u$ and $v^* = v$.

4. Apply the BC for u, v, u^*, v^*, and p. Then, the velocity and pressure at the present time instant are considered as the previous time-instant values $\phi_{old\,i,j} = \phi_{i,j}$ (for all the grid points), as the computations are to be continued further for the computation of their values $\phi_{i,j}$ for the present time instant.

5. Calculate the time step from the CFL and grid Fourier number criteria, and use the smaller value as the time step (Eqs. 9.34–9.36).

6. Prediction of the u and v velocities and the mass flux:

   (a) Compute mass, advection and diffusion fluxes at all the face centers in the domain, for the u-CVs (Eqs. 9.20 and 9.21) and the v-CVs (Eqs. 9.24 and 9.25).

   (b) Compute total out-flux and then predict the velocity and mass flux for the next time step (Eq. 9.22 for u^* and m_x, and Eq. 9.26 for v^* and m_y) at the internal grid points for the velocity CVs.

   (c) For non-Dirichlet BC, apply the BC to update the u^*/v^* and compute m^*_x/m^*_y (Eqs. 9.23 and 9.27) at the boundary grid points of the velocity CVs.

7. Compute the mass source $S^*_{m,P}$ from the predicted mass flux (Eq. 9.28). If $S^*_{m,P} < \epsilon$, then the predicted velocity, predicted mass flux, and the previous time-instant value of the pressure become the present time-instant values, and go to step 14 else continue.

8. Apply the initial condition for the pressure correction as $p' = 0$.

9. Apply the appropriate boundary conditions for the pressure-correction.

10. Compute the pressure-correction $p'^{,N+1}_{i,j}$ (Eq. 9.29), using the latest available pressure-correction of the neighboring grids points in the Gauss-Seidel method.

11. Compute the mass-flux correction $m'^{,N+1}_{x/y\,i,j}$ and the updated value of the predicted mass flux $m^{*,N+1}_{x/y\,i,j}$ and the predicted velocity $\overrightarrow{u}^{*,N+1}_{i,j}$ (Eq. 9.30).

12. Compute the updated value of mass source $S^{*,N+1}_{m\,i,j}$ and pressure $p'^{,N+1}_{i,j}$ (Eq. 9.31). If $S^{*,N+1}_{m\,i,j} \leq 0$ then continue else go to step 9.

13. The updated values of pressure, velocity, and mass flux correspond to the present time instant (Eq. 9.32).

14. Calculate a stopping parameter—steady-state convergence criterion (L.H.S of Eq. 9.37)—called here as unsteadiness. If the unsteadiness is greater than steady-state tolerance ϵ_{st} (Eq. 9.37), go to step 4 for the next time-instant solution.

15. Otherwise, stop and post-process the steady-state results for the analysis of the CFD problem. In an inherently unsteady problem, the condition in previous step will never be satisfied and a different stopping parameter is used.

9.5 Initial and Boundary Conditions

For fluid dynamics and heat transfer, although the initial and boundary conditions were presented in Chap. 3, more about it in the context of the CFD is presented here. The present book is on the methods for unsteady heat and fluid flow problems, with timewise accurate transient results, except the finite difference method-based steady-state solution, presented in Chap. 4. However, the present approach is often used to solve steady-state problems, using a false-transient approach. The approach starts with an initial condition for the velocity and pressure fields and proceeds with time to obtain the steady-state solutions.

9.5.1 Initial Condition

For such false-transient problem or in an inherently unsteady problem, the initial condition for the velocity is mostly known but not for the pressure. A compatible initial condition for the pressure is computed from the initial velocity field by solving a pressure Poisson equation, given as

$$\nabla^2 p = -\rho \nabla \cdot \left(\vec{u} \cdot \nabla \vec{u} \right) \tag{9.38}$$

This equation is obtained by the divergence operation on the momentum equation and applying the continuity equation, with suitable boundary conditions (Sharma and Eswaran 2003). Using an arbitrary instead of the compatible initial condition for the pressure results in a pseudo-pressure field, without affecting the correctness of the velocity field.

The pseudo- and the actual pressure field differs by an arbitrary constant. Thus, the pseudo-pressure field correctly computes those engineering parameters which requires the relative differences (not the absolute value) of the pressure, such as the lift (and drag) force for an external flow and pressure drop (and gradient) in an internal flow problem. Since the determination of the compatible initial condition for the pressure is complicated, most often an initial condition of zero is used. Then, after obtaining the velocity field, the pressure Poisson equation (Eq. 9.38) is solved to get the actual absolute value of the pressure if needed. Mostly, the velocity field starts with the quiescent (zero) conditions and the pressure may be assumed to be constant, often atmospheric, with a gage pressure of zero.

9.5.2 Boundary Condition

For an external or internal heat and fluid fluid flow problem (Fig. 2.7b,c), the boundary conditions for the velocity, pressure, and pressure-correction are as follows:

- **No-Slip BC:** For the mostly encountered stationary walls, the BC is given as

$$\vec{u} = 0, \; \frac{\partial p}{\partial \eta} = \mu \frac{\partial^2 u_n}{\partial \eta^2} \approx 0, \; \frac{\partial p'}{\partial \eta} = 0$$

  where η is the coordinate normal to the boundary.

- **Inlet BC:** The BC corresponds to a specified velocity u_{in} at the inlet boundary and is given as

$$u_\eta = u_{in}, \; u_t = 0, \; \frac{\partial p}{\partial \eta} = - \left[\frac{\partial (\rho u_\eta)}{\partial t} + \nabla \cdot (\rho \vec{u} \, u_\eta) - \mu \nabla^2 u_\eta \right] \approx 0, \; \frac{\partial p'}{\partial \eta} = 0$$

  where u_t is the velocity tangential to the boundary and u_η is the normal velocity. The u_{in} is specified as a constant or as a function of the coordinate along the boundary, and may be given a function of time, for a time-varying BC.

- **Free slip or symmetric BC:** The free-slip BC is mostly applied to the cross-stream boundary, for an external flow problem (Fig. 2.7b). This is an artificial boundary condition, to limit the size of the computational domain, for the unbounded physical domain, and is given as

$$u_\eta = 0, \; \frac{\partial u_t}{\partial \eta} = 0, \; \frac{\partial p}{\partial \eta} = \mu \frac{\partial^2 u_n}{\partial \eta^2} \approx 0, \; \frac{\partial p'}{\partial \eta} = 0$$

- **Outlet BC:** The BC corresponds to a boundary where the flow is leaving the domain, and is probably the most challenging BC in CFD. For an external flow over a body, a better outlet BC allows the vortices in the flow to exit the domain smoothly and has a negligible effect of the BC on the flow near the body. In this case also, the outlet BC is an artificial boundary to limit the streamwise length of the unbounded physical domain, without affecting the flow near the body. For the velocity at the outlet, a Neumann BC (corresponding to the fully developed flow, refer Fig. 2.7b,c) is the simplest and a convective BC is a better BC. Both the Neumann and convective BCs for velocity, along with the BC for pressure and pressure-correction, are given as

$$\frac{\partial \vec{u}}{\partial \eta} = 0, \; \frac{\partial \vec{u}}{\partial t} + U_c \frac{\partial \vec{u}}{\partial \eta} = 0, \; p = 0, \; p' = 0$$

Convective as compared to Neumann BC is demonstrated as a better BC, for flow across a cylinder, by Sohankar et al. (1998). They showed that the convective BC decreases the number of iterations per time step and the number of time step to reach a periodic flow, and allows the usage of a smaller streamwise length of the domain.

Note that the velocity BCs, corresponding to the no slip, free slip, and fully developed flow, are the *natural BC*, whereas both pressure and pressure-correction BCs are the *derived BCs*. Since the normal velocity is prescribed in all the above BC (except the outlet BC), the BC for the pressure-correction is the normal gradient equal to zero (Eq. 9.33). The pressure BC above corresponds to a simplified form of the momentum equation for the normal velocity u_η. However, they are commonly approximated as the Neumann BC, with the normal gradient equal to zero $\partial p/\partial \eta = 0$, except at the outlet boundary, where the pressure is atmospheric $p = 0$ (Dirichlet BC). From the above BCs, it is worth noting that the pressure as well as pressure-correction BCs are Neumann, if the BC for the normal velocity u_η is Dirichlet, and vice versa. Also note that both initial and boundary conditions are much more complicated for the pressure as compared to that for the velocity.

Example 9.1: CFD study for a 2D Lid Driven Cavity (LDC) flow.
The LDC flow is probably the most commonly used problem for the testing of an in-house Navier-Stokes (NS) solver in a 2D Cartesian grid. This is shown in Fig. 9.11, as a *square* cavity with the left, right, and bottom wall as stationary. The top wall, called here as the *lid*, acts like a long conveyor belt and is moving horizontally with a constant velocity u_0.

The motion results in a lid-driven recirculating flow inside the cavity. The cavity is represented by a closed 2D Cartesian square domain of size $L_1 = L_2$, with all the boundaries as the solid walls. Figure 9.11 also shows the initial and boundary conditions for the non-dimensional computational setup of the problem. The simplicity in the shape of the domain and the boundary conditions has led to the wide application of the LDC flow as the benchmark problem for a NS solver.

Using the flux-based solution methodology of the CFD development presented above for the *FOU scheme*, develop a computer program for the semi-explicit method-based 2D unsteady NS solver on a uniform staggered grid. The code should be written in a non-dimensional form, with the length of the cavity L_1 as the characteristic length, and the lid velocity u_0 as the characteristic velocity scale. They are shown in Fig. 9.11, along with the *governing parameter* for the isothermal flow as *Reynolds number* $Re = \rho u_0 L/\mu$. The Re is implemented in the code by a computational setup as $\rho = U_0 = L = 1$ and $\mu = 1/Re$. Use the developed and tested codes in the previous chapters (for conduction and advection heat transfer) as the *generic subroutines* in the present code development for the NS solver. After the CFD development, run the code with a convergence tolerance of $\epsilon_{st} = 10^{-3}$ for the steady state and $\epsilon = 10^{-8}$ for the mass conservation and perform the following study:

(a) *Steady-State flow patterns*: For a Reynolds number of 100 and a grid size of 42×42, present and discuss a figure for the velocity vector, pressure contour, and streamlines in the flow domain.

(b) *Grid-independence study*: For the Reynolds number of 100 and 4 *uniform* grid sizes (7×7, 12×12, 22×22, and 42×42), draw an overlap plot of the results (on all the grid sizes) as follows:

  (1) Variation of U-velocity along the *vertical* centerline.
  (2) Variation of V-velocity along the *horizontal* centerline.

  Discuss on the variation in the results with the change in the grid size.

(c) *Code-verification study*: Along with the results for the grid independence study, overlap the benchmark results reported by Ghia et al. (1982) and discuss the accuracy of the present results (on the grid size of 42×42) as compared to the benchmark results.

Solution:

(a) The flow properties obtained from the simulation are shown in Fig. 9.12a as velocity vector and Fig. 9.12 b as pressure contour. Figure 9.12b also shows an overlap plot of the streamlines as a flow visualization technique. The figure shows a clockwise recirculating flow and a primary vortex inside the cavity, with the pressure as largest on the top-right and smallest on the top-left corner of the cavity. The streamlines in the figure show a secondary, reactive, and counter-clockwise vortex on the bottom-left and bottom-right corners of the cavity.

(b) The steady-state results obtained by running the code on four different grid sizes are shown in Fig. 9.13, for the variation of U-velocity on the vertical and V-velocity on the horizontal centerlines of the cavity. Notice that the variation is in accordance with the flow patterns shown in Fig. 9.12. With grid refinement, the figure shows a much larger variation in the results among the coarser grids as compared to the variation among the finer grids. This indicates an asymptotic variation in the accuracy of the results, with grid refinement. Thus, the grid size of 42×42 is sufficiently fine for the grid-independent results, *i.e.*, there is almost no change in the results with further grid refinement.

(c) For the finest grid size of 42×42, Fig. 9.13 shows an excellent agreement between the present and the benchmark results. Since both the results are numerical, the agreement leads to the code verification (not code validation; refer Fig. 2.1).

Problems

9.1 Consider the example problem, *Example 9.1*, include the implementation of the QUICK scheme in the code. Using the QUICK scheme and a uniform grid size

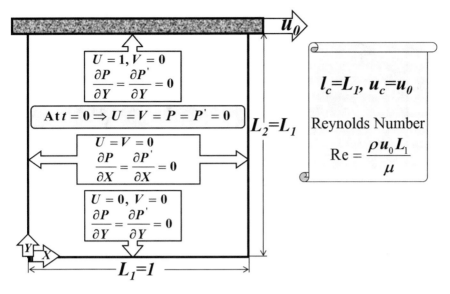

Fig. 9.11 Computational domain and boundary conditions for the lid-driven cavity flow

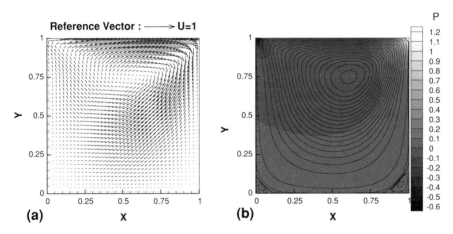

Fig. 9.12 CFD application of the semi-explicit method-based 2D Navier-Stokes solver on a uniform staggered Cartesian grid, for a lid-driven cavity flow at $Re = 100$ and a grid size of 42×42: steady-state results for (**a**) velocity vector and (**b**) pressure contour overlapped with the streamlines

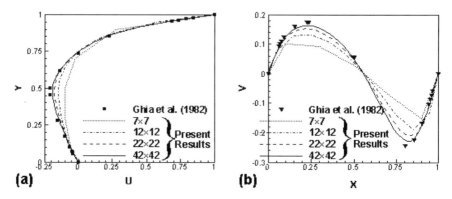

Fig. 9.13 For a Reynolds number of 100, variation of (**a**) U-velocity along the vertical centerline and (**b**) V-velocity along the horizontal centerline of a lid-driven cavity, considering the four different uniform grid sizes.

of 42×42, run the code for Reynolds number of 100, 400, and 1000. Similar to the example problem, present the flow patterns and code-verification study; refer (Ghia et al. 1982) for the benchmark results. Discuss the variation in the flow patterns and the agreement between the present and benchmark results, with the change in the Reynolds number.

9.2 Consider a rectangular cavity instead of the square cavity (Fig. 9.11) in *Example 9.1*, with (a) $L_2 = L_1/2$ and (b) $L_2 = 2L_1$. Run the FOU scheme-based code on a uniform grid size of 42×42, for the Reynolds number of 400. Finally, present the flow patterns and the centerline velocity profiles, similar to the example problem.

9.3 Consider the bottom wall along with the top wall of the cavity moving with the same velocity u_0 (Fig. 9.11) in *Example 9.1*; however, consider the direction of motion as opposite, *i.e.*, $u = -u_0$ at the bottom wall. Run the QUICK scheme-based code on a uniform grid size of 42×42, for the Reynolds number of 400. Finally, present the flow patterns and the centerline velocity profiles, similar to the example problem.

9.4 Consider a *developing flow in a plane channel*, shown in Fig. 9.14. The figure shows the non-dimensional computational setup, with $l_c = L_2$ and $u_c = u_\infty$. Modify the computational setup in the non-dimensional code (*Example 9.1*), and run the FOU scheme-based code on a uniform grid size of 82×22, for Reynolds number of 50. Finally, present the results as follows:

(a) Steady-state flow patterns: velocity vector, pressure contour, and streamlines in the flow domain.

(b) Variation of the streamwise U-velocity, along the horizontal centerline of the channel (at $Y = 0.5$).

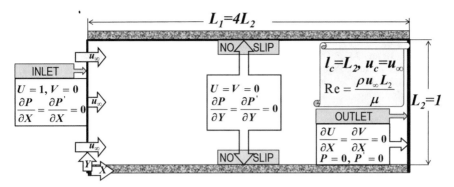

Fig. 9.14 Computational domain and boundary conditions for the developing flow in a plane channel

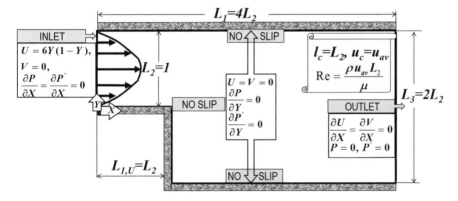

Fig. 9.15 Computational domain and boundary conditions for the backward facing step flow

   (c) Variation of the streamwise gradient of the cross-sectional averaged pressure, $\partial P_{av}/\partial X$ (non-dimensional value), along the non-dimensional streamwise X-direction.

9.5 Consider a *backward facing step flow*, with a fully developed flow at the inlet, shown in Fig. 9.15. Note from the figure that the cross-sectional average streamwise velocity u_{av} (equal to u_∞, from the previous problem) is considered here as the velocity scale. Modify the computational setup in the non-dimensional code (*Example 9.1*) and run the code for the both FOU and QUICK schemes, at a Reynolds number of 25 and uniform grid size of 82×22. Present the results (for both the schemes) similar to the previous problem, and compare the streamline pattern with the result presented by Biswas et al. (2004).

9.6 Solve the previous problem with $L_3 = 0.5L_2$ (Fig. 9.15), for the *forward facing step flow*.

References

Biswas, G., Breuer, M., & Durst, F. (2004). Backward-facing step flows for various expansion ratios at low and moderate Reynolds numbers. *ASME Journal of Fluids Engineering, 126,* 362–374.

Ghia, U., Ghia, K. N., & Shin, C. T. (1982). High-Re solutions for incompressible flow using the Navier-Stokes equations and a multigrid method. *Journal of Computational Physics, 48,* 387–411.

Harlow, F. H., & Welch, J. E. (1965). Numerical calculation of time-dependent viscous incompressible flow of fluid with free surface. *Physics of Fluids, 8,* 2182–2189.

Patankar, S. V. (1980). *Numerical heat transfer and fluid flow.* New York: Hemisphere Publishing Corporation.

Patankar, S. V., & Spalding, D. B. (1972). A calculation procedure for heat, mass and momentum transfer in three-dimensional parabolic flows. *International Journal of Heat and Mass Transfer, 15,* 1787–1806.

Sharma, A., & Eswaran, V. (2003). A finite volume method, Chapter 12. In K. Muralidhar & T. Sundararajan (Eds.), *Computational fluid flow and heat transfer* (2nd ed.). New Delhi: Narosa Publishing House.

Sohankar, A., Norberg, C., & Davidson, L. (1998). Low-Reynolds number flow across a square cylinder at incidence: Study of blockage, onset of vortex shedding and outlet boundary condition. *International Journal for Numerical Methods in Fluids, 26,* 39–56.

Chapter 10
Computational Fluid Dynamics on a Co-Located Grid

The solution methodology in the previous chapter, for the CFD on the *non-coinciding staggered grid points*, is extended to this chapter for a *coinciding* velocity, pressure, and temperature grid points—called as a *co-located grid*. The motivation is that the CFD on the co-located as compared to the staggered grid is much easier to implement, and computationally less expensive to use. This is more relevant for the non-orthogonal curvilinear grid-based CFD development on a complex geometry. However, the CFD on a co-located grid is first presented here for the simple geometry— on a non-uniform Cartesian grid; and will be presented later for the complex geometry in Chap. 12.

The previous chapter was presented with the flux-based solution methodology, and the present chapter will consider the coefficient of LAE-based methodology for the CFD development. Furthermore, the previous chapter was on a uniform grid, and the present chapter is on a non-uniform grid. The finite volume method for the present chapter was already presented in the Chap. 8; and the philosophy for the pressure-correction method (presented in the previous chapter) is also applicable to the present chapter.

However, to avoid the pressure-velocity decoupling (handled by the staggered grid in the previous chapter), the present chapter starts with the theory as well as the formulation for another method— called as momentum interpolation method; for the co-located grid system. Thereafter, using the coefficient of LAEs-based solution methodology on the non-uniform Cartesian grid, the semi-explicit as well as a semi-implicit method and their solution algorithm are presented. This is shown in Fig. 10.1 as the mind map for this chapter.

© The Author(s), under exclusive license to Springer Nature Switzerland AG 2022
A. Sharma, *Introduction to Computational Fluid Dynamics*,
https://doi.org/10.1007/978-3-030-72884-7_10

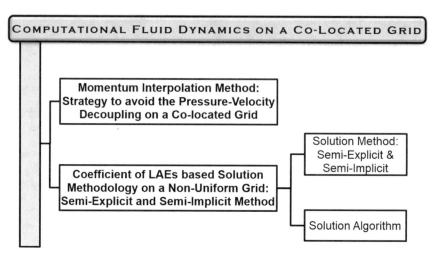

Fig. 10.1 Mind map for Chap. 10

10.1   Momentum Interpolation Method: Strategy to Avoid the Pressure-Velocity Decoupling on a Co-Located Grid

In the previous chapter, the *linear interpolation* (*II*-approximation) for both *mass flux and pressure* was shown to lead to the *pressure-velocity decoupling*; resulting in the physically unrealistic oscillatory pressure and velocity field. One of the first strategy to avoid the decoupling was the staggered grid, where the grid points for the velocity and the pressure are spatially staggered such that these linear interpolations are completely avoided. Another strategy is to use a *momentum interpolation*, instead of the linear interpolation, as the *II approximation for the mass flux* on a co-located grid. Thus, for the co-located as compared to the staggered grid, the interpolation is not avoided; instead, it is modified from the linear to the momentum interpolation for the mass flux. Whereas, for the pressure, the linear interpolation is avoided during the momentum interpolation-based computation of the mass fluxes (at the face centers) but not during the computation of the velocity (at the cell centers); presented below.

For a non-uniform grid in the computational domain, Fig. 10.2 shows the coinciding co-located grid points for the velocity, pressure, and temperature; the grid points are represented by circles in the figure. However, as compared to the co-located grid points, the figures shows the grid points for the mass flux (represented at the face centers by square and triangles) are staggered. Note that there are *two velocity* in the momentum transport: *first, advected as well as diffused velocity* which is considered *co-located* with pressure and temperature; and *second, advecting velocity* (normal to the surface area of a CV) which is shown as the *staggered* grid/CVs in the figure. The first velocity acts like a passenger, and the second velocity as the driver (for the bulk motion) in the momentum and energy transport process.

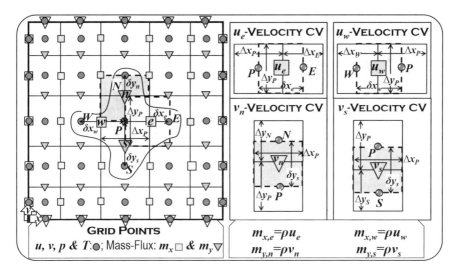

Fig. 10.2 Co-located grid in a computational domain for velocity, pressure, and temperature; and a representative staggered CV/grid for normal velocity/mass flux

A representative CV for the staggered normal velocity is shown in Fig. 10.2; separately for u_e, u_w, v_n, and v_s. The staggered CVs are shown here to demonstrate the formulation of the momentum interpolation, for the computation of the mass flux at the four face centers; however, note that the geometrical parameters corresponding to the staggered CVs will be avoided in the final equations (presented below). Thus, the role of the staggered grid here is limited to the formulation of the momentum interpolation by applying the already presented physical law-based FVM and semi-explicit method for CFD. Considering the staggered CV for the normal velocity u_e (Fig. 10.2), the resulting final discretized form of LAE for the prediction of the velocity (similar to Eq. 9.9) is given as

$$u_e^* = u_e^n + \frac{\Delta t}{\rho \Delta V_{u,e}} \left[\left(D_{u,e}^n - A_{u,e}^n \right) + \left(p_P^n - p_E^n \right) \Delta y_P \right]$$
$$= \tilde{u}_e^n + \frac{\Delta t}{\rho} \frac{\left(p_P^n - p_E^n \right)}{\delta x_e}$$

where $\tilde{u}_e^n = u_e^n + \dfrac{\Delta t}{\rho \Delta V_{u,e}} \left(D_{u,e}^n - A_{u,e}^n \right) \approx \dfrac{\Delta x_P \tilde{u}_E^n + \Delta x_E \tilde{u}_P^n}{\Delta x_P + \Delta x_E} \equiv \overline{\tilde{u}_E, \tilde{u}_P}$

$$(10.1)$$

where $\Delta V_{u,e} = \delta x_e \Delta y_P$ is the volume of the staggered CV for u_e; the width (δx_e and Δy_P) of the u_e−CV can be seen in Fig. 10.2. Furthermore, $\tilde{u}_e$ is a provisional velocity which corresponds to the predicted velocity without the source term for the pressure. The face center provisional velocity $\tilde{u}_e$ is approximated by a linear interpolation of the adjoining co-located cell center value of the provisional velocity $\tilde{u}_E$ and $\tilde{u}_P$. These velocities are predicted by the solution of an unsteady advection-diffusion equation (without the source term) for the x-momentum transport; will be presented below. Here, as shown in Eq. 10.1 for the provisional velocity, we define a "over-bar"

Table 10.1 Over-bar operator, at the various face centers, for the linear interpolation of the cell center values of a generic variable ϕ across a face f; to determine the face center value ϕ_f

$f \downarrow$		
	$\phi_f \approx \overline{\phi_{NB}, \phi_P}$	$\equiv (\Delta s_P \phi_{NB} + \Delta s_{NB} \phi_P) / (\Delta s_P + \Delta s_{NB})$
e	$\phi_e \approx \overline{\phi_E, \phi_P}$	$\equiv (\Delta x_P \phi_E + \Delta x_E \phi_P) / (\Delta x_P + \Delta x_E)$
w	$\phi_w \approx \overline{\phi_W, \phi_P}$	$\equiv (\Delta x_P \phi_W + \Delta x_W \phi_P) / (\Delta x_P + \Delta x_W)$
n	$\phi_n \approx \overline{\phi_N, \phi_P}$	$\equiv (\Delta y_P \phi_N + \Delta y_N \phi_P) / (\Delta y_P + \Delta y_N)$
s	$\phi_s \approx \overline{\phi_S, \phi_P}$	$\equiv (\Delta y_P \phi_S + \Delta y_S \phi_P) / (\Delta y_P + \Delta y_S)$

operator for the "linear interpolation". This is shown in Table 10.1 for the various face center in a 2D Cartesian CV.

Similar application of the FVM and the semi-explicit solution method for the staggered CVs (for u_w, v_n and v_s in Fig. 10.2) results in their predicted values. The predicted normal velocity as well as the corresponding mass flux are given for all the face centers as

Semi-Explicit Method:

$$u_e^* = \widetilde{u}_e^n + \frac{\Delta t}{\rho} \frac{(p_P^n - p_E^n)}{\delta x_e} \left\{ \widetilde{u}_e^n = \overline{\widetilde{u}_E, \widetilde{u}_P} \right\}; \quad m_{x,e}^* = \rho u_e^*$$

$$u_w^* = \widetilde{u}_w^n + \frac{\Delta t}{\rho} \frac{(p_W^n - p_P^n)}{\delta x_w} \left\{ \widetilde{u}_w^n = \overline{\widetilde{u}_W, \widetilde{u}_P} \right\}; \quad m_{x,w}^* = \rho u_w^* \qquad (10.2)$$

$$v_n^* = \widetilde{v}_n^n + \frac{\Delta t}{\rho} \frac{(p_P^n - p_N^n)}{\delta y_n} \left\{ \widetilde{v}_n^n = \overline{\widetilde{v}_N, \widetilde{v}_P} \right\}; \quad m_{y,n}^* = \rho v_n^*$$

$$v_s^* = \widetilde{v}_s^n + \frac{\Delta t}{\rho} \frac{(p_S^n - p_P^n)}{\delta y_s} \left\{ \widetilde{v}_s^n = \overline{\widetilde{v}_S, \widetilde{v}_P} \right\}; \quad m_{y,s}^* = \rho v_s^*$$

where the over-bar operator for the linear interpolation is shown in Table 10.1, with $\phi = \widetilde{u}/\widetilde{v}$. Furthermore, the width of the co-located u_P/v_P–CVs (Δx_P and Δy_P) used in the interpolation can be seen in Fig. 10.2.

Since the above method considers the application of the momentum conservation law to the staggered CVs and corresponds to the linear interpolation of the provisional velocity, it is called as the momentum interpolation method. However, note that it uses the co-located grid point value of the pressure at the face center of the staggered CVs (for the normal velocities in Fig. 10.2). Thus, the pressure difference between two *adjacent* grid points now becomes the driving force for the mass flux between those grid points. This is exactly as it is in the staggered grid, and this act as the remedy to the pressure-velocity decoupling (Sharma and Eswaran 2003).

Also note that the II-approximation for the pressure is avoided during the prediction of the normal velocity (Eq. 10.2) at the staggered CV, but not during the prediction of the velocity at the cell center of the co-located CV (will be presented below). The momentum interpolation method, as a remedy to the pressure-velocity decoupling, was first proposed by Rhie and Chow (1983). Later, around the same time, (Majumdar, 1988) and (Miller and Schmidt, 88) have removed the problem of under relaxation parameter dependency of the results, observed in Rhie and Chow's formula. Date (1996) proposed a more general momentum interpolation-based pressure-correction algorithm. Melaaen (1992) and Choi et al. (1994a, 1994b) have reported performance comparisons between the CFD development on the staggered and co-located grid arrangements.

10.2 Coefficients of LAEs-based Solution Methodology on a Non-Uniform Grid: Semi-Explicit and Semi-Implicit Method

The discretized form of momentum equations (Eq. 8.13) are given as

$$
\left.
\begin{aligned}
&\rho \Delta V_P \frac{u_P^{n+1} - u_P^n}{\Delta t} = D_{u,P}^\chi - A_{u,P}^\chi + S_{u,P}^{n+1} \\
&\text{where } S_{u,P} = (\overline{p_W,\, p_P} - \overline{p_E,\, p_P})\, \Delta y_P \\
&\rho \Delta V_P \frac{v_P^{n+1} - v_P^n}{\Delta t} = D_{v,P}^\chi - A_{v,P}^\chi + S_{v,P}^{n+1} \\
&\text{where } S_{v,P} = (\overline{p_S,\, p_P} - \overline{p_N,\, p_P})\, \Delta x_P
\end{aligned}
\right\}
\quad
\begin{aligned}
&\chi = n \text{ for Semi-Explicit} \\[2em]
&\chi = (n+1) \text{ for Semi-Implicit}
\end{aligned}
$$

$$\tag{10.3}$$

where the over-bar operator represents the linear interpolation for the pressure (Table 10.1, with $\phi = p$), to obtain $S_{u,P} = (p_w - p_e)\Delta y_P$ and $S_{v,P} = (p_s - p_n)\Delta x_P$) in Eq. 8.7. Here, the time level for the *advection and the diffusion* term are either *explicit or implicit* (depends on χ) and the *pressure* term ($S_{u,P}^{n+1}$ and $S_{v,P}^{n+1}$) is always taken as *implicit*. Thus, for $\chi = n$, the method is called $\underline{\text{S}}$emi-$\underline{\text{E}}$xplicit (SI) as all the terms except pressure is explicit. Whereas, $\chi = n + 1$ in the above equations results here a fully implicit method; however, this method becomes a $\underline{\text{S}}$emi-$\underline{\text{I}}$mplicit (SI) due to an approximation while deriving the velocity correction equation below.

Thus, the existing methods for the incompressible Navier-Stokes equations are neither fully explicit nor fully implicit, instead they are either semi-explicit or semi-implicit. For both the methods, the mass flux (in the mass conservation) and the pressure term are considered implicitly. Both advection and diffusion term are considered explicit in the semi-explicit and implicit in the semi-implicit method. For both the methods, the coupling of the mass conservation with the momentum conservation reduces the mass-conservation equation to a Poisson equation for the pressure corrections; and solved by an iterative method. The iterative method is also used for the solution of the other flow properties (velocity and temperature for non-isothermal flow) in the semi-implicit method; thus, the equation for *each* flow property is solved in a *sequential iteration*.

Both explicit and implicit methods have been extensively used for more than 40 years. *Explicit* methods have the advantage of *algorithm simplicity*, while the *implicit* ones that of greater *numerical stability*. For non-uniform grid arrangements with very fine grid size (near a solid surface, say), the use of reasonably large time steps

Table 10.2 Coefficient of the LAE given in Eq. 10.4, for 2D unsteady fluid dynamics. Note that $\phi = u$ for the x-momentum and $\phi = v$ for the y-momentum conservation.

$aE = \left(-m_{x,e}^{-,n} + \dfrac{\mu}{\delta x_e}\right)\Delta y_P$	$aW = \left(m_{x,w}^{+,n} + \dfrac{\mu}{\delta x_w}\right)\Delta y_P$
$aN = \left(-m_{y,n}^{-,n} + \dfrac{\mu}{\delta y_n}\right)\Delta x_P$	$aS = \left(m_{y,s}^{+,n} + \dfrac{\mu}{\delta y_s}\right)\Delta x_P$
$aP = \lambda\left(\sum_{NB} aNB + S_{m,P}\right) + aP^0,\ \sum_{NB} aNB = aE + aW + aN + aS$	
$b = \left[aP^0 - (1-\lambda)\left(\sum_{NB} aNB + S_{m,P}\right)\right]\phi_P^n - \lambda A_{\phi,P}^{d,N} - (1-\lambda)A_{\phi,P}^{d,n}$	
$aP^0 = \rho\Delta V_P/\Delta t,\ S_{m,P} = \left(m_{x,e}^n - m_{x,w}^n\right)\Delta y_P + \left(m_{y,n}^n - m_{y,s}^n\right)\Delta x_P \approx 0$	

$$A_{\phi,P}^d = \left[\left(m_{x,e}^{+,n}\phi_e^{d,+} + m_{x,e}^{-,n}\phi_e^{d,-}\right) - \left(m_{x,w}^{+,n}\phi_w^{d,+} + m_{x,w}^{-,n}\phi_w^{d,-}\right)\right]\Delta y_P$$
$$\left[\left(m_{y,n}^{+,n}\phi_n^{d,+} + m_{y,n}^{-,n}\phi_n^{d,-}\right) - \left(m_{y,s}^{+,n}\phi_s^{d,+} + m_{y,s}^{-,n}\phi_s^{d,-}\right)\right]\Delta x_P$$

where $\phi_{f=e,w,n,s}^{d,\pm} = \phi_f^{HOS,\pm} - \phi_f^{FOU,\pm} = \phi_f^{\pm} - \phi_U$

is allowed by the implicit methods, but not by the explicit methods; due to numerical instability. However, explicit methods are favored by parallel computers, which is now the preferred architecture of the fastest computers. Explicit computations can be made at thousands of grid points simultaneously on such computers.

The above Eq. 10.3 is the unsteady advection-diffusion algebraic equation, with the source term for the pressure. Thus, using the deferred correction approach, the coefficient of LAE-based methodology for the unsteady heat convection (Sect. 7.3) is applicable here; with an additional source term for $S_{u,P}/S_{v,P}$. Considering a generic variable ϕ which is equal to u for x-momentum and v for y-momentum equation, the LAE form of the Eq. 10.3 is given as

$$aP\,\phi_P^{n+1} = aE\left[\lambda\phi_E^{n+1} + (1-\lambda)\phi_E^n\right] + aW\left[\lambda\phi_W^{n+1} + (1-\lambda)\phi_W^n\right]$$
$$+aN\left[\lambda\phi_N^{n+1} + (1-\lambda)\phi_N^n\right] + aS\left[\lambda\phi_S^{n+1} + (1-\lambda)\phi_S^n\right] + b + S_{\phi,P}^{n+1}$$

$$(10.4)$$

where the coefficient of the above LAE is given in Table 10.2.

Substituting $\lambda = 0$ (for the semi-explicit method) and $\lambda = 1$ (for the semi-implicit method) in the above equation and Table 10.2, the final LAE form for the two methods as

$$\textbf{SE}: \begin{cases} aP\,\phi_P^{n+1} = \sum_{NB} aNB\phi_{NB}^n + b + S_{\phi,P}^{n+1} \\ \text{where } aP = aP^0,\ b = \left(aP^0 - \sum_{NB} aNB\right)\phi_P^n - A_{\phi,P}^{d,n} \\ \sum_{NB} aNB\phi_{NB} = aE\,\phi_E + aW\,\phi_W + aN\,\phi_N + aS\,\phi_S \end{cases} \quad (10.5)$$

$$\textbf{SI}: \begin{cases} aP\,\phi_P^{n+1} = \sum_{NB} aNB\phi_{NB}^{n+1} + b + S_{\phi,P}^{n+1} \\ \text{where } aP = \sum_{NB} aNB + aP^0 \text{ and } b = aP^0\phi_P^n - A_{\phi,P}^{d,N} \end{cases} \quad (10.6)$$

where aE, aW, aN, aS, aP^0, and $A_{\phi,P}^d$ are given in Table 10.2.

10.2.1 Predictor Step

Prediction of the Velocity at the Cell Center:

The prediction considers the pressure at the previous time level n, the modifies the source term as $S_{u,P}^n$ and $S_{v,P}^n$ in the *original proposition* for the semi-explicit method (Eq. 10.3). This results in the predicted velocity at the co-located grid point as

$$\left.\begin{array}{l} \rho \Delta V_P \dfrac{u_P^* - u_P^n}{\Delta t} = D_{u,P}^\chi - A_{u,P}^\chi + S_{u,P}^n \\[3mm] \rho \Delta V_P \dfrac{v_P^* - v_P^n}{\Delta t} = D_{v,P}^\chi - A_{v,P}^\chi + S_{v,P}^n \end{array}\right\} \quad \begin{array}{l} \chi = n \text{ for } \mathbf{SE} \\[3mm] \chi = * \text{ for } \mathbf{SI} \end{array} \qquad (10.7)$$

where χ is the previous time level value for the semi-explicit method, and the *predicted* value for the *semi-implicit* method. The above equation is given in a LAE form for the general variable ϕ (u for x-momentum and v for y-momentum equation) as

$$\mathbf{SE:} \quad \left\{\begin{array}{l} aP \, \phi_P^* = \sum_{NB} aNB \, \phi_{NB}^n + b + S_{\phi,P}^n \\[3mm] \phi_P^* = \tilde{\phi}_P + \dfrac{S_{\phi,P}^n}{a_P} \left\{ \tilde{\phi}_P = \dfrac{\sum_{NB} aNB\phi_{NB}^n + b}{a_P} \right\} \end{array}\right. \qquad (10.8)$$

$$\mathbf{SI:} \quad \left\{\begin{array}{l} aP \, \phi_P^* = \sum_{NB} aNB \, \phi_{NB}^* + b + S_{\phi,P}^n \\[3mm] \phi_P^* = \tilde{\phi}_P + \dfrac{S_{\phi,P}^n}{a_P} \left\{ \tilde{\phi}_P = \dfrac{\sum_{NB} aNB\phi_{NB}^* + b}{a_P} \right\} \end{array}\right. \qquad (10.9)$$

Prediction of the Mass Flux at the various Face Centers:

The already presented momentum interpolation method-based prediction of the mass flux for the semi-explicit method (Eq. 10.1) is indeed applicable to the semi-implicit method here, with a change in the advection and diffusion term from explicit to predicted values, *i.e.*, $A_{u,e}^*$ and $D_{u,e}^*$, respectively. This leads to a LAE (similar to Eq. 10.8 and 10.9 for $\phi_P = u_e$; instead of $\phi_P = u_P$) to predict the normal velocity at the various face center as

$$\mathbf{SE:} \quad \left\{\begin{array}{l} ae \, u_e^* = \sum_{NB} aNB \, u_{NB}^n + b + \left(p_P^n - p_E^n\right) \Delta y_P \\[3mm] u_e^* = \tilde{u}_e + \dfrac{\left(p_P^n - p_E^n\right) \Delta y_P}{ae} \left\{ \tilde{u}_e = \dfrac{\sum_{NB} aNB \, u_{NB}^n + b}{ae} \right\} \\[3mm] = \overline{\tilde{u}_E, \tilde{u}_P} + \dfrac{\Delta t}{\rho} \dfrac{\left(p_P^n - p_E^n\right)}{\delta x_e} \left\{ a_e = \dfrac{\rho \delta x_e \Delta y_P}{\Delta t} \right\} \end{array}\right. \qquad (10.10)$$

$$
\mathbf{SI:} \left\{
\begin{array}{l}
ae\, u_e^* = \sum_{NB} aNB\, u_{NB}^* + b + \left(p_P^n - p_E^n\right) \Delta y_P \\[2mm]
u_e^* = \widetilde{u}_e + \dfrac{\left(p_P^n - p_E^n\right)\Delta y_P}{ae} \quad \left\{ \widetilde{u}_e = \dfrac{\sum_{NB} aNB\, u_{NB}^* + b}{ae} \right\} \\[4mm]
= \overline{\widetilde{u}_E, \widetilde{u}_P} + \left(p_P^n - p_E^n\right)\Delta y_P \times \overline{\dfrac{1}{aE}, \dfrac{1}{aP}}
\end{array}
\right. \qquad (10.11)
$$

where the provisional velocity at the east face center $\widetilde{u}_e$ is obtained from the linear interpolation of the cell center values across the face $\widetilde{u}_P$ and $\widetilde{u}_E$. They are obtained from Eq. 10.8 for the semi-explicit method, and Eq. 10.9 for the semi-implicit method; considering the coefficients for the LAEs in Table 10.2. Furthermore, the linear interpolation at the east face center is also needed for $1/ae$ in the semi-implicit but not in the semi-explicit method (Eq. 10.10 and 10.11). Finally, considering Eq. 10.5 and 10.6 along with their coefficients in Table 10.2, the coefficients presented in the above equations are given as

$$
\mathbf{SE:} \left\{ ae = ae^0 = \frac{\rho \Delta V_{e,u}}{\Delta t} = \frac{\rho \delta x_e \Delta y_P}{\Delta t} \right. \qquad (10.12)
$$

$$
\mathbf{SI:} \left\{
\begin{array}{l}
\dfrac{1}{ae} = \overline{\dfrac{1}{aE}, \dfrac{1}{aP}} \\[2mm]
aP = \left(\sum_{NB} aNB\right)_P + \rho \Delta V_P / \Delta t \\[2mm]
aE = \left(\sum_{NB} aNB\right)_E + \rho \Delta V_E / \Delta t
\end{array}
\right. \qquad (10.13)
$$

where the over-bar operator for the linear interpolation is given in Table 10.1, with ϕ equal to the inverse of the coefficient of the LAE for a grid point.

The equations for the prediction of the normal velocity and mass flux, at the various face center, are already given in Eq. 10.2 for the semi-explicit method. For the semi-implicit method, the equations are given as

$$
\mathbf{SI:} \left\{
\begin{array}{ll}
u_e^* = \overline{\widetilde{u}_E, \widetilde{u}_P} + \left(p_P^n - p_E^n\right)\Delta y_P \times \overline{\dfrac{1}{aE}, \dfrac{1}{aP}}; & m_{x,e}^* = \rho u_e^* \\[2mm]
u_w^* = \overline{\widetilde{u}_W, \widetilde{u}_P} + \left(p_W^n - p_P^n\right)\Delta y_P \times \overline{\dfrac{1}{aW}, \dfrac{1}{aP}}; & m_{x,w}^* = \rho u_w^* \\[2mm]
v_n^* = \overline{\widetilde{v}_N, \widetilde{v}_P} + \left(p_P^n - p_N^n\right)\Delta x_P \times \overline{\dfrac{1}{aN}, \dfrac{1}{aP}}; & m_{y,n}^* = \rho v_n^* \\[2mm]
v_s^* = \overline{\widetilde{v}_S, \widetilde{v}_P} + \left(p_S^n - p_P^n\right)\Delta x_P \times \overline{\dfrac{1}{aS}, \dfrac{1}{aP}}; & m_{y,s}^* = \rho v_s^*
\end{array}
\right. \qquad (10.14)
$$

Prediction of the Mass Source at the Cell Center:

The predicted mass flux is used to obtain the predicted *mass source* as

$$
S_{m,P}^* = \left(m_{x,e}^* - m_{x,w}^*\right)\Delta y_P + \left(m_{y,n}^* - m_{y,s}^*\right)\Delta x_P \qquad (10.15)
$$

10.2.2 Corrector Step

Correction of the Velocity at the Cell Center:

Since the original proposition is considered as the sum of the predictor and the corrector step, the equations for the *corrector step* is obtained by subtracting Eq. 10.8 from Eq. 10.5 for semi-explicit and Eq. 10.9 from Eq. 10.6 for semi-implicit method. Furthermore, substituting $\phi^{n+1} - \phi^* = \phi'$ (u' for x-momentum and v' for y-momentum) and $p^{n+1} - p^n = p'$, the velocity correction is given as

$$
\begin{aligned}
&\textbf{SE:} && \left\{ aP\,\phi'_P = S'_{\phi,P} \text{ where } aP = \rho\Delta V_P/\Delta t \right. \\
&\textbf{SI:} && \left\{ \begin{aligned} & aP\,\phi'_P = \sum_{NB} aNB\,\phi'_{NB} + S'_{\phi,P} \\ & \qquad\qquad \approx S'_{\phi,P} \text{ (fully implicit to semi-implicit)} \\ & \text{where } aP = \sum_{NB} aNB + \rho\Delta V_P/\Delta t \end{aligned} \right. \\
&&& S'_{u,P} = \left(p'_w - p'_e\right)\Delta y_P = \left(\overline{p'_W,\,p'_P} - \overline{p'_E,\,p'_P}\right)\Delta y_P, \\
&&& S'_{v,P} = \left(p'_s - p'_n\right)\Delta x_P = \left(\overline{p'_S,\,p'_P} - \overline{p'_N,\,p'_P}\right)\Delta x_P
\end{aligned}
$$

(10.16)

where the $\sum_{NB} aNB\,\phi'_{NB}$ is dropped in the equation for the semi-implicit method. This neglected term represents the influence of the pressure corrections at the neighboring grid points on the neighboring velocity, leading to the velocity correction at the grid point under consideration (u'_P and v'_P). This approximation leads to a change in the method from the fully implicit (Eq. 10.17 for the SI method) to semi-implicit. This is to avoid a velocity correction equation, which will involve the pressure correction at all the grid points in the computational domain (Patankar 1980); and continue with the solution strategy used in CFD—the solution of the LAEs with the sparse coefficient matrix. The approximation ensures that the velocity correction leads to a pressure-correction equation (presented below), where the pressure correction at a particular grid point is a function of the correction at the four neighboring (not all) grid points in the domain.

Correction of the Mass Flux at the various Face Centers:

Similar to Eq. 10.3 for u_P/v_P in the co-located CV, the discretized form of momentum conservation for u_e and v_n in the staggered CVs (Fig. 10.2) are given as

$$
\left.\begin{aligned}
\rho\Delta V_e \frac{u_e^{n+1} - u_e^n}{\Delta t} &= D_{u,e}^\chi - A_{u,e}^\chi + \left(p_P^{n+1} - p_E^{n+1}\right)\Delta y_P \\
\rho\Delta V_n \frac{v_n^{n+1} - v_n^n}{\Delta t} &= D_{v,n}^\chi - A_{v,n}^\chi + \left(p_P^{n+1} - p_N^{n+1}\right)\Delta x_P
\end{aligned}\right\}
\begin{aligned}
&\chi = n \text{ for } \textbf{SE} \\
&\chi = (n+1) \text{ for}\textbf{SI}
\end{aligned}
$$

(10.17)

From the above original proposition for the normal velocity at the east and north face center, the predicted values are obtained as

$$
\left.\begin{aligned}
\rho\Delta V_e \frac{u_e^* - u_e^n}{\Delta t} &= D_{u,e}^\chi - A_{u,e}^\chi + \left(p_P^n - p_E^n\right)\Delta y_P \\
\rho\Delta V_n \frac{v_n^* - v_n^n}{\Delta t} &= D_{v,n}^\chi - A_{v,n}^\chi + \left(p_P^n - p_N^n\right)\Delta x_P
\end{aligned}\right\}
\begin{aligned}
&\chi = n \text{ for } \textbf{SE} \\
&\chi = * \text{ for}\textbf{SI}
\end{aligned}
$$

(10.18)

The equations for the *corrector step* is obtained by subtracting Eq. 10.18 from Eq. 10.17. Furthermore, substituting $u_e^{n+1} - u_e^* = u_e'$, $v_n^{n+1} - v_n^* = v_n'$ and $p^{n+1} - p^n = p'$, the corrections for the normal velocity are given as

$$
\text{SE:} \begin{cases} ae\, u_e' = \left(p_P' - p_E'\right)\Delta y_P \text{ where } ae = \rho\delta x_e \Delta y_P / \Delta t \\ an\, v_n' = \left(p_P' - p_N'\right)\Delta x_P \text{ where } an = \rho\delta y_n \Delta x_P / \Delta t \end{cases}
$$

$$
\text{SI:} \begin{cases} ae\, u_e' = \sum_{NB} aNB\, u_{NB}' + \left(p_P' - p_E'\right)\Delta y_P \approx \left(p_P' - p_E'\right)\Delta y_P \\ an\, v_n' = \sum_{NB} aNB\, v_{NB}' + \left(p_P' - p_N'\right)\Delta x_P \approx \left(p_P' - p_N'\right)\Delta x_P \\ \text{where } ae = 1 \Big/ \left(\dfrac{1}{aE}, \dfrac{1}{aP}\right) \text{ and } an = 1 \Big/ \left(\dfrac{1}{aN}, \dfrac{1}{aP}\right) \end{cases} \tag{10.19}
$$

where, for the semi-implicit method, the terms corresponding to the velocity corrections of the neighboring grid points ($\sum_{NB} aNB\, u_{NB}'$ and $\sum_{NB} aNB\, v_{NB}'$) are neglected; as discussed above. The above velocity correction is used to obtain mass-flux correction at the east and north face centers as

$$
m_{x,e}' = \rho u_e' = -\rho\frac{(p_E' - p_P')\Delta y_P}{ae}; m_{x,w}' = \rho u_w' = -\rho\frac{(p_P' - p_W')\Delta y_P}{aw}
$$

$$
m_{y,n}' = \rho v_n' = -\rho\frac{(p_N' - p_P')\Delta x_P}{an}; m_{y,s}' = \rho v_s' = -\rho\frac{(p_P' - p_S')\Delta x_P}{as} \tag{10.20}
$$

where the equation for the mass-flux correction is for both semi-explicit and semi-implicit method. As presented in the previous chapter, this equation is similar to Fourier's law of heat conduction; thus, the above equation is presented in a diffusion coefficient form as

$$
m_{x,e}' = -\Gamma_{p',e}\frac{(p_E' - p_P')}{\delta x_e} \quad ; \quad m_{x,w}' = -\Gamma_{p',w}\frac{(p_P' - p_W')}{\delta x_w}
$$

$$
m_{y,n}' = -\Gamma_{p',n}\frac{(p_N' - p_P')}{\delta y_n} \quad ; \quad m_{y,s}' = -\Gamma_{p',s}\frac{(p_P' - p_S')}{\delta y_s} \tag{10.21}
$$

where $\Gamma_{p',f}$ is the numerical diffusion coefficient for the pressure correction at the various face centers and is given as

$$
\text{SE:} \begin{cases} \Gamma_{p',e} = \Gamma_{p',w} = \Gamma_{p',n} = \Gamma_{p',s} = \Delta t \\[4pt] \end{cases}
$$

$$
\text{SI:} \begin{cases} \Gamma_{p',e} \equiv \dfrac{\rho\delta x_e \Delta y_P}{ae} \quad \Gamma_{p',w} \equiv \dfrac{\rho\delta x_w \Delta y_P}{aw} \\[10pt] \Gamma_{p',n} \equiv \dfrac{\rho\delta y_n \Delta x_P}{an} \quad \Gamma_{p',s} \equiv \dfrac{\rho\delta y_s \Delta x_P}{as} \end{cases} \tag{10.22}
$$

$$
\frac{1}{ae} = \frac{1}{aE}, \frac{1}{aP}, \frac{1}{aw} = \frac{1}{aW}, \frac{1}{aP}, \frac{1}{an} = \frac{1}{aN}, \frac{1}{aP} \text{ and } \frac{1}{as} = \frac{1}{aS}, \frac{1}{aP}
$$

where the coefficients for the various face center are obtained by the linear interpolation (Table 10.1, with ϕ equal to the inverse of the coefficient), for the semi-implicit method. The above equation for the semi-implicit method is general, as it results in the equation for the semi-explicit method by substituting $ae = \rho\delta x_e \Delta y_P/\Delta t$, $aw = \rho\delta x_w \Delta y_P/\Delta t$, $an = \rho\delta y_n \Delta x_P/\Delta t$ and $as = \rho\delta y_s \Delta y_P/\Delta t$. For the semi-explicit method, substituting Δt for the diffusion coefficients in Eq. 10.21, note that the resulting equation for the mass-flux correction for the co-located grid is same as Eq. 9.13 for the staggered grid.

Pressure Correction:

The discretized form of mass conservation (Eq. 9.2 in terms of mass flux), for the mass source, is given as

$$S_{m,P}^{n+1} = \left(m_{x,e}^{n+1} - m_{x,w}^{n+1}\right) \Delta y_P + \left(m_{y,n}^{n+1} - m_{y,s}^{n+1}\right) \Delta x_P = 0$$

Substituting $m_{x/y,f}^{n+1} = m_{x/y,f}^{*} + m_{x/y,f}^{'}$, we get

$$\left(m_{x,e}^{'} - m_{x,w}^{'}\right) \Delta y_P + \left(m_{y,n}^{'} - m_{y,s}^{'}\right) \Delta x_P$$
$$= -\left[\left(m_{x,e}^{*} - m_{x,w}^{*}\right) \Delta y_P + \left(m_{y,n}^{*} - m_{y,s}^{*}\right) \Delta x_P\right]$$

For the various faces of a CV, substituting the mass-flux correction from Eq. 10.21, the above equation is given in a LAE form as

$$aP\, p_P^{'} = aE\, p_E^{'} + aW\, p_W^{'} + aN\, p_N^{'} + aS\, p_S^{'} + b$$
$$\text{where } aE = \Gamma_{p',e} \Delta y_P/\delta x_e,\ aW = \Gamma_{p',w} \Delta y_P/\delta x_w,$$
$$aN = \Gamma_{p',n} \Delta x_P/\delta y_n,\ aS = \Gamma_{p',s} \Delta x_P/\delta y_s,\ aP = \sum_{NB} aNB$$
$$b = -S_{m,P}^{*} = -\left[\left(m_{x,e}^{*} - m_{x,w}^{*}\right) \Delta y_P + \left(m_{y,n}^{*} - m_{y,s}^{*}\right) \Delta x_P\right] \tag{10.23}$$

where the diffusion coefficient $\Gamma_{p'}$ is given in Eq. 10.22. The above steady-state mass-conservation equation for the pressure correction is solved at each time step by an iterative method. This involves finding the correct pressure-correction iteratively such that the converged pressure-correction-based mass flow rate corrections (Eq. 10.21) when added to the predicted mass flow rate in (Eqs. 10.2 and 10.14) obeys the mass-conservation law. For the iterative solution of the pressure-correction equation, the convergence criterion considered is $\left| p_{i,j}^{',N+1} - p_{i,j}^{',N} \right|_{max} \leq \epsilon$; ϵ is the convergence tolerance.

The converged pressure correction p' is used to obtain velocity correction at the cell center (Eq. 10.16); and mass-flux correction at the face center (Eqs. 10.21 and 10.22). Then, for the present time step $n + 1$, the pressure, velocity, and mass fluxes are given as

$$p_P^{n+1} = p_P^{*} + p_P^{'};\ u_P^{n+1} = u_P^{*} + u_P^{'};\ v_P^{n+1} = v_P^{*} + v_P^{'}$$
$$m_{x/y,f=e,w,n,s}^{*,n+1} = m_{x/y,f}^{*} + m_{x/y,f}^{'} \tag{10.24}$$

Table 10.3 Characteristic velocity as well as non-dimensional diffusion coefficient and source term, for the various types of non-dimensional convection heat transfer study; with a characteristic temperature difference $\Delta T_c = T_H - T_C$. Here, Re is the Reynolds number, Pr is the Prandtl number, Gr is the Grashof number, Pe is the Peclet number, Ra is the Rayleigh number, and Ri is the Richardson number

Convection	Characteristic	Diffusion coefficient		Source term	
	velocity	$\Gamma^*_{U/V}$	Γ^*_θ	S^*_U	S^*_V
Forced	$u_c = u_0$	$1/Re$	$1/Pe$	0	0
Mixed	$u_c = u_0$	$1/Re$	$1/Pe$	0	$Ri \times \theta$
Natural	$u_c = \alpha/l_c$	Pr	1	0	$Ra\,Pr \times \theta$

where $Re = u_c l_c/\nu$, $Pr = \nu/\alpha$, and $Gr = g\beta\Delta T_c l_c^3/\nu^2$

Forced/Mixed: $Pe = Re\,Pr$	Mixed: $Ri = Gr/Re^2$	Natural: $Ra = Gr\,Pr$

The initial as well as boundary condition and stability criterion (for semi-explicit method) as well as stopping criterion (for the unsteady solution), presented in the previous chapter is also applied for the solution on a co-located grid here. The fluid dynamics results, obtained from the methodology presented above, are used to obtain heat transfer results. The coupling of the methodology for the fluid dynamics and the heat transfer is one-way for the forced convection, and two-way for the mixed and natural convection heat transfer. This is because the heat transfer induced buoyancy force is considered negligible in the forced convection, and finite in the mixed/natural convection heat transfer problems. Thus, the fluid flow is independent of the heat transfer for the forced convection and becomes dependent for the mixed/natural convection. This is also reflected in the governing equation for the various convective heat transfer problems, presented in non-dimensional form as

$$\frac{\partial \overrightarrow{U}}{\partial \tau} + \nabla \cdot \left(\overrightarrow{U}\,\overrightarrow{U} \right) = -\nabla P + \Gamma^*_{U/V}\nabla^2 \overrightarrow{U} + S^*_{U/V}$$
$$\frac{\partial \theta}{\partial \tau} + \nabla \cdot \left(\overrightarrow{U}\,\theta \right) = \Gamma^*_\theta \nabla^2 \theta \qquad (10.25)$$
$$\text{where } \overrightarrow{X} = \frac{\overrightarrow{x}}{l_c}, \ \tau = \frac{t\,u_c}{l_c}, \ \overrightarrow{U} = \frac{\overrightarrow{u}}{u_c}, \ P = \frac{p}{\rho u_c^2}, \text{ and } \theta = \frac{T - T_C}{T_H - T_C}$$

where the u_c, $\Gamma^*_{U/V/\theta}$ and $S^*_{U/V}$ are presented in Table 10.3.

For a *forced* convection heat transfer problem, the mass conservation obeying mass flux at the new time level m_f^{n+1} is used to compute the temperature at the cell center T_P^{n+1}; using the solution methodology already presented in the Chap. 7 on computational heat convection. Whereas, for the *mixed/natural* convection problem, the solution at each time step is mostly obtained first for temperature and then for velocity and pressure (Sharma, 2004). Thus, the previous time level mass flux m_f^n is used to obtain the temperature field T_P^{n+1} which is considered in the computation of the temperature-based source term in the $y-momentum$ equation. The source term $S^*_V = f(\theta)$, shown in Eq. 10.25 and Table 10.3, represents the heat transfer induced buoyancy force modeled as a body force.

10.2.3 Solution Algorithm

For the semi-explicit/semi-implicit method and the coefficient of LAEs-based solution methodology on a non-uniform grid, presented above, the algorithm for 2D unsteady-state CFD is as follows:

1. User input:

   (a) Thermo-physical property of the fluid: enter density, viscosity, specific heat, and thermal conductivity.

   (b) Geometrical parameters: enter the length of the domain L_1 and L_2.

   (c) Grid: enter the maximum number of grid points $imax$ and $jmax$.

   (d) IC (initial condition) for the velocity, pressure, and temperature.

   (e) BCs (Boundary conditions): enter the parameter corresponding to the BCs; such as lid velocity and the temperature of the walls, for the example problem shown in Fig. 10.3.

   (f) Governing parameter: enter the parameter in a non-dimensional form (such as Reynolds, Prandtl, and Richardson number in the example problem below).

   (g) Convergence tolerance: enter the value of ϵ_{st} (practically zero defined by the user) for the steady-state convergence. For the iterative solution of the pressure-correction equation and the velocity prediction in the semi-implicit method, also enter the iteration convergence tolerance ϵ.

2. Non-uniform Cartesian grid generation: Determine the coordinates of vertices of the CVs, using an equation for the non-uniform grid generation. Thereafter, compute the non-uniform width of the CVs $\Delta x_i / \Delta y_j$ and distance between the cell centers $\delta x_i / \delta y_j$; using the procedure, introduced in Chap. 5.

3. Apply the IC for u, v and p; and also for the predicted velocity with $u^* = u$ and $v^* = v$.

4. Apply the BC for u, v, u^*, v^*, and p. Then, the velocity and pressure at the present time instant is considered as the previous time-instant values $\phi_{old\,i,j} = \phi_{i,j}$ (for all the grid points), as the computations is to be continued further for the computation of their values $\phi_{i,j}$ for the present time instant.

5. Calculate the time step from the CFL and grid Fourier number criteria, and use the smaller value as the time step (Eqs. 9.34–9.36).

6. Predict the mass flux at the various face center from Eq. 10.2 for the semi-explicit method, and Eq. 10.14 for the semi-implicit method, using the initial values.

7. For the *mixed and natural convection problems*, solve the energy equation (Eq. 10.25 and Table 10.3) using the previous time step mass flux m_f^n and the algorithm presented in Chap. 7 for the heat convection. The resulting temperature field for the present time step is used in the solution of Y-momentum equation.

8. *Prediction of the cell center velocity and the face center mass flux*: compute the provisional and predicted velocity for the semi-explicit and semi-implicit method, using Eqs. 10.8 and 10.9 (solved iteratively), respectively. Use linear interpolation to obtain the provisional velocity and pressure at the face center;

and also for the inverse of the coefficient for the semi-implicit method. The interpolated values are used to predict the mass flux, using Eq. 10.2 for the semi-explicit method and Eq. 10.14 for the semi-implicit method. Finally, using the predicted mass flux, compute the mass source $S_{m,P}^*$ (Eq. 10.15).

9. Apply the initial condition as $p' = 0$ and appropriate boundary conditions for the pressure correction.

10. Compute the *pressure correction* (Eq. 10.23) iteratively, using the mass source $S_{m,P}^*$ and the latest available pressure correction of the neighboring grids points in the Gauss-Seidel method. Apply the pressure-correction BC after every iteration.

11. *Corrections and new time level value for pressure, velocity and mass flux*: From the converged solution of the pressure correction, compute the cell center velocity correction (Eq. 10.16) and face center mass-flux correction (Eqs. 10.21 and 10.22). Finally, these corrections are used to compute the new time level value of the pressure, velocity, and mass flux from Eq. 10.24.

12. Apply the BC for u, v, u^*, v^*, p, and T.

13. For the *forced convection* problems, solve the energy equation (Eq. 10.25 and Table 10.3) using the mass fluxes at the present time step m_f^{n+1}.

14. Calculate a stopping parameter—steady-state convergence criterion (L.H.S of Eq. 9.37)— called here as unsteadiness. If the unsteadiness is greater than steady state tolerance ϵ_{st} (Eq. 9.37), then the velocity and pressure at present time instant becomes the previous time-instant values as the computations is to be continued further. Go to step 4, for the next time-instant solution.

15. Otherwise, stop and post-process the steady-state results for analysis of the CFD problem. In an inherently unsteady problem, the condition in previous step will never be satisfied and a different stopping parameter is used.

Example 10.1: CFD study for a 2D Lid Driven Cavity (LDC) flow, with Forced/Mixed Convection

For the lid driven cavity flow, introduced in the previous chapter, a buoyancy-induced flow is generated due to a difference in the temperature of the top as compared to other walls of the cavity. This is shown in Fig. 10.3, with the temperature as T_H at the top and T_C at the other walls. The combined heat and fluid flow correspond to forced convection for extremely small buoyancy-induced flow as compared to the forced flow (induced by the lid motion). Whereas, considerable buoyancy-induced flow results in mixed convection heat transfer.

The figure also shows the initial and boundary conditions for the non-dimensional computational setup of the problem. It can be seen that the non-dimensional temperature, $\theta = (T - T_C)/(T_H - T_C)$, is equal to 1 at the top wall and 0 at the other walls of the cavity. Finally, the figure shows the various non-dimensional governing parameters, where the Gr as well as Ri are equal to zero for the forced convection and non-zero for the mixed convection.

Using the coefficient of LAEs-based methodology of CFD development presented above for the FOU scheme, develop a computer program for the semi-explicit method-based 2D unsteady NS solver on a *uniform* co-located grid. The code should be written in a non-dimensional form, with the governing parameters Re, Pr, and Ri implemented in the code as $\rho = c_p = U_0 = L_1 = L_2 = 1$, $\mu = 1/Re$, $k = 1/(RePr)$ and $S_V^* = Ri\theta$. Use the developed and tested codes in the previous chapters (for conduction and advection heat transfer) as the generic subroutines in the present code development. After the development of the NS solver, run the code for various Richardson number $Ri = 0$ and 100, at a constant $Re = 100$ and $Pr = 1$. Use a grid size of 42 × 42, with a convergence tolerance of $\epsilon_{st} = 10^{-3}$ for the steady state and $\epsilon = 10^{-8}$ for the pressure-correction equation. Report the results as follows:

(a) *Flow patterns*: For the mixed convection cases, present and discuss a figure for the velocity vector, streamlines, and temperature contour in the flow domain.
(b) *Code verification study*: For both forced and mixed convection cases, plot the variation of the non-dimensional temperature θ along the vertical centerline—overlapped with the benchmark results reported by Torrance et al. (1972). Discuss the accuracy of the present as compared to the benchmark results. Furthermore, present and discuss the figures as follows:

    (1) Variation of U-velocity on the vertical centerline.
    (2) Variation of V-velocity on the horizontal centerline.

Solution:

(a) The flow properties obtained from the simulation are shown in Fig. 10.4a for velocity vector and streamlines, and Fig. 10.4b for temperature contour. Figure 10.4a,c show twin vortex for the mixed as compared to a single vortex for the forced convection. Figure 10.4b,d shows temperature variation restricted near the upper wall for the mixed as compared to the force convection.
(b) Fig. 10.5 shows a good agreement between the present and published results for the variation of the non-dimensional temperature along the vertical centerline of the cavity. This as well as Fig. 10.6a shows that the temperature and U-velocity variation is restricted near the top wall for the mixed as compared to the forced convection. The formation of twin vortex for the mixed convection (Fig. 10.4a) results in a negative V-velocity near the left as well as right wall, shown in Fig. 10.6b.

Problems

10.1 Consider the example problem on 2D lid driven square cavity, *Example 9.1*, run the co-located grid-based NS solver (developed for *Example 10.1*) for various advection schemes (FOU, SOU, and QUICK), Reynolds numbers (100 and 1000) and uniform grid sizes (12 × 12 and 52 × 52). Thereafter, for *each* Reynolds number and grid size, draw an overlap plot for the centerline velocity

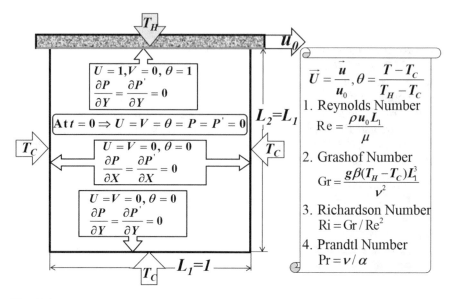

Fig. 10.3 Computational domain and boundary conditions for the lid driven cavity flow, with forced/mixed convection heat transfer

profiles (similar to Fig. 9.13) obtained by the FOU, SOU, and QUICK scheme; along with the benchmark results (Ghia et al. 1982). Discuss the accuracy of the various advection scheme by comparing with the benchmark results; similar study reported by (Sharma and Eswaran, 2002).

10.2 Solve the example problem, *Example 10.1*, for the CWT BC at the right wall (T_H) and left wall (T_C); and the other (top and bottom) walls as insulated.

10.3 For natural convection, the flow is only due to buoyancy with no forced flow. Thus, the lid is also taken as stationary here. Thus, the physical situation corresponds to a buoyancy-induced flow in a differentially heated closed square cavity. The computational setup for this problem is shown in Fig. 10.7, with the thermal BC similar to the previous problem and all the walls as stationary.

Using the non-dimensional governing equation (Eq. 10.25) and the corresponding variables in Table 10.3 for the natural convection, modify the code (developed for *Example 10.1*) to simulate the natural convective flow, on a grid size of 42×42, for $Pr = 0.71$ and $Ra = 10^3$. Use a convergence tolerance of $\epsilon_{st} = 10^{-3}$ for the steady state and $\epsilon = 10^{-8}$ for the pressure-correction equation. Report the results as follows:

(a) *Flow patterns*: Present and discuss the figures for streamlines, heatlines, and temperature contours in the flow domain.

(b) *Code verification study*: Plot and discuss the variation of V-velocity and temperature along the *horizontal* centerline of the cavity; overlapped with the benchmark results, reported by Deng and Tang (2002).

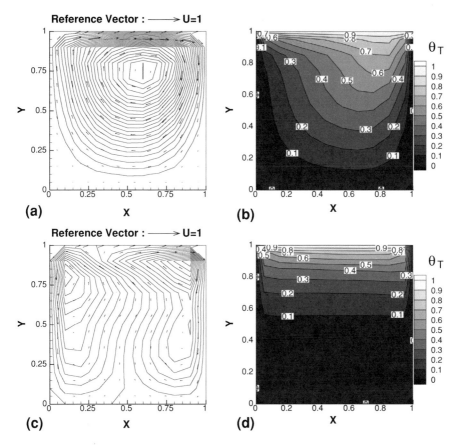

Fig. 10.4 CFD application of the semi-explicit method-based 2D code on a uniform co-located Cartesian grid, for a lid driven cavity flow with force/mixed convection heat transfer. The steady-state results are obtained on a grid size of 42×42 for a Richardson number of (**a,b**) 0 and (**c** and **d**) 100, at $Re = 100$ and $Pr = 1$: (**a** and **c**) velocity vector overlapped with the streamlines and (**b** and **d**) temperature contour

10.4 Modify the code (developed for the *Example 10.1*) for a simulation of free-stream forced convection flow across a cylinder; of square cross section. The computational setup for the problem is shown in Fig. 10.8. The figure shows both the types of thermal BC at the cylinder surface: CWT (constant wall temperature) and UHF (uniform heat flux). For both the thermal BCs, using a uniform grid size of 102×102 and FOU advection scheme, run the code for a Reynolds number of 40. Present the steady-state streamlines and isotherms; and compare with the flow patterns reported by Sharma and Eswaran (2004).

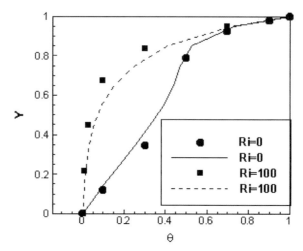

Fig. 10.5 Comparison of present (lines) and published (symbols; Torrance et al. 1972) results for the variation of temperature along the vertical centerline of a lid driven cavity, for various Ri at $Re = 100$ and $Pr = 1$

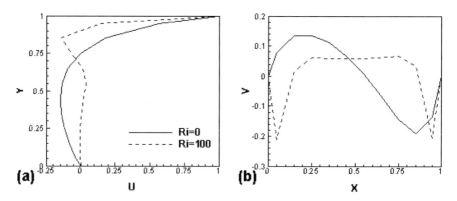

Fig. 10.6 Variation of (**a**) U-velocity along the vertical centerline and (**b**) V-velocity along the horizontal centerline of a lid driven cavity, for various Ri at $Re = 100$ and $Pr = 1$

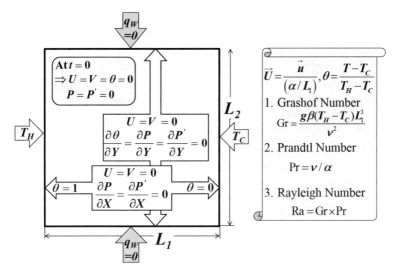

Fig. 10.7 Computational domain and boundary conditions for natural convection in a differentially heated square cavity

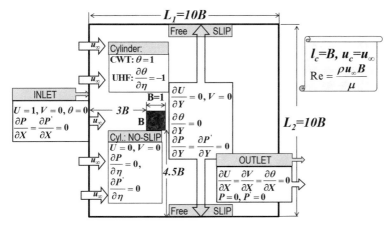

Fig. 10.8 Computational domain and boundary conditions for a free-stream flow across a square cylinder

References

Choi, S. K., Yun, N. H., & Cho, M. (1994a). Use of staggered and non-staggered grid arrangements for incompressible flow calculations on non orthogonal grids. *Numerical Heat Transfer B, 25,* 193–204.

Choi, S. K., Yun, N. H., & Cho, M. (1994b). Systematic comparison of finite volume methods with staggered and nonstaggered grid arrangements. *Numerical Heat Transfer B, 25,* 205–221.

Date, A. W. (1996). Complete pressure correction algorithm for solution of incompressible Navier-Stokes equations on a nonstaggered grid. *Numerical Heat Transfer B, 29,* 441–458.

Deng, Q. H., & Tang, G. F. (2002). Numerical visualization of mass and heat transport for conjugate natural convection/conduction by streamline and heatline. *International Journal of Heat and Mass Transfer, 45,* 2373–2385.

Ghia, U., Ghia, K. N., & Shin, C. T. (1982). High-Re solutions for incompressible flow using the Navier-Stokes equations and a multigrid method. *Journal of Computational Physics, 48,* 387–411.

Majumdar, S. (1988). Role of Under-relaxation in momentum interpolation for calculation of flow with nonstaggered grids. *Numerical Heat Transfer, 13,* 125–132.

Melaaen, M. C. (1992). Calculation of fluid flows with staggered and non-staggered curvilinear non-orthogonal grid − a comparison. *Numerical Heat Transfer B, 21,* 21–39.

Miller, T. F., & Schmidt, F. W. (1988). Use of a pressure-weighted interpolation method for the solution of incompressible Navier-Stokes equations on a non-staggered grid system. *Numerical Heat Transfer, 14,* 213–233.

Patankar, S. V. (1980). *Numerical heat transfer and fluid flow.* New York: Hemisphere Publishing Corporation.

Rhie, C. M., & Chow, W. L. (1983). A numerical study of the turbulent flow past an isolated airfoil with trailing edge separation. *AIAA Journal, 21,* 1525–1532.

Sharma, A., & Eswaran, V. (2002). Conservative and consistent implementation and systematic comparison of SOU and QUICK Scheme for recirculating flow computation on a non-staggered Grid. In: *Proc. 2nd Int. and 29th Nat. FMFP Conf.*, Roorkee, India, pp. 351–358.

Sharma, A., & Eswaran, V. (2003). A finite volume method, Chapter 12. In: K. Muralidhar & T. Sundararajan (Eds.), *Computational fluid flow and heat transfer* (2nd ed.). New Delhi: Narosa Publishing House.

Sharma, A., & Eswaran, V. (2004). Heat and Fluid Flow across a Square Cylinder in the Two-Dimensional Laminar Flow Regime. *Numerical Heat Transfer A, 45*(3), 247–269.

Torrance, K., Davis, R., Eike, K., Gill, P., Gutman, T., Hsui, A., et al. (1972). Cavity flow driven by buoyancy and shear. *Journal of Fluid Mechanics, 51*(2), 221–231.

Part III
CFD for a Complex-Geometry

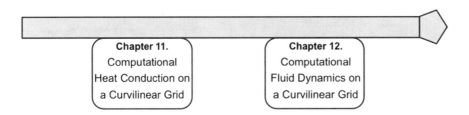

Chapter 11.
Computational
Heat Conduction on
a Curvilinear Grid

Chapter 12.
Computational
Fluid Dynamics on
a Curvilinear Grid

After building the foundation for a course in CFD (in the Part I), the CFD is presented first for the simpler Cartesian geometry (in the Part II) and then for the mostly encountered complex-geometry in the present Part III. The Part III consists of two chapters, starting with heat-conduction and ending with the fluid-dynamics. Both staggered and co-located grid based CFD development are presented in the previous part; whereas, in this part, CFD will be presented only on the colocated grid. This is because the staggered grid (with non-coinciding pressure and velocity CVs) is difficult to implement, and computationally expensive to use, in complex geometries. Furthermore, CFD on both uniform and non-uniform Cartesian grid systems are presented in the previous part; whereas, in this part, CFD for complex-geometry will be presented on non-uniform and non-orthogonal curvilinear grid. As compared to the previous part on two types of methodology (flux and coefficient of LAEs) and solution-method (explicit and implicit method), this part present the CFD development for the flux-based methodology and explicit-method. For many years to come, probably the three parts are sufficient for an introductory course in CFD.

Chapter 11
Computational Heat Conduction on a Curvilinear Grid

The numerical methodology for the heat conduction (presented in Chap. 5 for the Cartesian and simple geometry) is extended here for the complex geometry on a body-fitted curvilinear grid. The mind map for this chapter is shown in Fig. 11.1. The figure shows that the chapter starts with the algebraic and elliptic-PDE-based methods for a curvilinear grid generation. Thereafter, the physical law-based FVM, computation of geometrical properties for a curvilinear grid, and the flux-based solution methodology are presented.

11.1 Curvilinear Grid Generation

The curvilinear grid generation involves a mathematical equations-based discretization, of the continuous spatial Cartesian coordinates, in a complex physical-domain; leading to the coordinates of certain fixed number of discrete points. The points are joined by piecewise linear segments, such that they correspond to the vertices of the resulting CVs. The piecewise linear segments lead to the curvilinear grid lines; thus, called as the curvilinear grid. The coordinates of the vertices are sufficient enough to obtain the various other geometrical properties of the CVs. The geometrical properties are used in the FVM and solution methodology for the CFD development on a complex geometry; presented below. A curvilinear grid is presented here first from an algebraic equation and then from an elliptic partial differential equation, called as the algebraic and elliptic grid generation, respectively. Moreover, this is a much more detailed subject with exclusive books on grid generation; such as the classical book by Thompson et al. (1985) and a handbook by Thompson et al. (1988).

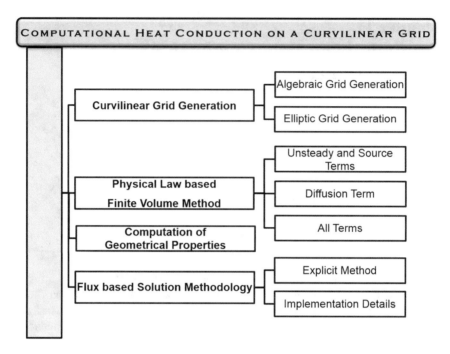

Fig. 11.1 Mind map for Chap. 11.

11.1.1 Algebraic Grid Generation

The algebraic grid generation is presented here for an example problem of a 2D heat transfer in a plate, shown in Fig. 11.2b. Figure 11.2 shows two different domains: computational in the ξ–ζ coordinate system, and physical in the Cartesian x–y coordinate system. Note that the computational-domain is square shaped, with a unit dimension ($\xi_{max} = \zeta_{max} = 1$). Furthermore, it can be seen that the uniform grid generation presented earlier for the Cartesian-geometry is used here for the grid generation in the computational-domain. This involves certain number of equi-spaced horizontal (ζ =constant) and vertical (ξ =constant) lines, and the determination of the ξ and ζ coordinates of the vertices of the resulting CVs. These ξ and ζ coordinates are used to determine the x– and y-coordinates of the corresponding vertices of the resulting CVs in the physical-domain, using an algebraic equation shown in Fig. 11.2b; thus, called as the algebraic grid generation. Note that this method is also used for the non-uniform Cartesian grid generation, presented in Prog. 5.5.

Note from the Fig. 11.2 a one-to-one correspondence between the computational and physical-domain, for the grid points (at the vertices) as well as the grid lines (ξ =constant and ζ =constant). Also note that the grid points are defined here at the vertices, for the grid generation methodology; as compared to the centroid in the FVM-based solution methodology. Finally, note from the figure that the grid

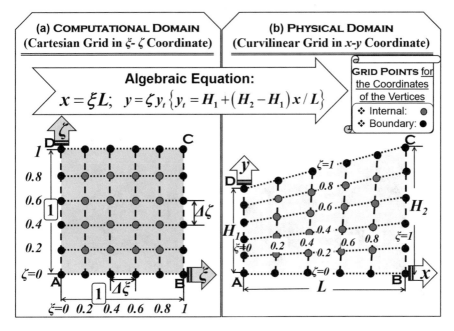

Fig. 11.2 Schematic of an algebraic equation-based generation of a boundary-fitted curvilinear grid, representing 2D heat transfer in a plate with a complex geometry.

lines are always horizontal and vertical in the computational-domain, and seen here as the inclined lines in the physical-domain. However, in general for a complex geometry, they are *piecewise linear segments-based curvilinear grid* lines in the physical-domain, presented below in Fig. 11.3b.

There are lots of *similarity* between the 2D *Cartesian* grid (presented earlier for the simple geometry) and the 2D *curvilinear* grid; presented here for the complex geometry. The 2D Cartesian grid is identified by a constant value of x-coordinate (and y-coordinate) based vertical (horizontal) grid lines. Similarly, a 2D curvilinear grid is represented in a physical-domain by a constant value of the curvilinear (ξ and ζ) coordinate-based grid lines. However, each of the coordinate-based grid lines is linear for the Cartesian grid and is piecewise linear segments-based curves for the curvilinear grid. Each segment of the physical-domain boundary is described by a constant value of one of the Cartesian coordinate (x/y) lines for the Cartesian simple geometry. Similarly, for a complex geometry, the boundary of the physical-domain is represented by a constant value of one of the curvilinear coordinate (ξ/ζ) lines; thus, it is called as the boundary or *body-fitted grid*. This is shown in Fig. 11.2, with $\xi = 0$ for the west-boundary, $\xi = 1$ for the east-boundary, $\zeta = 0$ for the south-boundary, and $\zeta = 1$ for the north-boundary of the physical-domain.

Another similarity between the *Cartesian* and the *curvilinear* grid is that both are *structured grid*, i.e., neighboring cell center information is in-built for a grid point (i, j): $(i + 1, j)$ for the east neighbor, $(i - 1, j)$ for the west neighbor, $(i, j + 1)$

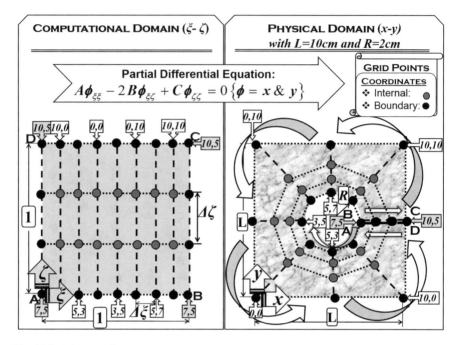

Fig. 11.3 Schematic for an elliptic PDE-based generation of a body-fitted curvilinear grid, representing 2D heat transfer in a plate with a hole. The marked arrow shows the (x, y) coordinates in *cm* for the various boundary grid points in the computation domain, obtained from the corresponding point in the physical-domain; for the hole radius $R = 2\,cm$ and dimension of the square plate $L = 10\,cm$.

for the north neighbor, and $(i, j - 1)$ for the south neighbor. This is because, for a Cartesian grid, the total number of intersection of the various x =constant lines with any y =constant line remains constant (and vice-versa); similarly, for the intersection of ξ =constant and ζ =constant lines in the curvilinear grid. In contrast to the uniform and orthogonal grid in the ξ–ζ computational-domain, it can be seen from Fig. 11.2 that the grid in the $x - y$ physical-domain are non-uniform and non-orthogonal.

11.1.2 Elliptic Grid Generation

The grid generation is presented here for an example problem of a 2D heat transfer a square plate with a hole, shown in Fig. 11.3b. This is a more complex geometry as compared to the plate in Fig. 11.2. Furthermore, along with the outer boundary for the plate, the hole in the plate here results in an inner boundary. Figure 11.3 shows that the inner circular-boundary and outer square-boundary of the physical-domain are not simply connected; whereas, the boundary of the square-shaped computational-domain is simply connected. There should be one-to-one correspondence between the boundary of the physical-domain and one or more of the curvilinear coordinates lines at the boundary of the computational-domain ($\xi = 0, \xi = 1, \zeta = 0$ and $\zeta = 1$)—to ensure the body-fitted grid. Thus, for the mapping of the two disconnected boundary of the complex physical-domain to the simply connected boundary of the square-shaped computational-domain, a cut is required—called as a *branch cut* and is shown as BC and AD (the two edges resulting from the cut) in Fig. 11.3b. The cut BC is seen as the east boundary, and AD as the west-boundary, of the computational-domain in Fig. 11.3a. Furthermore, it can be seen that the inner circular-boundary AB and outer square-boundary CD of the plate are mapped with the south and north-boundary of the computational-domain, respectively.

Considering the outer dimension of the plate as $L = 10$ cm and the radius of the hole as $R = 2$ cm, subdivide the inner boundary AB, branch cut BC/AD and outer boundary CD into certain equal division as seen in Fig. 11.3b. Using the geometrical information and considering an equi-spaced subdivision of the branch cut, square-boundary, and circular-boundary, the x–y coordinates of the resulting boundary grid points (shown as black color-filled circles in Fig. 11.3b) can be determined. These x–y coordinates in the physical-domain are transformed and correspond to the respective boundary grid points of the computational-domain, at the east-, west-, north-, and south-boundary in Fig. 11.3a. The value of the x–y coordinates are shown in Fig. 11.3, at the certain boundary grid points in both physical and computational domain. The transformed x–y coordinates (of the boundary grid points) act as the boundary conditions for a finite difference method based *numerical solution* of a partial differential equation in the *computational-domain*; presented below. Since there is one-to-one correspondence of the vertices in the computational-domain to that in the physical-domain, the solution results in the x–y coordinates at the vertices of the CVs in the physical-domain. Using the coordinates of the vertices (as the input) in a graphical software, any two consecutive vertices are joined by straight lines and results in the curvilinear grid (Fig. 11.3b).

Note that the methodology for the curvilinear grid generation involves determination of x–y coordinates at the vertices of the physical-domain, from the ξ–ζ coordinates at the vertices of the computational-domain, using an algebraic or partial differential equation. The differential equations considered here is the inverse transformation of an elliptic equation. The elliptic equation is given as

$$\nabla^2 \xi = 0 \text{ and } \nabla^2 \zeta = 0$$

Since the ξ and ζ coordinates are the dependent variables, and the x– and y-coordinates are the independent variables, an inverse transformation of the above equations is used in the elliptic grid generation; given (Hoffmann and Chiang 2000) as

$$\left.\begin{array}{l} A\dfrac{\partial^2 x}{\partial \xi^2} - 2B\dfrac{\partial^2 x}{\partial \xi \partial \zeta} + C\dfrac{\partial^2 x}{\partial \zeta^2} = 0 \\[3mm] A\dfrac{\partial^2 y}{\partial \xi^2} - 2B\dfrac{\partial^2 y}{\partial \xi \partial \zeta} + C\dfrac{\partial^2 y}{\partial \zeta^2} = 0 \end{array}\right\} \quad \begin{array}{l} A = \left(\dfrac{\partial x}{\partial \zeta}\right)^2 + \left(\dfrac{\partial y}{\partial \zeta}\right)^2 \\[3mm] B = \dfrac{\partial x}{\partial \xi}\dfrac{\partial x}{\partial \zeta} + \dfrac{\partial y}{\partial \xi}\dfrac{\partial y}{\partial \zeta} \\[3mm] C = \left(\dfrac{\partial x}{\partial \xi}\right)^2 + \left(\dfrac{\partial y}{\partial \xi}\right)^2 \end{array} \tag{11.1}$$

The above non-linear partial differential equation is solved using a finite difference method, presented in Sect. 4.1. The application of central difference-based FDM is given for a general variable ϕ (corresponding to x and y) as

$$\phi_{i,j} = \frac{A_{i,j}\Delta\zeta^2(\phi_{i+1,j} + \phi_{i-1,j}) + C_{i,j}\Delta\xi^2(\phi_{i,j+1} + \phi_{i,j-1}) - 2B_{i,j}\Delta\xi^2\Delta\zeta^2 d_{i,j}}{2\left(A_{i,j}\Delta\zeta^2 + C_{i,j}\Delta\xi^2\right)}$$

$$\text{where } d_{i,j} = \frac{\phi_{i+1,j+1} + \phi_{i-1,j-1} - \phi_{i-1,j+1} - \phi_{i+1,j-1}}{4\Delta\xi\Delta\zeta} \tag{11.2}$$

Furthermore,

$$A_{i.j} = \left(\frac{x_{i,j+1} - x_{i,j-1}}{2\Delta\zeta}\right)^2 + \left(\frac{y_{i,j+1} - y_{i,j-1}}{2\Delta\zeta}\right)^2; \tag{11.3}$$

$$C_{i.j} = \left(\frac{x_{i+1,j} - x_{i-1,j}}{2\Delta\xi}\right)^2 + \left(\frac{y_{i+1,j} - y_{i-1,j}}{2\Delta\xi}\right)^2;$$

$$B_{i.j} = \left(\frac{x_{i+1,j} - x_{i-1,j}}{2\Delta\xi}\right)\left(\frac{x_{i,j+1} - x_{i,j-1}}{2\Delta\zeta}\right)$$

$$+ \left(\frac{y_{i+1,j} - y_{i-1,j}}{2\Delta\xi}\right)\left(\frac{y_{i,j+1} - y_{i,j-1}}{2\Delta\zeta}\right)$$

Solution algorithm, for the FDM and Gauss-Seidel method based *elliptic grid generation* is as follows:

1. Enter the values of maximum number of grid points (and lines) in the ξ and ζ direction: $(imax - 1)$ and $(jmax - 1)$; and the convergence tolerance ϵ (say 10^{-4}). Note that the grid points here for the vertices are one less than the maximum number of cell centers ($imax$ and $jmax$); during the solution for flow properties, presented in the previous chapters.
2. Calculate the grid size in the computational-domain: $\Delta\xi = 1/(imax - 2)$ and $\Delta\zeta = 1/(jmax - 2)$; note that the domain is square, with a dimensionless size of unity.
3. Enter the initial guess and boundary conditions (for x and y; similar to the BCs shown in Fig. 11.3, at certain boundary grid points) in the computational-domain; and start the Gauss-Seidel iteration.
4. Assign $x_{old\,i,j} = x_{i,j}$ and $y_{old\,i,j} = y_{i,j}$, for all the i and j.

5. Using the $x_{i,j}$ and $y_{i,j}$, first compute $A_{i,j}$, $B_{i,j}$, $C_{i,j}$, $d_{i,j}$ and then obtained the new iterative value of $x_{i,j}$ as well as $y_{i,j}$. This is done for $i = 2$ to $imax - 2$ and $j = 2$ to $jmax - 2$, using Eqs. 11.2 and 11.3.

6. Check the convergence, using the convergence criterion $max\left(\left|x_{i,j} - x_{old\,i,j}\right|_{max}\right.$, $\left.\left|y_{i,j} - y_{old\,i,j}\right|_{max}\right) \le \epsilon$. If the convergence criterion is not satisfied, go to Step 4 and continue till convergence.

The elliptic grid presented in Fig. 11.3b is called as *O-type*; this is because the $\zeta =$ constant lines in the physical-domain of the figure are the enclosing curves which gradually change from circular to square. Moreover, there are other types of branch cut and the mapping of the boundary from the physical to the computational-domain. They are shown in Fig. 11.4 as *C-type* and *H-type*; along with the *O*-type of curvilinear grid. As compared to the *horizontal* branch cut (AD and BC) for the *O*-type of grid, inner circular-boundary is *also* considered in the branch cut A-A'-B'-B for the *C*-type of grid. For the *C*-type and the *H*-type of grid, the branch cut A-A'-B'-B corresponds only to the bottom boundary of the *computational-domain*; as compared to the branch cut BC for the east-boundary and AD for the west-boundary in case of the *O*-type grid. During the generation of *H*-type of grid, note from the figure that the computational-domain and the solution of the elliptic PDE results in the grid for the upper half of the physical-domain; thereafter, the grid for the lower half is easily obtained as it is symmetric about the branch cut A-A'-B'-B.

Example 11.1: Elliptic Grid Generation.
Consider a parallelogram shaped physical-domain, with all the sides as $L_1 = L_2 = 1$ m, shown in Fig. 11.5. Using the elliptic grid generation method, generate a curvilinear grid in the domain with $imax = jmax = 12$. For an accurate implementation of BCs for the flow properties (discussed later) present the formulation of the BCs (for the Cartesian coordinates) to ensure that the grid line intersecting a boundary of the physical-domain is normal at the boundary.

Solution:
For the orthogonality of the grid lines at the boundary, the condition is given (Sundararajan 2003) as

$$B = \frac{\partial x}{\partial \xi}\frac{\partial x}{\partial \zeta} + \frac{\partial y}{\partial \xi}\frac{\partial y}{\partial \zeta} = 0 \qquad (11.4)$$

where B is the coefficient of the cross-derivative term in the governing equation (Eq. 11.1) for the elliptic grid generation.

Applying the above equation for the north- and south-boundaries of the physical-domain (Fig. 11.5), since the y-coordinate remains constant along the horizontal boundary, $\partial y/\partial \xi = 0$; refer Fig. 11.6. Furthermore, since the figure shows that the variation of the x-coordinate with increasing ξ, at the north- and south-boundary, is a non-zero constant value c; thus, $\partial x/\partial \xi = c$. Thus, for the *horizontal boundary*, Eq. 11.4 results in the BCs as

$$\frac{\partial x}{\partial \zeta} = 0 \text{ at the north- and south-boundary} \tag{11.5}$$

whereas, for the inclined east and west-boundary of the physical-domain (Fig. 11.5), since the variation of both x and y coordinate is a non-zero constant value with increasing ζ, $\partial x / \partial \zeta = c_1$ and $\partial y / \partial \zeta = c_2$ (Fig. 11.6). Furthermore, since the angle of inclination is 45° for the plate, the variation of both x and y coordinate with ζ is same ($c_1 = c_2$), *i.e.*, $\partial x / \partial \zeta = \partial y / \partial \zeta$. Thus, for the inclined boundary, Eq. 11.4 results in the BCs as

$$\frac{\partial x}{\partial \xi} + \frac{\partial y}{\partial \xi} = 0 \text{ at the east and west-boundary} \tag{11.6}$$

The central difference discretization of Eq. 11.5 and Eq. 11.6 results in the BCs as

$$
\left.
\begin{aligned}
& x_{i,1} = x_{i,2} \} \text{ south-boundary} \\
& x_{imax-1,j} = x_{imax-2,j} \} \text{ north-boundary} \\
& \left. \begin{aligned} x_{1,j} &= x_{2,j} + (y_{2,j} - y_{1,j}^{old}) \\ y_{1,j} &= y_{2,j} + (x_{2,j} - x_{1,j}^{old}) \end{aligned} \right\} \text{ west-boundary} \\
& \left. \begin{aligned} x_{imax-1,j} &= x_{imax-2,j} - (y_{imax-1,j}^{old} - y_{imax-2,j}) \\ y_{imax-1,j} &= y_{imax-2,j} - (x_{imax-1,j}^{old} - x_{imax-2,j}) \end{aligned} \right\} \text{ east-boundary}
\end{aligned}
\right. \tag{11.7}
$$

where the superscript *old* corresponds to the previous iterative values.

Here, $imax$ and $jmax$ are the maximum number of grid points in the x- and y-direction, respectively. As discussed above, these grid points correspond to the cell centers in the FVM-based solution for flow properties. Whereas, for the vertices here in the curvilinear grid generation, the maximum number of grid points for the FDM-based solution is equal to ($imax - 1$) and ($jmax - 1$).

Considering 11×11 as the grid size for the vertices, the *non-Dirichlet* BCs (Eq. 11.7) are applied at the boundary of the computational-domain. This is done at every iteration of the solution algorithm, with x^{old} and y^{old} as the previous iterative values in Eq. 11.7. Using the formulation and solution algorithm presented above, for the elliptic grid generation, a computer program is developed. The results obtained from the program is presented, as the curvilinear grid, in Fig. 11.6b.

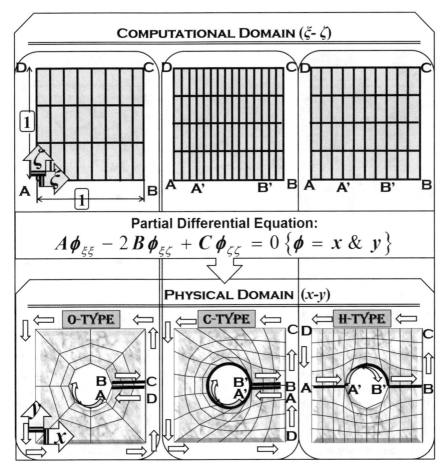

$$A\phi_{\xi\xi} - 2B\phi_{\xi\zeta} + C\phi_{\zeta\zeta} = 0\,\{\phi = x \,\&\, y\}$$

Fig. 11.4 Schematic of the various types of elliptic grid generation, for a physical-domain in-between a square and a circular region.

Fig. 11.5 Schematic of a parallelogram shaped physical-domain

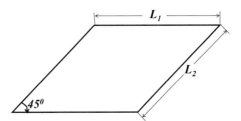

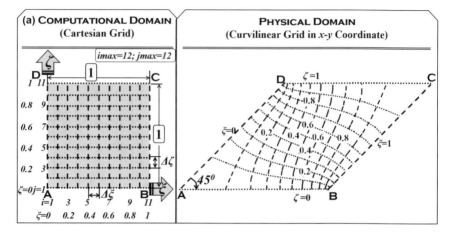

Fig. 11.6 Schematic for a curvilinear grid generation (with grid lines almost orthogonal at the boundary of the physical-domain) in a parallelogram shaped physical-domain, on a grid size of 11×11.

11.2   Physical Law-based Finite Volume Method

Using the curvilinear grid presented above, consider a representative CV for heat conduction in a complex geometry. As compared to the Cartesian CV for simple geometries, presented earlier and also shown in Figs. 11.7a, b shows that a *complex CV* for a complex geometry problem has two major differences. First, the figures show that the *faces for the CV* are not orthogonal to each other and are not inclined along the direction of the Cartesian coordinates. Second, it can also be seen that the *lines joining the cell center P and neighboring cell centers* are also not inclined along the direction of the Cartesian coordinates, do not pass through the corresponding face centers, and are not normal to the face. The complex CVs are considered to present the physical law-based (PDE free) FVM below, for 2D conduction in a plate of complex geometry.

11.2.1   Unsteady and Source Term

For the unsteady and volumetric heat generation term, the two approximations for the FVM already presented in Chap. 5 is applicable—with a modification in the expression for the volume of the CV. The finally approximated algebraic formulation for the rate of change of enthalpy stored inside the CV $\Delta \mathcal{H} / \Delta t$, and generated from the CV Q_{gen}, are given as

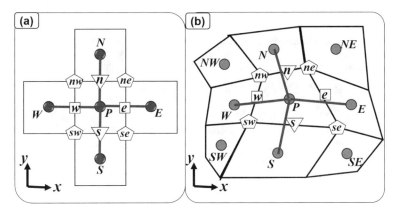

Fig. 11.7 A representative control volume (CV) for (**a**) simple geometry and (**b**) complex geometry (with non-uniform, non-orthogonal, and curvilinear grid arrangement)

$$\frac{\Delta \mathcal{H}}{\Delta t} \approx \rho c_p \Delta V_P \frac{T_P^{n+1} - T_P^n}{\Delta t} \tag{11.8}$$

$$Q_{gen,P} \approx \overline{Q}_{gen,P} \Delta V_P \tag{11.9}$$

where ΔV_P is the volume of the representative complex CV P (Fig. 11.7b); the expression will be presented below.

11.2.2 Diffusion Term

For the diffusion term in a complex CV, corresponding to the net conduction heat transfer rate into a CV Q_{cond}^{mean}, the two approximations are shown in Fig. 11.8. It can be seen that the approximations for the 2D complex CV are similar to that for the simple Cartesian CV (refer Fig. 5.4). However, the conduction fluxes (q_x and q_y) are in the orthogonal Cartesian coordinates directions for the Cartesian CV (Fig. 5.4); this is not the case for the formulation presented here for the fluxes in a complex CV (Fig. 11.8). The figure shows that the fluxes as $q_{s,f}$ and $q_{t,f}$ at the various face centers f ($= e$, w, n and s); using a *local non-orthogonal coordinate* system $s - t$. Moreover, it can be seen in the figure that both q_s and q_t components acts on each of the four surface of the complex CV, as compared to either q_x or q_y components acting on a surface of the Cartesian CV.

In general, the conduction flux acts in a direction normal to a surface and is represented as q_η; the subscript η represents a coordinate which is normal to a surface. Since the direction of the η−coordinate varies from the one to the other surface of a complex CV, its definition is local to a surface; thus, η−*coordinate* is called here as a *local coordinate*. Now, the *Cartesian coordinate* is called as the *global coordinate*.

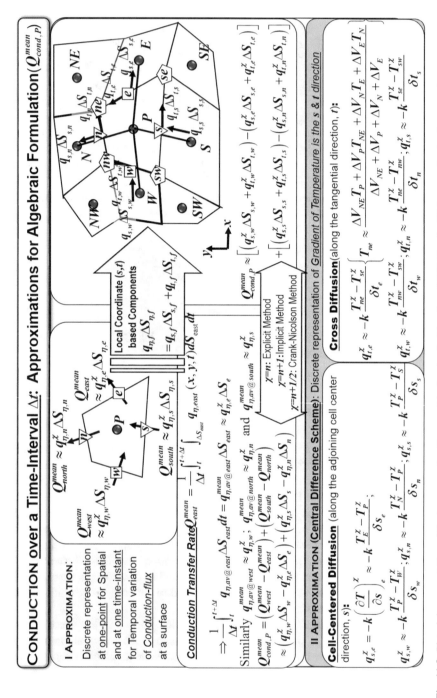

Fig. 11.8 Pictorial representation of the two approximations used in the physical law-based FVM for the 2D heat conduction in a complex geometry.

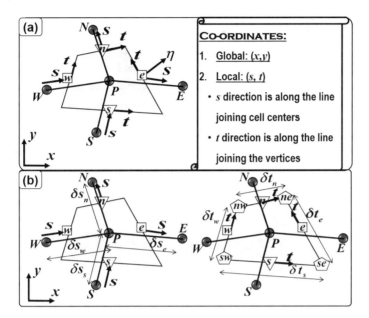

Fig. 11.9 Schematic representation of the (**a**) global and local coordinates, and (**b**) distances involved in the second approximations of the FVM, for a representative complex CV

For a 2D problem here, at each of the four surface, the local η−coordinate consists of two non-orthogonal components: s−$coordinate$, along the line joining the cell centers across the surface; and t−$coordinate$, along the direction tangential to the surface. The global (x, y) and the local (s, t) coordinate are shown in Fig. 11.9a. Since the curvilinear grid generation presents the results in the Cartesian coordinates (discussed in the previous section), the global *Cartesian coordinates* are considered here for the computation of the various *geometrical parameters* of all the complex CVs. Whereas, the *local coordinate* are considered during an approximated *algebraic formulation* of the conduction heat transfer rate, at the various surfaces of the complex CVs; presented below.

First Approximation:
The first approximation corresponds to a discrete representation of the time and space-averaged conduction flux at a surface $q_{\eta,av@face}^{mean}$ ($face = east/west/north/ south$)—one point for the spatial variation, and one time instant for the temporal variation. As seen from Fig. 11.8, the first approximation for the conduction flux q_η in a complex CV is similar to that for q_x/q_y in the 2D Cartesian coordinate (presented in Chap. 5). The first approximation, during the determination of the conduction heat transfer at the east surface Q_{east}^{mean}, is given as

$$Q_{east}^{mean} = \frac{1}{\Delta t} \int_t^{t+\Delta t} \int_{\Delta S_{east}} q_{\eta,east} d\,S_{east} dt = q_{\eta,av@east}^{mean} \Delta S_{east}$$

$$q_{\eta,av@east}^{mean} \begin{cases} \approx q_{\eta,e}^n \Delta S_e & \text{for Explicit Method} \\ \approx q_{\eta,e}^{n+1} \Delta S_e & \text{for Implicit Method} \\ \approx q_{\eta,e}^{n+1/2} \Delta S_e & \text{for Crank - Nicolson Method} \end{cases} \tag{11.10}$$

where n, $n+1$ and $n+1/2$ are the time level corresponding to the time instant t, $t+\Delta t$ and $t+\Delta t/2$, respectively. Note that $q_{\eta,e}^{n+1/2} \approx \left(q_{\eta,e}^n + q_{\eta,e}^{n+1}\right)/2$ for the Crank-Nicolson method. Furthermore, the subscript av in $q_{\eta,av@face}^{mean}$ corresponds to the space-averaged value at a surface, and the superscript $mean$ corresponds to the time-averaged value. Using the equations similar to Eq. 11.10, for the other faces of the representative CV P, the conduction heat transfer rate is given by the first approximation (Fig. 11.8) as

$$\begin{aligned} Q_{east}^{mean} \approx q_{\eta,e}^\chi \Delta S_e, \quad Q_{west}^{mean} \approx q_{\eta,w}^\chi \Delta S_w, \\ Q_{north}^{mean} \approx q_{\eta,n}^\chi \Delta S_n, \quad \& Q_{south}^{mean} \approx q_{\eta,s}^\chi \Delta S_s \end{aligned} \tag{11.11}$$

where $\chi = n, n+1$ and $n+1/2$ for explicit, implicit, and Crank-Nicolson method, respectively. Furthermore, $\Delta S_{f=e,w,n,s}$ are the surface area of the various surfaces of the complex CV; their expressions will be presented below.

Considering the local $s-t$ components of the conduction flux $\vec{q}_{\eta,f}(= q_{s,f}\hat{s} + q_{t,f}\hat{t})$ as well as the surface area $\overrightarrow{\Delta S}_f(= \Delta S_{s,f}\hat{s} + \Delta S_{t,f}\hat{t})$, the conduction heat transfer rate at a surface of the complex CV is given as

$$Q_{face}^{mean} = \vec{q}_f \cdot \overrightarrow{\Delta S}_f \approx q_{\eta,f}^\chi \Delta S_f \approx q_{s,f}^\chi \Delta S_{s,f} + q_{t,f}^\chi \Delta S_{t,f} \tag{11.12}$$

where $f = e, w, n$, and s for the $surface = east, west, north$, and $south$ surfaces of the complex CV, respectively

Second Approximation:
The second approximation for the computational heat conduction, corresponding to the discrete representation of the conduction flux, can also be seen in Fig. 11.8. The figure shows the approximation for the both the fluxes at each face f $(e/w/n/s)$: $q_{s,f}^\chi$ in the $s-$direction, and $q_{t,f}^\chi$ in the $t-$direction. The figure shows that the $q_{s,f}^\chi$ involves the temperature at the cell centers adjoining the face f, and the approximation for $q_{t,f}^\chi$ involves the temperature at the associated vertices (to the ends of a face f) of the CVs. Using these temperatures, the *second-order central difference* approximation results in the local $s-t$ components of the conduction flux as

$$q_{s,e}^{\chi} \approx -k\frac{T_E^{\chi} - T_P^{\chi}}{\delta s_e}, \quad q_{t,e}^{\chi} \approx -k\frac{T_{ne}^{\chi} - T_{se}^{\chi}}{\delta t_e}$$

$$q_{s,n}^{\chi} \approx -k\frac{T_N^{\chi} - T_P^{\chi}}{\delta s_n}, \quad q_{t,n}^{\chi} \approx -k\frac{T_{ne}^{\chi} - T_{nw}^{\chi}}{\delta t_n} \tag{11.13}$$

$$q_{s,w}^{\chi} \approx -k\frac{T_P^{\chi} - T_W^{\chi}}{\delta s_w}, \quad q_{t,w}^{\chi} \approx -k\frac{T_{nw}^{\chi} - T_{sw}^{\chi}}{\delta t_w}$$

$$q_{s,s}^{\chi} \approx -k\frac{T_P^{\chi} - T_S^{\chi}}{\delta s_s}, \quad q_{t,s}^{\chi} \approx -k\frac{T_{se}^{\chi} - T_{sw}^{\chi}}{\delta t_s}$$

where $\delta s_{f=e,w,n,s}$ are the distances between the adjoining cell centers, and $\delta s_{f=e,w,n,s}$ are the distances between the associated vertices δt_f; shown in Fig. 11.9b.

The temperature at the vertices of the complex CV, involved in computation of cross-diffusion $q_{t,f}^{\chi}$ (Eq. 11.13), is approximated by a volume weighted interpolation of the temperature at the four adjoining cell centers as

$$T_{ne} \approx \frac{\Delta V_{NE}T_P + \Delta V_P T_{NE} + \Delta V_N T_E + \Delta V_E T_N}{\Delta V_{NE} + \Delta V_P + \Delta V_N + \Delta V_E} \tag{11.14}$$

$$T_{se} \approx \frac{\Delta V_{SE}T_P + \Delta V_P T_{SE} + \Delta V_S T_E + \Delta V_E T_S}{\Delta V_{SE} + \Delta V_P + \Delta V_S + \Delta V_E}$$

$$T_{nw} \approx \frac{\Delta V_{NW}T_P + \Delta V_P T_{NW} + \Delta V_N T_W + \Delta V_W T_N}{\Delta V_{NW} + \Delta V_P + \Delta V_N + \Delta V_W}$$

$$T_{sw} \approx \frac{\Delta V_{SW}T_P + \Delta V_P T_{SW} + \Delta V_S T_W + \Delta V_W T_S}{\Delta V_{SW} + \Delta V_P + \Delta V_S + \Delta V_W}$$

Using the two approximation for the diffusion term above, now the algebraic equation for the mean value (within a time interval Δt) of the total heat gained by conduction $Q_{cond,P}$ (refer Fig. 11.8) is presented as

$$Q_{cond,P}^{mean} = Q_{west}^{mean} - Q_{east}^{mean} + Q_{south}^{mean} - Q_{north}^{mean}$$

Applying the I−approximation (Eq. 11.11), the above equation is given as

$$Q_{cond,P}^{mean} \approx \left(q_{\eta,w}^{\chi}\Delta S_w - q_{\eta,e}^{\chi}\Delta S_e\right) + \left(q_{\eta,s}^{\chi}\Delta S_s - q_{\eta,n}^{\chi}\Delta S_n\right)$$

Substituting $q_{\eta,f}^{\chi}\Delta S_f$ (with face center $f = e, w, n$, and s) from Eq. 11.12, the above equation is given as

$$Q_{cond,P}^{mean} \approx Q_{s,P}^{mean} + Q_{t,P}^{mean} \tag{11.15}$$

$$\text{where} \quad Q_{s,P}^{mean} \approx \left(q_{s,w}^{\chi}\Delta S_{s,w} - q_{s,e}^{\chi}\Delta S_{s,e}\right) + \left(q_{s,s}^{\chi}\Delta S_{s,s} - q_{s,n}^{\chi}\Delta S_{s,n}\right)$$

$$Q_{t,P}^{mean} \approx \left(q_{t,w}^{\chi}\Delta S_{t,w} - q_{t,e}^{\chi}\Delta S_{t,e}\right) + \left(q_{t,s}^{\chi}\Delta S_{t,s} - q_{t,n}^{\chi}\Delta S_{t,n}\right)$$

where $Q_{s,P}^n$ and $Q_{t,P}^n$ are the total heat gained by the adjoining cell centered and the cross-diffusion, respectively. Applying the II−approximation (Eq. 11.13), the above equation is given as

$$
\begin{aligned}
Q_{cond,P} = {} & \left(k\frac{T_E^\chi - T_P^\chi}{\delta s_e}\Delta S_{s,e} - k\frac{T_P^\chi - T_W^\chi}{\delta s_w}\Delta S_{s,w} \right) \\
&+ \left(k\frac{T_N^\chi - T_P^\chi}{\delta s_n}\Delta S_{s,n} - k\frac{T_P^\chi - T_S^\chi}{\delta s_s}\Delta S_{s,s} \right) \\
&+ \left(k\frac{T_{ne}^\chi - T_{sw}^\chi}{\delta t_e}\Delta S_{t,e} - k\frac{T_{nw}^\chi - T_{sw}^\chi}{\delta t_w}\Delta S_{t,w} \right) \\
&+ \left(k\frac{T_{ne}^\chi - T_{nw}^\chi}{\delta t_n}\Delta S_{t,n} - k\frac{T_{se}^\chi - T_{sw}^\chi}{\delta t_s}\Delta S_{t,s} \right) + \overline{Q}_{gen,P}\Delta V_P
\end{aligned}
\tag{11.16}
$$

11.2.3 All Terms

Using Eqs. 11.8, 11.9, 11.15 and 11.16, the final governing linear algebraic equations for heat conduction are given as

$$
\begin{aligned}
\frac{\Delta \mathcal{H}}{\Delta t} &= Q_{s,P}^n + Q_{t,P}^n + Q_{gen,P} \\
\rho c_p \frac{T_P^{n+1} - T_P^n}{\Delta t}\Delta V_P &= \left[(q_{s,w}^\chi \Delta S_{s,w} - q_{s,e}^\chi \Delta S_{s,e}) + (q_{s,s}^\chi \Delta S_{s,s} - q_{s,n}^\chi \Delta S_{s,n}) \right] \\
&\quad + \left[(q_{t,w}^\chi \Delta S_{t,w} - q_{t,e}^\chi \Delta S_{t,e}) + (q_{t,s}^\chi \Delta S_{t,s} - q_{t,n}^\chi \Delta S_{t,n}) \right] + \overline{Q}_{gen,P}\Delta V_P
\end{aligned}
\tag{11.17}
$$

$$
\begin{aligned}
\rho c_p \frac{T_P^{n+1} - T_P^n}{\Delta t}\Delta V_P = {} & \left(k\frac{T_E^\chi - T_P^\chi}{\delta s_e}\Delta S_{s,e} - k\frac{T_P^\chi - T_W^\chi}{\delta s_w}\Delta S_{s,w} \right) \\
&+ \left(k\frac{T_N^\chi - T_P^\chi}{\delta s_n}\Delta S_{s,n} - k\frac{T_P^\chi - T_S^\chi}{\delta s_s}\Delta S_{s,s} \right) \\
&+ \left(k\frac{T_{ne}^\chi - T_{sw}^\chi}{\delta t_e}\Delta S_{t,e} - k\frac{T_{nw}^\chi - T_{sw}^\chi}{\delta t_w}\Delta S_{t,w} \right) \\
&+ \left(k\frac{T_{ne}^\chi - T_{nw}^\chi}{\delta t_n}\Delta S_{t,n} - k\frac{T_{se}^\chi - T_{sw}^\chi}{\delta t_s}\Delta S_{t,s} \right) + \overline{Q}_{gen,P}\Delta V_P
\end{aligned}
\tag{11.18}
$$

11.3   Computation of Geometrical Properties

After obtaining the *Cartesian coordinates* of the *vertices* of the complex CVs in the curvilinear grid, using the methods presented in the previous section, the coordinates of the *cell centers* are computed as

$$
\vec{x}_P = \frac{\vec{x}_{ne} + \vec{x}_{nw} + \vec{x}_{sw} + \vec{x}_{se}}{4}
$$

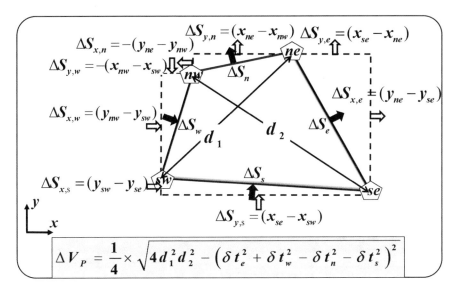

Fig. 11.10 Schematic representation of the surface areas and volume, for a representative complex CV. Also note that the normal direction (shown as the black color-filled arrows) for a surface area ΔS_f is outward for the positive (east and north) surface, and inward for the negative (west and south) surface. Considering the Cartesian coordinates of the relative (left/right and top/bottom) positions of the two adjoining vertices of the CV in the figure, the above equations for the *components* $\Delta S_{x/y,f=e,w,n,s}$ of the surface area are such that the *term inside the round bracket is positive*. Furthermore, outside the round bracket, the sign is seen above as positive and negative if the direction of $\Delta S_{x/y,f}$ (shown as the unfilled arrows) is in the positive and negative x/y direction, respectively

The coordinates of the cell centers and the vertices are used to compute distance $\delta s_{f=e,w,n,w}$ and $\delta t_{f=e,w,n,w}$ (Fig. 11.9), respectively. Thereafter, the distances δt_f and the diagonal (d_1 and d_2) of the CV, are used to determine the volume ΔV_P of the CVs, shown in Fig. 11.10. The figure also shows that the coordinates of the vertices are used to compute the various surface area ($\Delta S_{x,f=e,w,n,w}$ and $\Delta S_{x,f=e,w,n,w}$) of the CVs.

Since the areas shown in the figure correspond to the global Cartesian components and not the local $s - t$ components (needed in the final algebraic equation; Eq. 11.18), the Cartesian components of the various surface area are used to obtain the local components. This is done using a vector analysis, presented in Fig. 11.11; for the east face of a CV. The surface area vector, shown in the Fig. 11.11a, is expressed in terms of the Cartesian and local components as

$$\Delta \vec{S}_e = \Delta S_{x,e}\hat{i} + \Delta S_{y,e}\hat{j} = \Delta S_{s,e}\hat{s} + \Delta S_{t,e}\hat{i}$$

where the figure shows that the Cartesian components ($\Delta S_{x,e}$ and $\Delta S_{y,e}$) are orthogonal, and the local components ($\Delta S_{s,e}$ and $\Delta S_{t,e}$) are non-orthogonal; however, it is

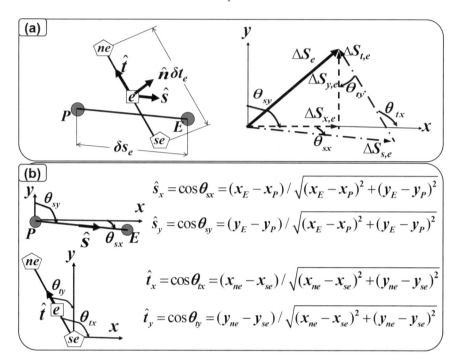

Fig. 11.11 For the east face of representative complex CV, schematic representation of (**a**) surface area vector and its components (Cartesian and local), and (**b**) unit vector along the adjoining cell centers and the associated vertices

interesting to note from the figures that ΔS_e *and* $-\Delta S_{t,e}$ *are orthogonal components of* $\Delta S_{s,e}$. The local components of the surface area and their direction cosines leads to the Cartesian components of the surface area (refer Fig. 11.11a) as

$$\Delta S_{x,e} = \Delta S_{s,e} \cos\theta_{sx} + \Delta S_{t,e} \cos\theta_{tx}$$
$$\Delta S_{y,e} = \Delta S_{s,e} \cos\theta_{sy} + \Delta S_{t,e} \cos\theta_{ty}$$

where the above direction cosines are the Cartesian components of the unit vector of the $\hat{s}$ and $\hat{t}$, shown in Fig. 11.11b. Solving the above equations, the local components of the area of the east surface are given as

$$\Delta S_{s,e} = \frac{\Delta S_{x,e} \cos\theta_{ty} - \Delta S_{y,e} \cos\theta_{tx}}{\cos\theta_{sx} \cos\theta_{ty} - \cos\theta_{sy} \cos\theta_{tx}} \qquad (11.19)$$

$$\Delta S_{t,e} = -\frac{\Delta S_{x,e} \cos\theta_{sy} - \Delta S_{y,e} \cos\theta_{sx}}{\cos\theta_{sx} \cos\theta_{ty} - \cos\theta_{sy} \cos\theta_{tx}}$$

Similarly, the local components of the area of the other (north, west, and south) surfaces can be obtained.

11.4   Flux-based Solution Methodology

For the 2D conduction, the flux-based unsteady-state solution methodology in a complex geometry is quite similar to that for the simple Cartesian-geometry, presented in Chap. 5. For a complex geometry with the curvilinear grid, the methodology is presented here for a representative problem, shown in Fig. 11.12a. The figure shows 2D conduction in parallelogram plate (with complex CVs) as the physical-domain, and the computational-domain (with square/Cartesian CVs) is shown in Fig. 11.12b. The figures for both the domains show a grid size of $imax \times jmax = 7 \times 7$. The purpose of showing the computational-domain here is to bring more clarity in the implementation details while presenting the solution methodology for the complex geometry.

Figure 11.12a also shows FDM-based discretized form of the BCs for the two cases: (a) Dirichlet BC at all the boundaries and (b) a different BC at each of the boundary. Note the curvilinear grid shown in the figure ensure that the grid lines intersecting a boundary are normal at the boundary. The grid orthogonality is ensured

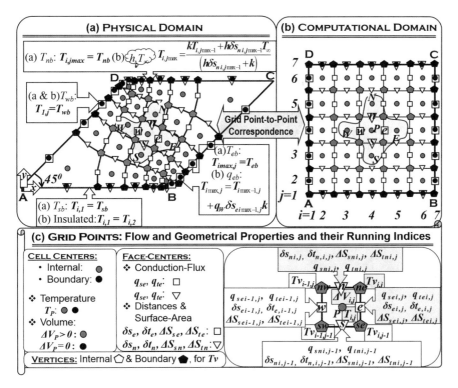

Fig. 11.12 Physical and computational-domain, discretized BCs, and different types of grid points and their running indices, for two different cases (**a** and **b**) of a 2D heat conduction problem in a complex geometry

by using appropriate BCs (Eq. 11.7) during the curvilinear grid generation; presented above in Example 11.1. This leads to an accurate algebraic implementation of the non-Dirichlet BCs (involves normal gradient at the boundary), as the line joining the boundary and the associated border grid points are almost normal to the boundary. For example, at the bottom boundary, it can be seen in Fig. 11.12a that the line (joining a boundary grid point and the just above border grid point) is almost in the vertical direction for the horizontal surface; similar orthogonality can be seen at the other boundaries.

Figure 11.12a, b shows the grid points for temperature as filled symbols; circles at the centroids and pentagons at the vertices of the CVs. Whereas, the grid points for the conduction flux are represented by the unfilled symbols; squares for $\overrightarrow{q}_e$ and triangles for $\overrightarrow{q}_n$. The different symbols and colors used here later help to clearly present the implementation of the algebraic formulation. Furthermore, since there is one-to-one correspondence of various types of grid points as well as CVs in the physical to that in the computational-domain, *an implementation of the above formulation is easier by considering the running indices (i, j) of the grid points in the computational-domain.* However, note that the geometrical parameter still corresponds to that in the physical $x–y$ coordinate.

For the running indices $i = j = 4$ in both physical and computational-domain, a representative CV with grid point P is shown in Fig. 11.12a, b. Figure 11.12c shows the CV separately, with a single running index (i, j) for the various flow and geometrical properties seen appropriately at the cell center, face center, and vertices of the CV. The single running index (for all the variables) require a convention—for defining the indices of the variables computed at the face center (q_{sf}, q_{tf}, δs_f, δs_f, ΔS_{sf}, and ΔS_{tf}) and vertices (Tv), in terms of the indices of the variables computed at the cell center (T). Since a face is common to two CVs, the convention presented earlier is also followed here—the running indices (i, j) for a heat flux corresponds to the indices of that CV for which it is at the positive (east and north) faces. Furthermore, since a vertex is common to the adjoining four CVs, the running indices of the temperature at the vertices Tv corresponds to the indices of that CV for which it is the *north-east* vertex. Appropriately, running indices for the other face centers and vertices can also be seen in Fig. 11.12c.

11.4.1 Explicit Method

Using the physical law-based FVM for the 2D unsteady-state heat conduction with *uniform* heat generation in a representative complex CV (Fig. 11.8), the explicit method-based LAE for the conduction flux (Eq. 11.13) and temperature (Eq. 11.17) are given as

$$q_{s,e}^n \approx -k\frac{T_E^n - T_P^n}{\delta s_e}, \quad q_{t,e}^n \approx -k\frac{T_{ne}^n - T_{se}^n}{\delta t_e}$$

$$q_{s,n}^n \approx -k\frac{T_N^n - T_P^n}{\delta s_n}, \quad q_{t,n}^n \approx -k\frac{T_{ne}^X - T_{nw}^X}{\delta t_n}$$

$$q_{s,w}^n \approx -k\frac{T_P^n - T_W^n}{\delta s_w}, \quad q_{t,w}^n \approx -k\frac{T_{nw}^n - T_{sw}^n}{\delta t_w} \qquad (11.20)$$

$$q_{s,s}^n \approx -k\frac{T_P^n - T_S^n}{\delta s_s}, \quad q_{t,s}^n \approx -k\frac{T_{se}^n - T_{sw}^n}{\delta t_s}$$

$$T_P^{n+1} = T_P^n + \frac{\Delta t}{\rho c_p \Delta V_P}\left[Q_{s,P}^n + Q_{t,P}^n + \overline{Q}_{gen}\Delta V_P\right]$$

$$\text{where } Q_{s,P}^n = (q_{s,w}^n \Delta S_{s,w} - q_{s,e}^n \Delta S_{s,e}) + (q_{s,s}^n \Delta S_{s,s} - q_{s,n}^n \Delta S_{s,n}) \qquad (11.21)$$

$$Q_{t,P}^n = (q_{t,w}^n \Delta S_{t,w} - q_{t,e}^n \Delta S_{t,e}) + (q_{t,s}^n \Delta S_{t,s} - q_{t,n}^n \Delta S_{t,n})$$

11.4.2 Implementation Details

Similar to the details presented in Chap. 5 (for 2D conduction in the Cartesian-geometry), and using the definition of the running indices (for the geometrical as well as flow properties) seen in Fig. 11.12c, the two-step flux-based implementation for the complex geometry is as follows:

First step: Calculate the conduction flux at the previous time level n (Eq. 11.20) as

$$q_{se\,i,j}^n = -k\frac{T_{i+1,j}^n - T_{i,j}^n}{\delta s_{e\,i,j}}, \quad q_{te\,i,j}^n = -k\frac{Tv_{i,j}^n - Tv_{i,j-1}^n}{\delta t_{e\,i,j}}$$

$$\implies \text{for } i = 1 \text{ to } imax - 1 \text{ for } j = 2 \text{ to } jmax - 1 \qquad (11.22)$$

$$q_{sn\,i,j}^n = -k\frac{T_{i,j+1}^n - T_{i,j}^n}{\delta s_{n\,i,j}}, \quad q_{tn\,i,j}^n = -k\frac{Tv_{i,j}^n - Tv_{i-1,j}^n}{\delta t_{n\,i,j}}$$

$$\implies \text{for } i = 2 \text{ to } imax - 1 \text{ for } j = 1 \text{ to } jmax - 1 \qquad (11.23)$$

where the temperature at the vertices of the CVs are given as

$$Tv_{i,j} \approx \frac{\Delta V_{i+1,j+1}T_{i,j} + \Delta V_{i,j}T_{i+1,j+1} + \Delta V_{i,j+1}T_{i+1,j} + \Delta V_{i+1,j}T_{i,j+1}}{\Delta V_{i+1,j+1} + \Delta V_{i,j} + \Delta V_{i,j+1} + \Delta V_{i+1,j}}$$

$$\implies \text{for } i = 2 \text{ to } imax - 1 \text{ for } j = 2 \text{ to } jmax - 1 \qquad (11.24)$$

Note that the above equation is used to obtain the temperature at those vertices of the CVs, which are inside the domain. Whereas, for the boundary vertices, shown by the black color-filled pentagon in Fig. 11.12b, a distance-based linear interpolation (of the temperature at the two adjoining boundary grid points) is used to obtain the vertex values Tv. Alternatively, for the boundary vertices, the above equation can be used with the volume of all the boundary CVs as zero, *i.e.*, $\Delta V_{1,j} = 0$, $\Delta V_{imax,j} = 0$, $\Delta V_{i,1} = 0$, and $\Delta V_{i,jmax} = 0$. Note that the above computation of the temperature at the boundary grid points is needed for the non-Dirichlet BC but not for the Dirichlet BC; as the boundary temperature can be directly used at the vertices.

Second Step: Calculate the total heat in-flux/gained by conduction at the internal grid points, and then the temperature (Eq. 11.21) as

$$Q^n_{s\,i,j} = (q^n_{s\,i-1,j}\Delta S_{se\,i-1,j} - q^n_{s\,i,j}\Delta S_{se\,i,j})$$
$$+(q^n_{s\,i,j-1}\Delta S_{sn\,i,j-1} - q^n_{s\,i,j}\Delta S_{sn\,i,j})$$
$$Q^n_{t\,i,j} = (q^n_{t\,i-1,j}\Delta S_{te\,i-1,j} - q^n_{t\,i,j}\Delta S_{te\,i,j})$$
$$+(q^n_{t\,i,j-1}\Delta S_{tn\,i,j-1} - q^n_{t\,i,j}\Delta S_{tn\,i,j}) \tag{11.25}$$
$$T^{n+1}_{i,j} = T^n_{i,j} + \frac{\Delta t}{\rho c_p \Delta V_P}(Q^n_{s,i,j} + Q^n_{t,i,j} + \overline{Q}_{gen}\Delta V_P)$$
$$\Longrightarrow \text{for } i = 2 \text{ to } imax - 1 \text{ for } j = 2 \text{ to } jmax - 1$$

The computational stencil and the solution algorithm for the complex geometry are similar to that presented for the Cartesian-geometry (Chap. 5). However, note that the dimension of the data structure of the variables for the geometrical properties of the CVs increases for the complex geometry as compared to the Cartesian-geometry, For example, the surface area of the CV is represented by one-value Δx (and Δy) for the uniform Cartesian grid, 1D matrix Δx_i (and Δy_j) for the non-uniform Cartesian grid, and 2D matrix $\Delta S_{i,j}$ here for the complex geometry. Also note that the computation of geometrical properties are mathematically much more involved for the complex CVs as compared to the simple CVs. Finally, for both the CVs, note that both geometrical and fluid properties correspond to the coefficients in the algebraic formulation, and the flow properties correspond to the unknowns in the LAEs.

Example 11.2: Testing of the formulation for complex geometry on a simple geometry problem.

Consider the 2D heat conduction in a square-shaped *long* stainless-steel plate, presented in Example 5.2. The Cartesian-geometry problem is considered here as a complex geometry problem by a 45° clockwise rotation of the plate (Fig. 5.10), without rotating the horizontal x-coordinate and vertical y-coordinate. The resulting physical as well as computational-domain and the CVs are shown in Fig. 11.13.

Using the formulation as well as the solution methodology, presented above, develop a computer program for the 2D conduction in a complex geometry. Run the code for the above problem (Fig. 11.13), with *no volumetric heat generation*, grid size of $imax \times jmax = 12$, and the steady-state convergence tolerance as $\epsilon_{st} = 10^{-4}$.

Plot the steady-state temperature distribution (obtained from the complex geometry formulation) in the rotated plate and compare it with the result for the non-rotated plate (Example 5.2), as the purpose of this problem is to demonstrate that the results for the above complex geometry degenerate to the results for the simple geometry problem.

Solution:
The grid generation here involves computation of the vertices of the CVs; and then the determination of the various geometrical properties needed for a complex geometry problem (presented above).

After developing the computer program (not presented here), the code is run for the present computational setup and the resulting temperature contour is shown in Fig. 11.14a. Figure 11.14b also shows the result for the non-rotated plate, presented earlier in Fig. 5.12a. The figure shows good agreement between the two results, confirming that the results for a complex geometry formulation degenerates to that for the simple geometry problem.

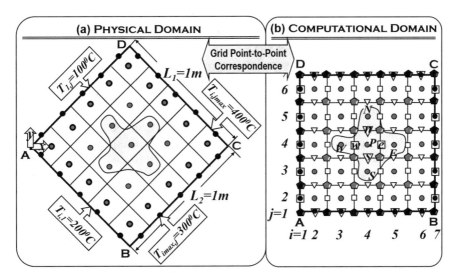

Fig. 11.13 Schematic of (**a**) physical and (**b**) computational-domain, for a square plate (Fig. 5.10) rotated at angle of 45° from the horizontal x-direction

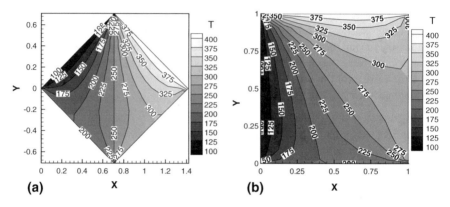

Fig. 11.14 Steady state temperature contour in (**a**) rotated and (**b**) non-rotated plate, obtained from the complex and Cartesian-geometry formulations, respectively

---

Example 11.3: CFD application for 2D heat conduction in a parallelogram plate.

Consider the 2D conduction in the plate, with the two cases of the BCs, shown in Fig. 11.12; with $L_1 = L_2 = 1$ m. For the case (a) in the figure, consider the BCs as $T_{wb} = 100\,°C$, $T_{sb} = 200\,°C$, $T_{eb} = 300\,°C$ and $T_{nb} = 400\,°C$. Whereas, for the case (b), consider $T_{wb} = 100\,°C$, $q_W = 10\,KW/m^2$, $h = 1000\,W/m^2K$ and $T_\infty = 30\,°C$. For both the cases, consider no heat generation, the plate is initially at a uniform ambient temperature of $30\,°C$, and is made up of stainless steel ($\rho = 7750\,kg/m^3$, $c_p = 500\,J/kg\,K$ and $k = 16.2\,W/m\,K$). Also consider a grid size of $imax = jmax = 32$ and the steady-state convergence tolerance as $\epsilon_{st} = 10^{-4}$. Generate and present the curvilinear grid, using the method presented in the *Example 11.1*. Using the code developed and tested for the previous problem, simulate the two cases on the curvilinear grid and present the steady-state temperature contour in the plate.

Solution:

Using the method and the program presented in Example 11.1, the curvilinear grid for physical-domain is shown in Fig. 11.15— on a grid size of 32×32. As discussed above, note from the figure that the grid lines are orthogonal at the boundary of the plate.

Using the curvilinear grid and the program used in the previous problem, a CFD simulation is done for the two cases of BCs (Fig. 11.12); and the resulting steady-state temperature contour is shown in Fig. 11.16. For the case (a) of BCs, comparing Fig. 11.16a (for the complex geometry) and Fig. 11.14 (for the Cartesian-geometry), it is interesting to note that one of the diagonal for both the plates is at $250\,°C$; mean of the boundary conditions of $100\,°C$, $200\,°C$, $300\,°C$, and $400\,°C$. Whereas, for the case (b) of BCs, since the heat flux is zero on the

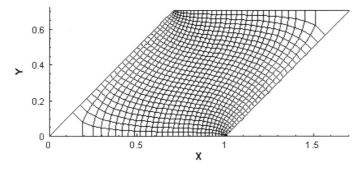

Fig. 11.15 Curvilinear grid in a parallelogram plate, on a grid size of 32×32

bottom boundary and a constant non-zero value at the right boundary, Fig. 11.16b shows that all the isotherms are intersecting the bottom boundary at 90°, and the right boundary with the same constant angle. However, the constant angle at the right boundary is different for the complex geometry here as compared to that for the simple geometry (Fig. 5.18a).

Problems

11.1 Solve the example problem, *Example 11.3*, considering a trapezoidal plate (Fig. 11.17; with $L_1 = L_2 = 1$ m) instead of the parallelogram plate (Fig. 11.15). Present the results similar to the example problem, using the grid generation method similar to *Example 11.1*.

11.2 Consider a square plate with a circular hole, shown in Fig. 11.3; with dimension of the plate $L = 10$ m and radius of the hole $R = 2$ m. Using the elliptic grid generation method and the Dirichlet BC (for x and y coordinates) in the computational-domain, modify the computer-program developed for *Example 11.1* and generate the three types of curvilinear grid: (a) O-type, (b) C-type, and (c) H-type. Take the number of grid points (vertices of CVs) for the grid generation as shown in Fig. 11.4; and plot similar figures for the 2D curvilinear grid.

11.3 Consider 2D heat conduction in the *long* square plate (made up of stainless steel; $\rho = 7750$ kg/m^3, $c_p = 500$ J/kg K and $k = 16.2$ W/m K) with a circular hole. Both the circular hole and square-boundary of the plate are maintained at a constant wall temperature $- T_C = 0$ °C (at the hole) and $T_H = 100$ °C (at the square-boundary); with an initial condition of the plate as 100 °C. Considering the computational-domain and the O-type grid (generated in previous problem), modify the code (developed for *Example 11.3*) to implement the present domain and boundary conditions. Use periodic boundary condition at the branch cut, *i.e.*, left and right boundary (AD and BC) of the computational-

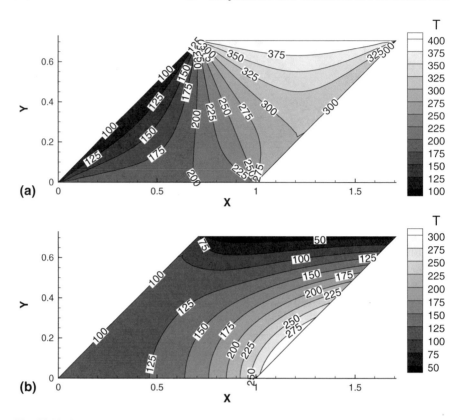

Fig. 11.16 Steady-state temperature contour in a parallelogram plate, for the two cases of BCs (Fig. 11.12): (a) Dirichlet BC on all the boundaries, and (b) a different BC in each of the boundary

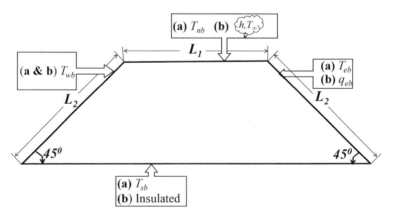

Fig. 11.17 Computational domain and boundary conditions, for 2D heat conduction in a trapezoidal plate

domain (Fig. 11.3). Run the code for the steady-state convergence tolerance $\epsilon_{st} = 10^{-4}$, and plot the steady state temperature contours in the plate.

References

Thompson, J. F., Warsi, Z. U. A., & Mastin, C. W. (1985). *Numerical grid generation: Foundations and applications*. North-Holland.

Thompson, J. F., Soni, B. K., & Wealtherill, N. P. (Eds.). (1988). *Handbook of grid generation*. CRC press.

Hoffmann, K. A., & Chiang, S. T. (2000). *Computational fluid dynamics: Volume 1, 2 and 3*, Engineering Education System, Kansas.

Sundararajan, T. (2003). Automatic grid generation, Chapter 13 (pp. 511). In: K. Muralidhar & T. undararajan (Eds.), *Computational fluid flow and heat transfer* (2nd ed.). New Delhi: Narosa Publishing House.

Chapter 12
Computational Fluid Dynamics on a Curvilinear Grid

The computational heat conduction on a curvilinear grid, presented in the previous chapter for a complex geometry, is extended in this chapter for computational fluid dynamics. The curvilinear grid generation and the computation of geometrical parameters, presented in the previous chapter, are also applicable for this chapter. The mind map for this chapter is shown in Fig. 12.1. The figure shows that the present chapter starts with physical law-based FVM, and ends with the semi-explicit method-based solution methodology for a body-fitted curvilinear grid. The approximated algebraic formulation in the FVM is presented separately from the mass and momentum/energy conservation, and the solution methodology is presented as a two-step predictor-corrector method.

12.1 Physical Law-based Finite Volume Method

The earlier presented (Chap. 8) application of the conservations laws for a Cartesian CV (over a time interval Δt) also corresponds to the complex geometry here; however, instead of the Cartesian CV earlier, a complex CV is considered here; the Cartesian and Complex CV are shown in Fig. 11.7. The earlier Cartesian componentwise application of the momentum conservation law is also considered here, as the flow field for the complex geometry corresponds to the Cartesian components of the velocity.

For the 2D Cartesian CV, the direction of a *surface area vector* is either in the x or y direction; thus, the normal component of a flux vector $\vec{f}$ at a surface is either f_x or f_y. Whereas, for a 2D complex CV, a surface in a complex CV can be in any arbitrary direction. Since the surface area $\overrightarrow{\Delta S} (= \hat{\eta} \Delta S)$ is a vector represented by a direction η normal to the surface, the *normal component of the flux* f_η is considered here for all the faces of a complex CV. Accordingly, using the *I-approximation*, the

© The Author(s), under exclusive license to Springer Nature Switzerland AG 2022   323
A. Sharma, *Introduction to Computational Fluid Dynamics*,
https://doi.org/10.1007/978-3-030-72884-7_12

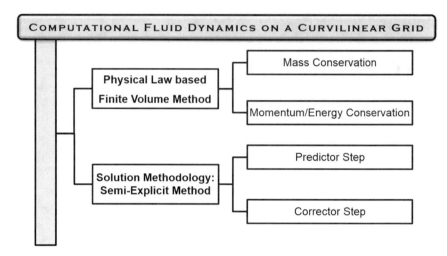

Fig. 12.1 Mind map for Chap. 12

approximated value of the *mean* (over the interval Δt) mass flow rate, advection flow
rate, diffusion at a surface of a complex CV are given as

$$
\begin{aligned}
M_{face}^{mean} &= \overrightarrow{m}_f^{n+1} \cdot \overrightarrow{\Delta S}_f = \rho \left(u_f^{n+1} \Delta S_{x,f} + v_f^{n+1} \Delta S_{y,f} \right) \equiv m_{\eta,f}^{n+1} \Delta S_f \\
A_{\phi,face}^{mean} &= \overrightarrow{a}_f^{\chi} \cdot \overrightarrow{\Delta S}_f = C m_{\eta,f}^n \phi_f^{\chi} \Delta S_f \equiv a_{\phi\eta,f} \Delta S_f \\
D_{\phi,face}^{mean} &= \overrightarrow{d}_f^{\chi} \cdot \overrightarrow{\Delta S}_f = \Gamma_\phi \left[\left(\frac{\partial \phi}{\partial s} \right)_f^{\chi} \hat{s} + \left(\frac{\partial \phi}{\partial t} \right)_f^{\chi} \hat{t} \right] \cdot \left(\Delta S_{s,f} \hat{s} + \Delta S_{t,f} \hat{t} \right) \\
&= \Gamma_\phi \left[\left(\frac{\partial \phi}{\partial s} \right)_f^{\chi} \Delta S_{s,f} + \left(\frac{\partial \phi}{\partial t} \right)_f^{\chi} \Delta S_{t,f} \right] \equiv d_{\phi\eta,,f} \Delta S_f
\end{aligned}
$$

$$(12.1)$$

where, at the centroid f of the surface, $\overrightarrow{m}_f^{n+1} \left(= \rho \overrightarrow{u}_f^{n+1} \right)$ is the mass flux, $\overrightarrow{a}_f^{\chi}$
$\left(= \rho C \overrightarrow{u}_f^n \phi_f^{\chi} \right)$ is the advection flux, and $\overrightarrow{d}_f^{\chi} \left(= \Gamma_\phi \overrightarrow{\nabla} \phi^{\chi} \right)$ is the diffusion flux.
Also the normal component of respective flux is represented above as $m_{\eta,f}$, $a_{\phi\eta,f}$,
and $d_{\phi\eta,f}$.

Note from the above equation that the 2D components of the mass flux are considered in the global $x-y$ coordinate direction; whereas, the diffusion flux is represented
in the local $s-t$ coordinate directions. The local η-coordinate, defined at a face in
the previous chapter (Fig. 11.9), consist of components along the s-direction (line
joining the cell centers across the face) and the t-direction (along the tangent to the
face). The unit vector in the respective directions are represented as $\hat{s}$ and $\hat{t}$. The 2D
components of the vector quantities in the above equation are given as

$$\text{Global} x - y : \vec{u}_f = u\hat{i} + v\hat{j}; \qquad \vec{\Delta S}_f = \Delta S_{x,f}\,\hat{i} + \Delta S_{y,f}\,\hat{j}$$

$$\text{Local} s - t : \quad \vec{\nabla\phi}_f = \left(\frac{\partial\phi}{\partial s}\right)_f \hat{s} + \left(\frac{\partial\phi}{\partial t}\right)_f \hat{t}; \quad \vec{\Delta S}_f = \Delta S_{s,f}\,\hat{s} + \Delta S_{t,f}\,\hat{t}$$

where the components for $\vec{u}_f$ and $\vec{\nabla\phi}_f$ are in the global and local coordinate system, respectively.

The *II-approximation* corresponds to the momentum interpolation for the mass flux $m_{\eta,f}$, advection schemes for the advection term, and central difference method for the diffusion term. The two approximations and the approximated algebraic formulation for the mass and momentum conservation are presented in separate subsections below.

12.1.1 Mass Conservation

Considering the application of the law of conservation of mass (over a time interval Δt) for a complex CV, Fig. 12.2 shows the first and the second approximation for the physical law-based FVM. After the first approximation, the figure shows the algebraic formulation for the mass source given as

$$
\begin{aligned}
S_{m,P}^{mean} &\approx \left(m_{\eta,e}^{n+1}\,\Delta S_e - m_{\eta,w}^{n+1}\,\Delta S_w\right) + \left(m_{\eta,n}^{n+1}\,\Delta S_n - m_{\eta,s}^{n+1}\,\Delta S_s\right) \\
&\approx \rho \left(u_{\eta,e}^{n+1}\,\Delta S_e - u_{\eta,w}^{n+1}\,\Delta S_w\right) + \rho \left(u_{\eta,n}^{n+1}\,\Delta S_n - u_{\eta,s}^{n+1}\,\Delta S_s\right) \quad (12.2)
\end{aligned}
$$

where $u_{\eta,f}$ is the normal velocity, and $m_{\eta,f}$ is the mass flux, at the various face centers $f = e, w, n, s$ of the CVs. The normal velocity is computed by the momentum interpolation method, as the co-located (not staggered) grid is used for the complex geometry. The momentum interpolation was introduced in Chap. 10 for 2D Cartesian-geometry and will be presented below for the complex geometry.

12.1.2 Momentum Conservation

The approximations and the resulting algebraic formulation are presented below; separately for the unsteady, advection, diffusion, and source terms.

12.1.2.1 Unsteady Term

For the unsteady term, the approximation and the algebraic formulation presented in Fig. 8.5 for the Cartesian-geometry is also applicable for the complex geometry; given as

Fig. 12.2 A representative fluid CV to demonstrate two approximations for algebraic representation of the balance statement in the mass conservation law for a complex CV. Here, $\vec{m}$ is the mass flux, M is the mass flow rate, and $S_{m,P}$ is the mass source

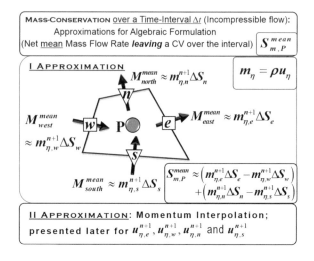

$$\frac{\Delta \mathcal{V}_\phi}{\Delta t} \approx \rho C \Delta V_P \frac{\phi_P^{n+1} - \phi_P^n}{\Delta t} \tag{12.3}$$

where the transported variable $\phi = u$, v, and T for x-momentum, y-momentum, and energy conservation, respectively; and $C = c_p$ for the energy conservation and $C = 1$ for the momentum conservation. Furthermore, ΔV_P is the volume of the representative complex CV, computed using the equation shown in Fig. 11.10.

12.1.2.2 Advection Term

For the advection term, Fig. 12.3 shows the first and the second approximation for the physical law-based FVM. The figure shows the advection flux $a_{\phi\eta}$ as x-momentum flux $a_{\phi u}$ $\left(= m_\eta u\right)$, y-momentum flux $a_{\phi v}$ $\left(= m_\eta v\right)$, and enthalpy flux $a_{\phi T}$ $\left(= m_\eta c_p T\right)$. The respective flux is used for the conservation of x-momentum, y-momentum, and energy. Furthermore, at the various faces of the complex CV, the *1-approximation* for the advection flux is shown as $a_{\phi\eta,av@face}^{mean} \approx a_{\phi\eta,f}^\chi$. Here, as discussed earlier for $a_{\phi\eta,av@face}^{mean}$, the time-averaged *mean* value of the flux is $\chi = n + 1$ for the implicit method, $\chi = n$ for the explicit method, and $\chi = n + 1/2$ for the Crank-Nicolson method; and the space-averaged $av@face$ value corresponding to the flux at the centroid f of the face. Furthermore, as discussed earlier, it can be seen that the mass flux is considered explicit (m_η^n) in the equation for the advection flux $a_{\phi\eta,f}^\chi = m_\eta^n C \phi^\chi$.

Advection scheme is used as the *II-approximation* for the advection term, shown in Fig. 12.3. The advection scheme was introduced earlier for the Cartesian-geometry (Chap. 6) and is presented here for the complex geometry; with the help of Fig. 12.4 and Table 12.1. As discussed in the Chap. 6, advection scheme is a direction of mass flux-based interpolation or extrapolation procedure. Thus, considering the positive

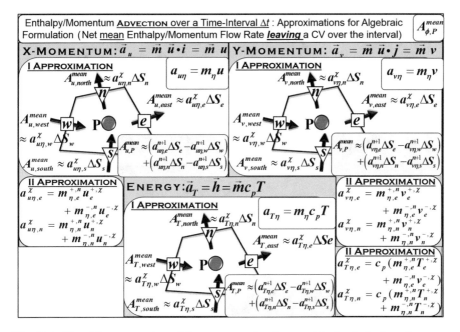

Fig. 12.3 Pictorial representation of the two approximations used in the derivation of governing LAE for the advection term of the law of conservation of momentum and energy. Here, the transported variable $\phi = u$ for x-momentum, $\phi = v$ for y-momentum and $\phi = T$ for energy conservation

or negative values of the mass flux at a face center $m_{\eta,f}$, ϕ_f is expressed in terms of cell center values which are downstream ϕ_D, upstream ϕ_U and further upstream ϕ_{UU} of the face center f.

Using the two approximations shown in Fig. 12.3 for $a^{\chi}_{\phi,f=e,w,n,s}$, the algebraic formulation for the advection term is given as

$$A^{mean}_{\phi,P} = A^{mean}_{west} - A^{mean}_{east} + A^{mean}_{south} - A^{mean}_{north}$$

$$A^{mean}_{\phi,P} \approx \left\{ \left[m^{+,n}_{\eta,e}\phi^{+,\chi}_e + m^{-,n}_{\eta,e}\phi^{-,\chi}_e \right] \Delta S_e - \left[m^{+,n}_{\eta,w}\phi^{+,\chi}_w + m^{-,n}_{\eta,w}\phi^{-,\chi}_w \right] \Delta S_w \right\} C$$
$$+ \left\{ \left[m^{+,n}_{\eta,n}\phi^{+,\chi}_n + m^{-,n}_{\eta,n}\phi^{-,\chi}_n \right] \Delta S_n - \left[m^{+,n}_{\eta,s}\phi^{+,\chi}_s + m^{-,n}_{\eta,s}\phi^{-,\chi}_s \right] \Delta S_s \right\} C$$
$$(12.4)$$

where $\Delta S_{f=e,w,n,s}$ is the surface area of the various surfaces of the CVs, computed using its Cartesian components $\Delta S_{x/y,f=e,w,n,s}$ (Fig. 11.10).

Since a face is common to two adjoining CVs, the advection scheme is presented in Table 12.1 for the north and east face center; in continuation with the solution methodology adopted in the previous chapter. The table shows a general expression for the product of the mass flux $m_{\eta,f}$ and the advected variable ϕ_f; needed for computation of for $A^{mean}_{\phi,P}$ in Eq. 12.4. Note from the table that the equation for ϕ^+_f correspond to positive flow ($m_{\eta,f} > 0$), and ϕ^-_f correspond to negative flow

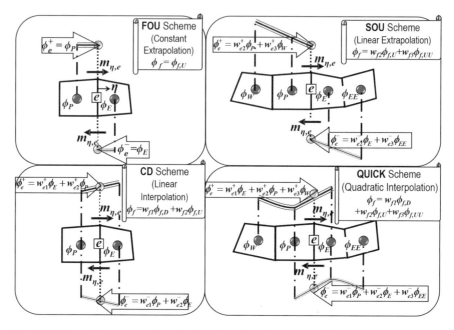

Fig. 12.4 Schematic representation of various advection schemes, to calculate the advected variable ϕ_e at the east face center; considering both positive/outward and negative/inward direction of the mass flux $m_{e,f}$ at the face

($m_{\eta,f} < 0$). This is demonstrated in the table for the east face center, where the equation for ϕ_e^+ and ϕ_e^- is obtained after substituting appropriate subscript D, U and UU in $\phi_{e,D}^\pm$, $\phi_{e,U}^\pm$ and $\phi_{e,UU}^\pm$; shown in Fig. 12.4, for the various advection schemes.

The figure and the table also show certain weights for the neighboring cell center values. Note that the equations for the weights in Table 12.1 are similar to that for the Cartesian-geometry (Table 6.2). However, here the equations correspond to a non-uniform volume of the complex CVs (instead of the non-uniform width of the earlier Cartesian CVs) weighted interpolation or extrapolation procedure in the various advection schemes. Note that the interpolation or extrapolation is volume weighted for the complex geometry instead of the distance weighted for the Cartesian-geometry; this is not only limited to advection scheme but used below during the linear interpolation for the various other terms (such as pressure) in the algebraic formulation.

12.1.2.3 Diffusion Term

For the diffusion term, Fig. 12.5 shows the first and the second approximation for the physical law-based FVM. The figure shows the diffusion flux $d_{\phi\eta}$ as viscous stress in the x-direction $d_{\phi u}\left(=\sigma_{\eta x}\right)$, viscous stress in the y-direction $d_{\phi v}\left(=\sigma_{\eta y}\right)$, and

Table 12.1 Advection schemes for the complex geometry problems (Sharma and Eswaran 2003): volume weighted interpolation or extrapolation-based equations for the product of mass flux $m_{\eta,f}$ ($\equiv \rho \vec{u}_f . \hat{n}$) and advected variable ϕ_f at the east and north face centers. The equations are given for both positive/outward and negative/inward direction of the mass flux at the face centers. Subscript D, U and UU correspond to the downstream, upstream, and upstream-of-upstream cell center of a face center f, respectively. The weights for the respective cell center value are represented as $w_{f,1}$, $w_{f,2}$ and $w_{f,3}$. Superscripts $+$ and $-$ correspond to $m_{\eta,f} > 0$ and $m_{\eta,f} < 0$, respectively

Advection Scheme: $\left. m_{\eta,f}\phi_f = m^+_{\eta,f}\phi^+_f + m^-_{\eta,f}\phi^-_f \right\} f = e, w, n, s$

$$\text{where } m_{\eta,f} = \vec{u}_f . \hat{n}, \quad \left. \begin{array}{l} m^+_{\eta,f} \equiv \max(m_{\eta,f}, 0) = +|m_{\eta,f}| \text{ if } m_{\eta,f} > 0 \\ m^-_{\eta,f} \equiv \min(m_{\eta,f}, 0) = -|m_{\eta,f}| \text{ if } m_{\eta,f} < 0 \end{array} \right\} \text{ otherwise } 0$$

$$\text{and } \phi^\pm_f = w^\pm_{f,1}\phi^\pm_{f,D} + w^\pm_{f,2}\phi^\pm_{f,U} + w^\pm_{f,3}\phi^\pm_{f,UU}$$

$\phi^+_e \text{ (for } m_{\eta,f} > 0) = w^+_{e,1}\phi_E + w^+_{e,2}\phi_P + w^+_{e,3}\phi_W$ and $\phi^-_e \text{ (for } m_{\eta,f} < 0) = w^-_{e1}T_P + w^-_{e,2}\phi_E + w^-_{e,3}\phi_{EE}$

$w_f \downarrow$	FOU	CD	SOU	QUICK
$w^+_{e,1}$	0	$\dfrac{\Delta V_E}{\Delta V_E + \Delta V_P}$	0	$\dfrac{\Delta V_P(2\Delta V_P + \Delta V_W)}{(\Delta V_E + \Delta V_P)(\Delta V_E + 2\Delta V_P + \Delta V_W)}$
$w^-_{e,1}$	0	$\dfrac{\Delta V_P}{\Delta V_P + \Delta V_E}$	0	$\dfrac{\Delta V_E(2\Delta V_E + \Delta V_{EE})}{(\Delta V_P + \Delta V_E)(\Delta V_P + 2\Delta V_E + \Delta V_{EE})}$
$w^+_{e,2}$	1	$\dfrac{\Delta V_P}{\Delta V_E + \Delta V_P}$	$\dfrac{2\Delta V_P + \Delta V_W}{\Delta V_P + \Delta V_W}$	$\dfrac{\Delta V_E(2\Delta V_P + \Delta V_W)}{(\Delta V_E + \Delta V_P)(\Delta V_P + \Delta V_W)}$
$w^-_{e,2}$	1	$\dfrac{\Delta V_E}{\Delta V_P + \Delta V_E}$	$\dfrac{2\Delta V_E + \Delta V_{EE}}{\Delta V_E + \Delta V_{EE}}$	$\dfrac{\Delta V_P(2\Delta V_E + \Delta V_{EE})}{(\Delta V_P + \Delta V_E)(\Delta V_E + \Delta V_{EE})}$
$w^+_{e,3}$	0	0	$-\dfrac{\Delta V_P}{\Delta V_P + \Delta V_W}$	$-\dfrac{\Delta V_E \Delta V_P}{(\Delta V_P + \Delta V_W)(\Delta V_E + 2\Delta V_P + \Delta V_W)}$
$w^-_{e,3}$	0	0	$-\dfrac{\Delta V_E}{\Delta V_E + \Delta V_{EE}}$	$-\dfrac{\Delta V_P \Delta V_E}{(\Delta V_E + \Delta V_{EE})(\Delta V_P + 2\Delta V_E + \Delta V_{EE})}$

The above equation for $\phi^\pm_e$ and $w^\pm_e$ is used to obtain similar equation for $\phi^\pm_n$ and $w^\pm_n$, by replacing EE, E, and W with NN, N, and S, respectively.

conduction flux $d_{\phi T}$ $(= -q_\eta)$. The respective flux is used for the conservation of x-momentum, y-momentum, and energy. Furthermore, at the various faces of the complex CV, the *I-approximation* for the diffusion flux is shown as $d^{mean}_{\phi\eta,av@face} \approx d^\chi_{\phi\eta,f}$; similar to presented above for the advection flux.

The figures also show that the conduction flux is represented by the normal gradient of temperature. Similarly, the figures shows the viscous stress in the x-direction (y-direction) represented by normal gradient of u-velocity (v-velocity); note that this generalized representation (from the viscous stress in Cartesian coordinate; refer Fig. 3.7) is applicable only for the incompressible flow. For the normal gradient of the velocity and temperature, the central difference method-based *II-approximation* results in the algebraic formulation for the diffusion flux. This was presented in the previous chapter for the conduction flux (Eq. 11.13), and extended here for the viscous stress; resulting in the generalized diffusion flux as

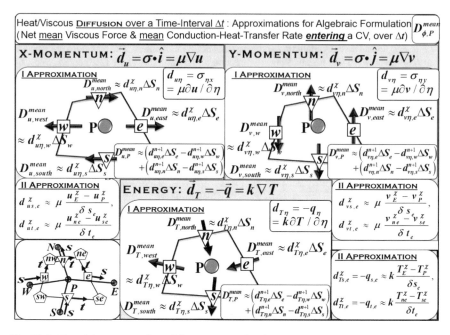

Fig. 12.5 Pictorial representation of the two approximations used in the derivation of governing LAE for the diffusion term of the law of conservation of momentum and energy. Here, the transported variable $\phi = u$ for x-momentum, $\phi = v$ for y-momentum and $\phi = T$ for energy conservation

$$
\begin{aligned}
&d^{\chi}_{\phi s,e} \approx \Gamma_\phi \frac{\phi^{\chi}_E - \phi^{\chi}_P}{\delta s_e}, \quad d^{\chi}_{\phi t,e} \approx \Gamma_\phi \frac{\phi^{\chi}_{ne} - \phi^{\chi}_{se}}{\delta t_e} \\[4pt]
&d^{\chi}_{\phi s,n} \approx \Gamma_\phi \frac{\phi^{\chi}_N - \phi^{\chi}_P}{\delta s_n}, \quad d^{\chi}_{\phi t,n} \approx \Gamma_\phi \frac{\phi^{\chi}_{ne} - \phi^{\chi}_{nw}}{\delta t_n} \\[4pt]
&d^{\chi}_{\phi s,w} \approx \Gamma_\phi \frac{\phi^{\chi}_P - \phi^{\chi}_W}{\delta s_w}, \quad d^{\chi}_{\phi t,w} \approx \Gamma_\phi \frac{\phi^{\chi}_{nw} - \phi^{\chi}_{sw}}{\delta t_w} \\[4pt]
&d^{\chi}_{\phi s,s} \approx \Gamma_\phi \frac{\phi^{\chi}_P - \phi^{\chi}_S}{\delta s_s}, \quad d^{\chi}_{\phi t,s} \approx \Gamma_\phi \frac{\phi^{\chi}_{se} - \phi^{\chi}_{sw}}{\delta t_s}
\end{aligned}
\tag{12.5}
$$

where $\phi = u$ (and $\phi = v$) and $\Gamma_\phi = \mu$ for the x-momentum (y-momentum) conservation; and $\phi = T$ and $\Gamma_\phi = k$ for the energy conservation. The above diffusion fluxes at the east face center are shown in Fig. 12.5, in terms of the *adjoining cell center* values for $d^{\chi}_{\phi s,e}$ and the *associated vertex values* for $d^{\chi}_{\phi t,e}$. Furthermore, the transported variable ϕ at the vertices of the CVs (Eq. 12.5) are computed by a volume-weighted interpolation of the adjoining four cell center values of ϕ; using Eq. 11.14, with the generic variable ϕ instead of T.

Using the two approximation for the diffusion term above, the algebraic equation for the mean value of total diffusion $D^{mean}_{\phi,P}$ (refer Fig. 12.5) is presented as

$$
D^{mean}_{\phi,P} = D^{mean}_{east} - D^{mean}_{west} + D^{mean}_{north} - D^{mean}_{south}
$$

Applying the *I-approximation* (Fig. 12.5), the above equation is given as

$$D_{\phi,P}^{mean} \approx \left(d_{\phi\eta,e}^{\chi}\Delta S_e - d_{\phi\eta,w}^{\chi}\Delta S_w\right) + \left(d_{\phi\eta,n}^{\chi}\Delta S_n - d_{\phi\eta,s}^{\chi}\Delta S_s\right)$$

Substituting the component of $d_{\eta,f}^{\chi}\Delta S_f$ (with face center $f = e, w, n$ and s) along the local s and t direction, $d_{\phi\eta,f}^{\chi}\Delta S_f = d_{\phi s,f}^{\chi}\Delta S_{\phi s,f} + d_{\phi t,w}^{\chi}\Delta S_{\phi t,f}$, the above equation is given as

$$D_{\phi,P}^{mean} \approx D_{\phi s,P}^{mean} + D_{\phi t,P}^{mean} \tag{12.6}$$

where
$$D_{\phi s,P}^{mean} \approx \left(d_{\phi s,e}^{\chi}\Delta S_{s,e} - d_{\phi s,w}^{\chi}\Delta S_{s,w}\right) + \left(d_{\phi s,n}^{\chi}\Delta S_{s,n} - d_{\phi s,s}^{\chi}\Delta S_{s,s}\right)$$
$$D_{\phi t,P}^{mean} \approx \left(d_{\phi t,e}^{\chi}\Delta S_{t,e} - d_{\phi t,w}^{\chi}\Delta S_{t,w}\right) + \left(d_{\phi t,n}^{\chi}\Delta S_{t,n} - d_{\phi t,s}^{\chi}\Delta S_{t,s}\right)$$

where $D_{\phi s,P}^{n}$ is the net cell-centered diffusion, and $D_{\phi t,P}^{n}$ is the net cross-diffusion; introduced in the previous chapter for conduction (Fig. 11.13). Applying the *II-approximation* (Eq. 12.5), the above equation is given as

$$D_{cond,P} = \left(\Gamma_\phi \frac{\phi_E^{\chi} - \phi_P^{\chi}}{\delta s_e}\Delta S_{s,e} - \Gamma_\phi \frac{\phi_P^{\chi} - \phi_W^{\chi}}{\delta s_w}\Delta S_{s,w}\right)$$
$$+ \left(\Gamma_\phi \frac{\phi_N^{\chi} - \phi_P^{\chi}}{\delta s_n}\Delta S_{s,n} - \Gamma_\phi \frac{\phi_P^{\chi} - \phi_S^{\chi}}{\delta s_s}\Delta S_{s,s}\right)$$
$$+ \left(\Gamma_\phi \frac{\phi_{ne}^{\chi} - \phi_{se}^{\chi}}{\delta t_e}\Delta S_{t,e} - \Gamma_\phi \frac{\phi_{nw}^{\chi} - \phi_{sw}^{\chi}}{\delta t_w}\Delta S_{t,w}\right)$$
$$+ \left(\Gamma_\phi \frac{\phi_{ne}^{\chi} - \phi_{nw}^{\chi}}{\delta t_n}\Delta S_{t,n} - \Gamma_\phi \frac{\phi_{se}^{\chi} - \phi_{sw}^{\chi}}{\delta t_s}\Delta S_{t,s}\right) \tag{12.7}$$

where $\Delta S_{s,f=e,w,n,s}$ and $\Delta S_{t,f=e,w,n,s}$ are the local components of the surface area $\Delta S_{f=e,w,n,s}$; obtained from Eq. 11.19, for the east face center.

12.1.2.4 Pressure as a Source Term

For the source term, Fig. 12.6 shows the approximations for the physical law-based FVM. Figure 12.6a,b shows that the pressure at all the four faces (instead of two surface for a Cartesian CV earlier) are multiplied by the Cartesian components of surface area, $\Delta S_{x,f}$ and $\Delta S_{y,f}$, to obtain the 2D Cartesian components of the net pressure force (acting on the CV) in respective direction. The figure shows the volume-weighted linear interpolation as the *II-approximation* for the pressure (similar to the central difference method in Fig. 12.4 and Table 12.1); represented by an *over-bar operator* for a general variable ϕ as

$$\phi_{f=e,w,n,s} = \overline{\phi_{NB}}, \quad \overline{\phi_P} = \frac{\Delta V_P \phi_{NB} + \Delta V_{NB} \phi_P}{\Delta V_P + \Delta V_{NB}} \tag{12.8}$$

where $NB = E, W, N$, and S for $f = e, w, n$, and s, respectively.

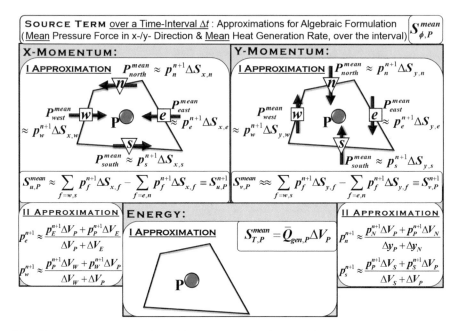

Fig. 12.6 Pictorial representation of the two approximations used in the derivation of governing LAE for source term for the law of conservation of momentum and energy. Here, S_u and S_v are the net pressure force in the x and y direction, respectively; and S_T is heat gain by volumetric heat generation

12.1.2.5 All Terms

The final algebraic formulation for 2D unsteady-state transport equation is given as

$$\frac{\Delta \mathcal{V}_\phi}{\Delta t} + A_{\phi,P}^{mean} = D_{\phi,P}^{mean} + S_{\phi,P}^{mean}$$

$$\rho \frac{u_P^{n+1} - u_P^n}{\Delta t} \Delta V_P + A_{u,P}^{\chi} = D_{u,P}^{\chi} + S_{u,P}^{n+1}$$

$$\rho \frac{v_P^{n+1} - v_P^n}{\Delta t} \Delta V_P + A_{v,P}^{\chi} = D_{v,P}^{\chi} + S_{v,P}^{n+1} \qquad (12.9)$$

$$\rho c_P \frac{T_P^{n+1} - T_P^n}{\Delta t} \Delta V_P + A_{T,P}^{\chi} = D_{T,P}^{\chi} + \overline{Q}_{gen,P} \Delta V_P$$

where

$$
\begin{aligned}
A^\chi_{\phi,P} &= \left\{\left[m^{+,n}_{\eta,e}\phi^{+,\chi}_e + m^{-,n}_{\eta,e}\phi^{-,\chi}_e\right]\Delta S_e - \left[m^{+,n}_{\eta,w}\phi^{+,\chi}_w + m^{-,n}_{\eta,w}\phi^{-,\chi}_w\right]\Delta S_w\right\}C \\
&\quad + \left\{\left[m^{+,n}_{\eta,n}\phi^{+,\chi}_n + m^{-,n}_{\eta,n}\phi^{-}_n\right]\Delta S_n - \left[m^{+,n}_{\eta,s}\phi^{+,\chi}_s + m^{-,n}_{\eta,s}\phi^{-,\chi}_s\right]\Delta S_s\right\}C \\
D^\chi_{\phi,P} &= \Gamma_\phi\left(\frac{\phi^\chi_E - \phi^\chi_P}{\delta s_e}\Delta S_{s,e} - \frac{\phi^\chi_P - \phi^\chi_W}{\delta s_w}\Delta S_{s,w} + \frac{\phi^\chi_N - \phi^\chi_P}{\delta s_n}\Delta S_{s,n}\right. \\
&\quad - \frac{\phi^\chi_P - \phi^\chi_S}{\delta s_s}\Delta S_{s,s} + \frac{\phi^\chi_{ne} - \phi^\chi_{se}}{\delta t_e}\Delta S_{t,e} - \frac{\phi^\chi_{nw} - \phi^\chi_{sw}}{\delta t_w}\Delta S_{t,w} \\
&\quad \left. + \frac{\phi^\chi_{ne} - \phi^\chi_{nw}}{\delta t_n}\Delta S_{t,n} - \frac{\phi^\chi_{se} - \phi^\chi_{sw}}{\delta t_s}\Delta S_{t,s}\right) \\
S^{n+1}_{u,P} &= -\left(\sum_{f=e,n} p^{n+1}_f\Delta S_{x,f} - \sum_{f=w,s} p^{n+1}_f\Delta S_{x,f}\right) \\
S^{n+1}_{v,P} &= -\left(\sum_{f=e,n} p^{n+1}_f\Delta S_{y,f} - \sum_{f=w,s} p^{n+1}_f\Delta S_{y,f}\right)
\end{aligned}
$$

$$(12.10)$$

where the transported variable at a face center $\phi^\pm_{f=e,w,n,s}$ is presented in Table 12.1; and the volume-weighted pressure at the various face center p_f is given in Eq. 12.8.

12.2 Solution Methodology: Semi-Explicit Method

For the semi-explicit method, the discretized form of momentum equations (Eq. 12.9) are given as

$$
\begin{aligned}
u^{n+1}_P &= u^n_P + \frac{\Delta t}{\rho\Delta V_P}\left[\left(D^n_{u,P} - A^n_{u,P}\right) + S^{n+1}_{u,P}\right] \\
v^{n+1}_P &= v^n_P + \frac{\Delta t}{\rho\Delta V_P}\left[\left(D^n_{v,P} - A^n_{v,P}\right) + S^{n+1}_{v,P}\right] \\
\text{where } S^{n+1}_{\phi,P} &= -\left(\sum_{f=e,n} p^{n+1}_f\Delta S_{\Phi,f} - \sum_{f=w,s} p^{n+1}_f\Delta S_{\Phi,f}\right)
\end{aligned}
$$

$$(12.11)$$

where the net advection $A^n_{\phi,P}$ and net diffusion $D^n_{\phi,P}$ are given in Eq. 12.10. Furthermore, for $S^{n+1}_{\phi,P}$ above, $\Phi = x$ for $\phi = u$ and $\Phi = y$ for $\phi = v$.

Similar to above equation for u_P and v_P, the discretized form of momentum conservation for the normal velocity at various face center (at the centroid of the staggered CVs, shown in Fig. 12.7) are given as

$$
u^{n+1}_{\eta,f=e,w,n,s} = u^n_{\eta,f} + \frac{\Delta t}{\rho\Delta V_{\eta,f}}\left[\left(D^n_{u_\eta,f} - A^n_{u_\eta,f}\right) + S^{n+1}_{u_\eta,f}\right]
$$
$$\text{where } S^{n+1}_{u_\eta,f} \approx -(\nabla p^{n+1}\cdot\hat{n})_f\Delta V_{u_\eta,f}$$

$$(12.12)$$

where the above formulation for $S^{n+1}_{u_\eta,f}$ is presented in Fig. 12.7. The figure shows that although the pressure is a flux term, it is converted into a volumetric term by using Gauss-divergence theorem. This is done to *avoid* the surface area of the staggered CVs in the algebraic formulation for the prediction of mass flux; presented below.

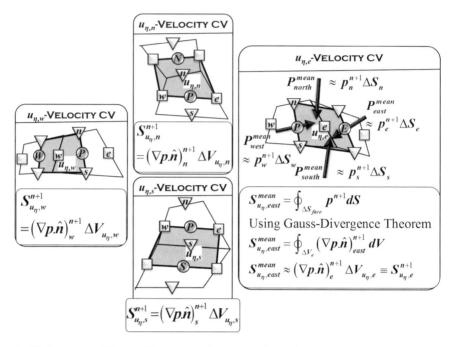

Fig. 12.7 For a curvilinear grid, representative staggered CVs for normal velocity at the various face centers $u_{\eta,f}$ and the associated staggered grid points

Note that the pressure term corresponds to the flux term in Eq. 12.11 and volumetric term in Eq. 12.12. The above equation is used for the momentum interpolation method-based computation of the mass flux; already introduced in Chap. 10 for the Cartesian-geometry.

12.2.1 Predictor Step

Prediction of the Velocity at the Cell Center:

From the original proposition for the semi-explicit method in Eq. 12.11, considering the pressure at the previous time level n results in the predicted velocity (at the co-located grid point) as

$$u_P^* = u_P^n + \frac{\Delta t}{\rho \Delta V_P} \left(D_{u,P}^n - A_{u,P}^n \right) + \frac{\Delta t}{\rho \Delta V_P} S_{u,P}^n \qquad (12.13)$$

$$= \tilde{u}_P - \frac{\Delta t}{\rho \Delta V_P} \left(\sum_{f=e,n} \overline{p_{NB}^n, \, p_P^n} \Delta S_{x,f} - \sum_{f=w,s} \overline{p_{NB}^n, \, p_P^n} \Delta S_{x,f} \right)$$

$$v_P^* = v_P^n + \frac{\Delta t}{\rho \Delta V_P} \left(D_{v,P}^n - A_{v,P}^n \right) + \frac{\Delta t}{\rho \Delta V_P} S_{v,P}^n$$

$$= \tilde{v}_P - \frac{\Delta t}{\rho \Delta V_P} \left(\sum_{f=e,n} \overline{p_{NB}^n, \ p_P^n} \Delta S_{y,f} - \sum_{f=w,s} \overline{p_{NB}^n, \ p_P^n} \Delta S_{y,f} \right)$$

where the over-bar operator for the pressure corresponds to the volume weighted linear interpolation, shown in Eq. 12.8. The *provisional* Cartesian velocities ($\tilde{u}_P$ and $\tilde{v}_P$) are given, as a general transported velocity $\tilde{\phi}_P$, as

$$\tilde{\phi}_P = \phi_P^n + \frac{\Delta t}{\rho \Delta V_P} \left(D_{\phi,P}^n - A_{\phi,P}^n \right) \tag{12.14}$$

Prediction of the Mass Flow Rate at the various Face Centers:

The already presented momentum interpolation method-based prediction of the mass flux for the Cartesian-geometry (Eq. 10.1) is indeed applicable for the complex geometry here; however, there is a modification the source term for pressure. This is because $u_{\eta,f}$ is equal to either of the Cartesian velocity (u_f for the vertical faces and v_f for the horizontal faces) for the Cartesian CV but not for the complex CV. Considering the pressure in Eq. 12.12 at the previous time level n, the predicted *normal* velocity (at the staggered grid points; Fig. 12.7) is given as

$$u_{\eta,f=e,w,n,s}^* = u_{\eta,f}^n + \frac{\Delta t}{\rho \Delta V_{u_\eta,f}} \left(D_{u_\eta,e}^n - A_{u_\eta,f}^n \right) - \frac{\Delta t}{\rho} (\nabla p \cdot \hat{n})_f^n$$

$$= \qquad \tilde{u}_{\eta,f}^n \qquad\qquad\qquad - \frac{\Delta t}{\rho} (\nabla p^n \cdot \hat{n})_f \tag{12.15}$$

where $\tilde{u}_{\eta,f}^n = u_{\eta,f}^n + \frac{\Delta t}{\rho \Delta V_{u_\eta,f}} \left(D_{u_\eta,f}^n - A_{u_\eta,f}^n \right)$

The predicted normal velocity leads to the prediction of the mass flow rate at the various face centers as

$$m_{\eta,f}^* \Delta S_f = \rho \tilde{u}_{\eta,f}^n \Delta S_f - \Delta t (\nabla p^n \cdot \hat{n})_f \Delta S_f$$

$$= \rho \left(\tilde{u}_f^n \Delta S_{x,f} + \tilde{v}_f^n \Delta S_{y,f} \right)$$

$$- \Delta t \left[(\nabla p^n \cdot \hat{s})_f \Delta S_{s,f} + (\nabla p^n \cdot \hat{t})_f \Delta S_{t,f} \right] \tag{12.16}$$

where $\tilde{u}_f^n = \overline{\tilde{u}_{NB}^n, \ \tilde{u}_P^n}$ and $\tilde{v}_f^n = \overline{\tilde{v}_{NB}^n, \ \tilde{v}_P^n}$

where the over-bar operator for the provisional velocity corresponds to the volume weighted linear interpolation (Eq. 12.8 with $\phi = \tilde{u}$ and $\phi = \tilde{v}$); and $-\Delta t (\nabla p^n \cdot \hat{s})_f \Delta S_{s,f}$ and $-\Delta t (\nabla p^n \cdot \hat{t})_f \Delta S_{t,f}$ are given in Eq. 12.5 (with $\phi = p, \Gamma_\phi = -\Delta t$). Note from the above equation that $\tilde{u}_{\eta,f}^n$ is the *provisional normal velocity*, and $\tilde{u}_f^n$ and $\tilde{v}_f^n$ are the *provisional Cartesian velocity*.

Prediction of the Mass Source at the Cell Center:

The predicted mass flow rate at the various face centers are used to obtain the predicted *mass source* as

$$S^*_{m,P} = \sum_{f=e,w,n,s} m^*_{\eta,f} \Delta S_f = \sum_{f=e,w,n,s} \rho u^*_{\eta,f} \Delta S_f \tag{12.17}$$

12.2.2 Corrector Step

Correction of the Velocity at the Cell Center:

The equations for the *velocity correction* are obtained by subtracting Eq. 12.13 from Eq. 12.11. Furthermore, substituting $u^{n+1} - u^* = u'$, $v^{n+1} - v^* = v'$, and $p^{n+1} - p^n = p'$, the velocity correction at the cell center is given (in terms of the pressure correction p') as

$$u'_P = -\frac{\Delta t}{\rho \Delta V_P} \left(\sum_{f=e,n} \overline{p'_{NB}, p'_P} \Delta S_{x,f} - \sum_{f=w,s} \overline{p'_{NB}, p'_P} \Delta S_{x,f} \right)$$

$$v'_P = -\frac{\Delta t}{\rho \Delta V_P} \left(\sum_{f=e,n} \overline{p'_{NB}, p'_P} \Delta S_{y,f} - \sum_{f=w,s} \overline{p'_{NB}, p'_P} \Delta S_{y,f} \right) \tag{12.18}$$

Correction of the Mass Flow Rate at the various Face Centers:

The equations for the *mass flow rate correction* are obtained by subtracting Eq. 12.15 from Eq. 12.12. Furthermore, substituting $u^{n+1}_{\eta,f} - u^*_{\eta,f} = u'_e$, $v^{n+1}_{\eta,f} - v^*_{\eta,f} = v'_n$ and $p^{n+1} - p^n = p'$, the correction for the normal velocity and mass flow rate at the various face center are given as

$$u'_{\eta,f=e,w,n,s} = -\frac{\Delta t}{\rho}(\nabla p' \cdot \hat{n})_f \tag{12.19}$$

$$m'_{\eta,f} \Delta S_f = -\Delta t (\nabla p' \cdot \hat{n})_f \Delta S_f \tag{12.20}$$

where $-\Delta t (\nabla p' \cdot \hat{n})_f$ is given in Eq. 12.5 (with $\phi = p'$, $\Gamma_\phi = -\Delta t$). As presented in the previous chapter, the above equation for the mass flux is similar to Fourier's law of heat conduction.

Pressure Correction:

Substituting $m^{n+1}_{\eta,f} = m^*_{\eta,f} + m'_{\eta,f}$ in the Eq. 12.2, we get

$$\sum_{f=e,n} m'_{\eta,f} \Delta S_f - \sum_{f=w,s} m'_{\eta,f} \Delta S_f = -\left(\sum_{f=e,n} m^*_{\eta,f} \Delta S_f - \sum_{f=w,s} m^*_{\eta,f} \Delta S_f \right)$$

$$= -S^*_{m,P}$$

For the various faces of a CV, substituting the mass flow-rate correction $m'_{\eta,f} \Delta S_f$ from Eq. 12.20, the above equation is given in a LAE form as

$$-\left(\sum_{f=e,n} \Delta t (\nabla p' \cdot \hat{n})_f \Delta S_f - \sum_{f=w,s} \Delta t (\nabla p' \cdot \hat{n})_f \Delta S_f \right) = -S^*_{m,P}$$

Substituting Eq. 12.7 (with $\phi = p'$, $\Gamma_\phi = -\Delta t$) for the LHS of the above equation, the pressure-correction equation is given as

$$
\begin{aligned}
-\Delta t &\left(\frac{p'_E - p'_P}{\delta s_e} \Delta S_{s,e} - \frac{p'_P - p'_W}{\delta s_w} \Delta S_{s,w} + \frac{p'_N - p'_P}{\delta s_n} \Delta S_{s,n} \right. \\
&- \frac{p'_P - p'_S}{\delta s_s} \Delta S_{s,s} + \frac{p'_{ne} - p'_{se}}{\delta t_e} \Delta S_{t,e} - \frac{p'_{nw} - p'_{sw}}{\delta t_w} \Delta S_{t,w} = -S^*_{m,P} \quad (12.21) \\
&\left. + \frac{p'_{ne} - p'_{nw}}{\delta t_n} \Delta S_{t,n} - \frac{p'_{se} - p'_{sw}}{\delta t_s} \Delta S_{t,s} \right)
\end{aligned}
$$

Similar to Eq. 9.17, the Gauss-Seidel method-based solution of the residual form of the above equation is given as

$$p'^{,N+1}_P = p'^{,N}_P - \frac{1}{a_P} \left(S^{*,N}_{m,P} + S'_{m,P} \right) \tag{12.22}$$

where $S^*_{m,P} = \sum_{f=e,n} m^*_{\eta,f} \Delta S_f - \sum_{f=w,s} m^*_{\eta,f} \Delta S_f$

$$
\begin{aligned}
S'_{m,P} = -\Delta t &\left(\frac{p'^{,N}_E - p'^{,N}_P}{\delta s_e} \Delta S_{s,e} - \frac{p'^{,N}_P - p'^{,N+1}_W}{\delta s_w} \Delta S_{s,w} + \right. \\
&+ \frac{p'^{,N}_N - p'^{,N}_P}{\delta s_n} \Delta S_{s,n} - \frac{p'^{,N}_P - p'^{,N+1}_S}{\delta s_s} \Delta S_{s,s} \\
&+ \frac{p'^{,N}_{ne} - p'^{,N}_{se}}{\delta t_e} \Delta S_{t,e} - \frac{p'^{,N}_{nw} - p'^{,N}_{sw}}{\delta t_w} \Delta S_{t,w} \\
&\left. + \frac{p'^{,N}_{ne} - p'^{,N}_{nw}}{\delta t_n} \Delta S_{t,n} - \frac{p'^{,N}_{se} - p'^{,N}_{sw}}{\delta t_s} \Delta S_{t,s} \right)
\end{aligned}
$$

$$a_P = \Delta t \left(\frac{\Delta S_{s,e}}{\delta s_e} + \frac{\Delta S_{s,w}}{\delta s_w} + \frac{\Delta S_{s,n}}{\delta s_n} + \frac{\Delta S_{s,s}}{\delta s_s} \right)$$

In the above equation, for the first iteration ($N + 1 = 1$), note that the previous iterative value of the predicted mass source $S^{*,0}_{m,P}$ is calculated from the predicted mass flow rate (Eq. 12.16) and the mass source correction $S'^{,0}_{m,P}$ is obtained from the initial condition for pressure correction.

After each iteration of the above equation, the pressure correction $p'^{,N+1}$ is substituted in Eq. 12.20 to obtain $m_f'^{,N+1} \Delta S_f$ which is used to obtained the updated value of the predicted mass flow rate as

$$m_{\eta,f=e,w,n,s}^{*,N+1} \Delta S_f = m_{\eta,f}^{*,N} \Delta S_f + m_{\eta,f}'^{,N+1} \Delta S_f \tag{12.23}$$

Furthermore, the updated (present iterative) value of the mass source and pressure is given as

$$S_{m,P}^{*,N+1} = \sum_{f=e,n} m_{\eta,f}^* \Delta S_f - \sum_{f=w,s} m_{\eta,f}^* \Delta S_f \tag{12.24}$$

$$p_P^{*,N+1} = p_P^{*,N} + p_P'^{,N+1}$$

The above mass source is checked for convergence (computationally zero defined by convergence criterion). If not converged, Eqs. 12.22 and 12.24 are solved sequentially; as per the philosophy of the pressure-correction method, presented in Chap. 9. Once converged, then the pressure correction is first used to obtain first the velocity correction (Eq. 12.18) and then the *correct* value of cell center velocity for the present time level as

$$u_P^{n+1} = u_P^* + u_P' \text{ and } v_P^{n+1} = v_P^* + v_P'$$

The initial as well as boundary condition and stability (for the semi-explicit method) as well as stopping criterion (for the unsteady solution), presented in Chap. 9 is also considered here. The solution algorithm for the complex geometry is similar to that in Chap. 10 for the Cartesian-geometry.

Example 12.1: CFD study for a 2D L̲id D̲riven C̲avity (LDC) flow, with a complex geometry for the cavity

The LDC flow was introduced in the Chap. 9, for a Cartesian-geometry. Here, for the complex geometry, Fig. 12.8(a1–c1) shows the LDC flow in a parallelogram, trapezoidal, and triangular cavity; with $L_1 = 1$. Using the FVM, the various types of advection schemes and the solution methodology (presented above), develop an in-house code for the N̲avier-S̲tokes (NS) solver on the complex geometry. Thereafter, use the code to solve the LDC flow in the various complex shaped cavities, using the BCs on the stationary and moving wall of the cavity (Fig. 12.8); similar to the BCs shown in Fig. 9.11. For a non-dimensional study, Fig. 12.8(a1–c1) shows that the length scale considered is the length L_1 for the parallelogram cavity; whereas, for the trapezoidal and triangular cavity, one-third of the height of the cavity $(L_2/3)$ is considered as the length scale. Furthermore, the velocity u_0 of the lid is taken as the velocity scale for all the cavities.

After the development of the NS solver, run the code at a Reynolds number $Re = 1000$ for the parallelogram cavity, and $Re = 500$ for both trapezoidal and triangular cavity. Use two different grid size (22×22 and 82×82), and the four advection schemes (FOU, CD, SOU and QUICK), with a convergence tolerance of $\epsilon_{st} = 10^{-4}$ (for the steady state) and $\epsilon = 10^{-8}$ (for the iterative solution of the pressure-correction equation). Report the results as follows:

(a) *Curvilinear grid generation*: Using the elliptic method presented in Section 11.1.2, generate a curvilinear grid for the parallelogram, trapezoidal, and triangular cavity
(b) *Flow patterns*: Present and discuss a figure for the streamlines in the various lid driven cavities, for the QUICK scheme on the grid size of 82×82.
(c) *Code verification study and comparison of the various advection schemes*: Using the FOU, CD, SOU and QUICK scheme on both the grid size of 22×22 and 82×82, draw an overlap plot of the results as follows:

   (1) Variation of U-velocity on the centerline AB of the cavity.
   (2) Variation of V-velocity on the centerline CD of the cavity.

   The centerline AB and CD can be seen in Fig. 12.8(a1–c1), for the various cavities. For the above variation in the figure (on the grid size of 82×82), include the benchmark results reported by Demirdzic et al. (1992) for the parallelogram shaped cavity, and Paramane and Sharma (2008) for the trapezoidal as well as triangular cavity. Discuss the accuracy of the present results as compared to the benchmark results. Also discuss the effect of various advection schemes on the accuracy of the results, for both the grid sizes.

Solution:

(a) The curvilinear grid generated for the various cavities are shown in Fig. 12.8(a1–c1).
(b) The streamlines in Fig. 12.8(a2–c2) show a bigger primary vortex and many smaller secondary vortices inside the various cavities. The primary vortex is seen in the central region and the secondary vortices are formed at the various corners of the cavities.
(c) Fig. 12.9(a2–c2) shows a excellent agreement between the present and published results for the variation of the non-dimensional velocity profiles, on the finer grid size, for all the advection schemes; except FOU scheme. Whereas, on the coarser grid size, the relative accuracy of the results obtained from the various advection schemes is more clearly seen in Fig. 12.9(a1–c1). The figure shows that the accuracy of the results is maximum for the QUICK and minimum for the SOU scheme. More detailed study and discussion on the qualitative as well as quantitative comparison of the various advection schemes for the various cavities was reported by Paramane and Sharma (2008).

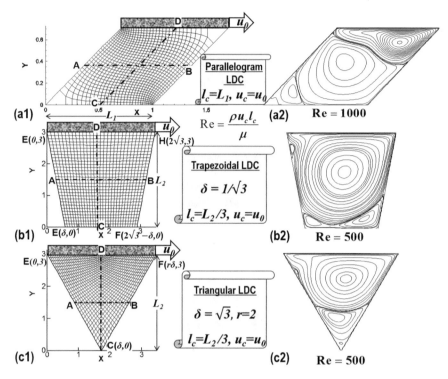

Fig. 12.8 (a1–c1) Computational-domain as well as curvilinear grid generation, and (a2–c2) streamlines for the lid driven flow; inside a (a1–a2) parallelogram, (b1–b2) trapezoidal, and (c1–c2) triangular cavity at the various Reynolds number

Problems

12.1 Solve the example problem, *Example 12.1*, for the non-isothermal forced and mixed convective flow. This corresponds to the computational setup shown in Fig. 12.8(a1–c1), along with the temperature as T_H at the top and T_C at the other walls. This results in the non-dimensional temperature, $\theta = (T - T_C)/(T_H - T_C)$, equal to 1 at the top wall and 0 at the other walls of the cavity. This leads to additional governing parameter as Prandtl number $Pr (= \nu/\alpha)$ for the forced convection, and also Richardson number $Ri (= Gr/Re^2)$ for the mixed convection. Using the grid and the governing parameter in the *Example 12.1*, along with $Ri = 0$ for forced convection and $Ri = 100$ for mixed convection, run the code (developed for *Example 12.1*) on the grid size of 82×82 for the QUICK scheme. Present and discuss the figures4 for the streamlines and temperature contours in the various lid driven cavities.

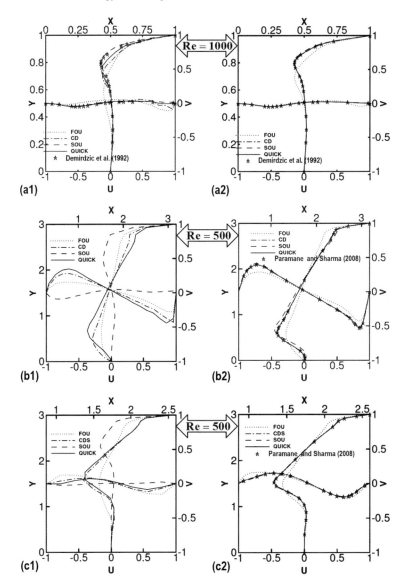

Fig. 12.9 Variation of U-velocity profile along centerline CD (Fig. 12.8)and V-velocity profile along centerline AB, obtained with the various advection schemes, on a grid size of (a1–c1) 22 × 22 and (a2–c2) 82 × 82 at the various Reynolds number

References

Demirdzic, I., Lilek, Z., & Peric, M. (1992). Fluid flow and heat transfer test problems for non-orthogonal grids: bench-mark solutions. *International Journal for Numerical Methods in Fluids, 15,* 329–354.

Paramane, S. B., & Sharma, A. (2008). Consistent implementation and comparison of FOU, CD, SOU and QUICK convection schemes on a square, skewed, trapezoidal and triangular lid driven cavity flow. *Numerical Heat Transfer B, 54,* 84–102.

Sharma, A., & Eswaran, V. (2003). A finite volume method, Chapter 12. In: K. Muralidhar & T. Sundararajan (Eds.), *Computational fluid flow and heat transfer* (2nd ed.). New Delhi: Narosa Publishing House.

Index

© The Author(s), under exclusive license to Springer Nature Switzerland AG 2022
A. Sharma, *Introduction to Computational Fluid Dynamics*,
https://doi.org/10.1007/978-3-030-72884-7

Printed in the United States
by Baker & Taylor Publisher Services